DRYING, ROASTING, AND CALCINING OF **MINERALS**

TMS2015

144th Annual Meeting & Exhibition

March 15-19, 2015 • Walt Disney World • Orlando, Florida, USA

DRYING, ROASTING, AND CALCINING OF MINERALS

Proceedings of a symposium sponsored by

The Minerals, Metals & Materials Society (TMS)

held during

TMS2015

144th Annual Meeting & Exhibition

March 15-19, 2015

Walt Disney World • Orlando, Florida, USA

EDITORS

Thomas P. Battle
Jerome P. Downey
Lawrence D. May
Boyd Davis
Neale R. Neelameggham
Sergio Sanchez-Segado
P. Chris Pistorius

Editors
Thomas P. Battle
Jerome P. Downey
Lawrence D. May
Boyd Davis
Neale R. Neelameggham
Sergio Sanchez-Segado
P. Chris Pistorius

ISBN 978-3-319-48600-0 ISBN 978-3-319-48245-3 (eBook)
DOI 10.1007/978-3-319-48245-3

Chemistry and Materials Science: Professional

Published by Springer International Publishers, Switzerland, 2016
Reprint of the original edition published by John Wiley & Sons, Inc., 2015, 978-1-119-07337-6

Printed on acid-free paper

This Springer imprint is published by Springer Nature
The registered company is Springer International Publishing AG
The registered company address is: Gewerbestrasse 11, 6330 Cham, Switzerland

TABLE OF CONTENTS
Drying, Roasting, and Calcining of Minerals

Roasting

Drying and Calcining

Fluidization, Reduction Roasting, and Microwave Treatment

Induration and Sintering

Sintering and Energy Use

PREFACE

Drying, roasting, and calcining are not the primary unit operations in extractive metallurgy. However, in most cases, the conversion of minerals into metals and metal compounds, whether by pyrometallurgy, hydrometallurgy, or electrometallurgy, involves one or more of these operations, generally on the feed material, sometimes on the product. These unit operations are rarely discussed in their own right, but usually as ancillaries to the core unit operations. In this symposium these processes are brought to the forefront, and discussed in the detail needed for practitioners to understand the state-of-the-art and also the importance of these technologies to extractive metallurgical flowsheets.

The symposium naturally divided itself into five sessions: Drying and Calcining; Fluidization, Reduction Roasting, and Microwave Treatment; Roasting; Induration and Sintering; and Sintering and Energy Use. While induration and sintering relate specifically to iron ores, the other papers cover a wide variety of materials, including copper chloride residue, salt and salt sludge, ammonium paratungstate, chromite ore fines, nickel laterites, petroleum coke, rare earth oxides, zinc oxide flue dust, red mud, and ferrochromium powder.

The technologies involved also cover a wide range. Using microwaves as a source of heat is considered in a number of the papers, while more traditional energy sources such as hot air and steam are discussed for drying. The greatest variety of papers is in the area of roasting. There are papers on the traditional route of roasting in air, where the high temperatures themselves are the main mechanism to modify the physical properties of the feeds, but even here, in sintering and induration the air used leads to complete oxidation of the iron units. A number of papers discuss reduction roasting, where the conditions are adjusted to reduce valence states on key elements in the starting minerals, and a few technologies result in major changes on the chemistry of the feeds, including sulfation, chlorination, bromination, and alkali roasting.

Exploring this book will provide a tour of the possibilities inherent in these unit operations—from old to new, potentially small scale to large, major commodities and specialty metals. It is clear that there is still much more to learn about these cornerstone operations in extractive metallurgy.

Thomas P. Battle, Midrex Technologies
Lead organizer

EDITORS

Thomas P. Battle

Thomas Battle is currently a Senior Metallurgist at Midrex Technologies in Pineville, North Carolina. He has undergraduate degrees in Materials Engineering and Astronomy from The University of Michigan, a Master's in Metallurgical Engineering from the Colorado School of Mines, and a Doctorate in Materials and Metallurgical Engineering from The University of Michigan. After a time as a Post-Doctoral Research Fellow in the Centre for Numerical Modelling and Process Analysis at Thames Polytechnic, he spent 18 years at various positions with the White Pigments and Mineral Products business at DuPont (now known at DuPont Titanium Technologies). He has spent the last six years as a senior metallurgist at Midrex Technologies, focusing on iron ore pelletizing technology and the direct reduction of iron.

Dr. Battle has been active with TMS for over 25 years, holding a number of volunteer positions, both technical and administrative, mainly for the Extraction and Processing Division. This culminated in a three-year term as chair of the division and a position on the TMS Board of Directors. He is a founding member of the North American Extractive Metallurgy Council.

Jerome P. Downey

Jerome (Jerry) Downey earned his Ph.D. in Metallurgical and Materials Engineering at Colorado School of Mines and his B.S. and M.S. degrees in Metallurgical Engineering at Montana Tech. Dr. Downey is a registered professional engineer with active licenses in Colorado and Montana. His experience base includes industrial operations, applied process research and development, and corporate management. Areas of technical expertise include chemical and metallurgical thermodynamics, thermal processing, materials synthesis and processing, and hazardous materials treatment.

Dr. Downey is presently Associate Professor, the Hazen Research Professor of Extractive Metallurgy, and Goldcorp Professor at Montana Tech. He is the campus Director of the Montana University System Materials Science Ph.D. program. Dr. Downey's research efforts are currently focused on the study of fundamental properties of slags, molten salts, and glasses; vapor phase extraction and refining of rare earth elements; and synthesis of non-oxide ceramic compounds for energy applications. Other research interests include pyro- and hydrometallurgical process development, polymer and biomass gasification, and wastewater treatment.

Lawrence D. May

Lawrence (Larry) May is a Vice President for Hazen Research, Inc., in Golden, Colorado. He leads the Thermal Processing group at Hazen and is also the Facility Manager. Mr. May has a Bachelor of Science degree in Chemical Engineering from the Colorado School of Mines and a Masters of Business Administration from the University of Colorado. He has over 28 years of experience at Hazen and his primary focus is the development of thermal processes for industrial clients interested in fluidized-bed, rotary kiln, and multiple hearth technologies. Mr. May has been a member of TMS since 2012, and this year is his second opportunity to co-chair a session in the TMS conference proceedings.

Boyd Davis

Boyd Davis is a principal of Kingston Process Metallurgy Inc., a contract research and development company based in Kingston, Ontario, Canada. He graduated with a Ph.D. in Chemical Metallurgy from Queen's University in Kingston. After consulting for a few years in applied thermochemistry, he formed KPM with his partner, Alain Roy. The company, currently with 20 employees, specializes in client-driven chemical process development—both at the lab and pilot scale—with techno-economic and process modeling in parallel. The company also develops its own processes in extractive metallurgy. Dr. Davis is active with both MetSoc-CIM (as its current president) and the TMS Extraction and Processing Division. He also volunteers his time as an adjunct at Queen's University teaching process design and has supervised over a dozen M.Sc. and Ph.D. students since 2002.

Neale R. Neelameggham

Neale R. Neelameggham is "The Guru" at IND LLC, involved in technology marketing and international consulting in the field of light metals and associated chemicals (boron, magnesium, titanium, lithium and alkali metals), rare earth elements, and battery and energy technologies, etc. He was a visiting expert at Beihang University of Aeronautics and Astronautics, Beijing, China.

He has over 38 years of expertise in magnesium production and was involved in process development of its startup company NL Magnesium through to the present US Magnesium LLC, UT from where he retired in 2011. He is developing thiometallurgical processes—a new concept of using sulfur as the reductant and/or fuel. He has published a heat transfer model for global anthropogenic warming based on thermal emissions independent of energy conversion source.

Dr. Neelameggham holds 16 patents and patent applications, and has published several technical papers. He has served in the Magnesium Committee of the Light Metals Division (LMD) since its inception in 2000, chaired it in 2005, and in 2007 he was made a permanent co-organizer for the Magnesium Technology symposium. He has been a member of the Reactive Metals Committee, Recycling Committee, Titanium Committee, and Programming Committee Representative of LMD and LMD council.

Dr. Neelameggham was the inaugural chair in 2008 when the Light Metals and Extraction and Processing Divisions created the Energy Committee, and has been a co-organizer of the Energy Technology symposium through the present. He received the LMD Distinguished Service Award in 2010. He is the chairman of Hydrometallurgy and Electrometallurgy Committee of EPD for the 2013-2015 term and is the organizer for the Rare Metal Technology 2014 and 2015 symposia. He is the co-editor of the *Essential Readings in Magnesium Technology* compendium of TMS papers published in 2014.

Sergio Sanchez-Segado

Sergio Sanchez-Segado obtained his Bachelor's Degree from Polytechnic University of Cartagena and Master of Chemical Engineering degree from the University of Murcia in Spain in 2002 and 2005, respectively. He was the recipient of Caja de Ahorros del Mediterráneo (CAM) Scholarship for pursuing his Ph.D. at Polytechnic University of Cartagena where he completed his Ph.D. degree in September 2010 in the area of bio-fuels production. After his Ph.D., he was appointed as a post-doctoral researcher at Polytechnic University of Cartagena for developing advanced separation processes of organic and inorganic compounds using ionic liquids. In October 2011, he joined the University of Leeds where he has been developing clean processes in extractive metallurgy of critical materials, area in which he was awarded with a Marie Curie Intra-European Fellowship in 2012.

P. Chris Pistorius

Chris Pistorius is POSCO Professor and Co-Director of the Center for Iron and Steelmaking Research in the Department of Materials Science and Engineering at Carnegie Mellon University. His research focuses on pyrometallurgy and electrochemistry. He was an associate professor (1991-1996) and professor (1997-2008) in the Department of Materials Science and Metallurgical Engineering, University of Pretoria, South Africa, and served as the chair of that department from May 2002 to June 2008. Dr. Pistorius has a Master's degree in Metallurgical Engineering from the University of Pretoria, and a Ph.D. from the University of Cambridge, United Kingdom.

SESSION CHAIRS

Roasting

Boyd Davis,
Kingston Process Metallurgy Inc.

Sergio Sanchez-Segado,
University of Leeds

Drying and Calcining

Neale R. Neelameggham,
IND LLC

James Sever,
Alpha / Omega Engineering

Fluidization, Reduction Roasting, and Microwave Treatment

Jerome P. Downey,
Montana Tech of the University of Montana

Lawrence D. May,
Hazen Research, Inc.

Induration and Sintering

Thomas P. Battle,
Midrex Technologies

P. Chris Pistorius,
Carnegie Mellon University

Sintering and Energy Use

Mingming Zhang,
ArcelorMittal Global R&D

Dean Gregurek,
RHI AG

DRYING, ROASTING, AND CALCINING OF

MINERALS

ROASTING

Session Chairs:

Boyd Davis

Sergio Sanchez-Segado

Drying, Roasting, and Calcining of Minerals
Edited by: Thomas P. Battle, Jerome P. Downey, Lawrence D. May, Boyd Davis, Neale R. Neelameggham, Sergio Sanchez-Segado, and P. Chris Pistorius
TMS (The Minerals, Metals & Materials Society), 2015

SULFATION ROASTING OF A BORNITE FLOTATION CONCENTRATE TO OPTIMIZE SILVER EXTRACTION IN A FERRIC CHLORIDE LEACH

Ryan J. Foy[1], Steve Lloyd[2], Brandon Steinborn[3], Jerome P. Downey[4]

[1]Montana Tech of the University of Montana; Graduate Student
Metallurgical and Materials Engineering Dept.; 1300 West Park St.; Butte, MT 59701, USA
[2]Mill Manager, Troy Mine, Inc.
Troy, MT 59355, USA
[3]Freeport-McMoRan Inc.; Metallurgical Engineer; Miami Operations-Smelter
Claypool, AZ 85532, USA
[4]Montana Tech of the University of Montana; Associate Professor
Metallurgical and Materials Engineering Dept.; 1300 West Park St.; Butte, MT 59701, USA

Keywords: Ferric chloride leaching, sulfation roasting, bornite, silver, copper

Abstract

Research was performed to evaluate copper and silver extraction from samples of bornite flotation concentrate. Sulfation roasting experiments were performed in a rotary tube furnace, a fluidized bed reactor, and a static bed muffle furnace. Roasting was plagued with rapid, low-temperature calcine sintering, apparently related to the presence of organic matter in the concentrate. Calcine entered a weak acid leaching process to remove copper, and silver was extracted from the weak acid leach residue in a ferric chloride and hydrochloric acid solution. Copper extraction efficiencies in excess of 90% were attained in the weak acid leach while 90% silver extraction was achieved in the ferric chloride leach, along with copper and lead extractions as high as 93% and 99% respectively. The two-stage leaching process resulted in overall extraction efficiencies of 99% copper and 90% silver.

Introduction

Revett Minerals, Inc. commissioned a project to investigate silver extraction from bornite flotation concentrate produced at the Troy Mine, which is located in northwest Montana. Results of preliminary experiments identified ferric chloride leaching (FCL) media as a promising method for silver extraction. Prior to the ferric chloride leach, copper contained in calcine produced by sulfation roasting of the concentrate was leached in weak sulfuric acid. Residue from the weak acid leaching (WAL) was the FCL process feed material.

X-ray diffraction analysis revealed the concentrate contains three copper-containing minerals: bornite, chalcopyrite, and djurleite. Elemental analysis (ICP) indicated the concentrate is sulfur deficient for the purpose of sulfation roasting. As shown in Equation 1, bornite, the principal copper-bearing mineral, is a net consumer of sulfur dioxide (R.Coussement, 1964). Chalcopyrite produces sulfur dioxide, as shown in Equation 2, but the amount of chalcopyrite in the flotation concentrate is insufficient to maintain the sulfur dioxide partial pressure necessary for complete sulfation of the copper minerals (R.S. Boorman, 1984).

$$2Cu_5FeS_{4(s)} + 2SO_{2(g)} + 39/2O_{2(g)} \rightarrow 10CuSO_{4(s)} + Fe_2O_{3(s)} \quad [1]$$

$$2CuFeS_{2(s)} + 15/2O_{2(g)} \rightarrow 2CuSO_{4(s)} + Fe_2O_{3(s)} + 2SO_{2(g)} \qquad [2]$$

Sulfation roasting of copper mineral concentrates can be effected over a wide temperature range from 550°C to 720°C (Griffith, 1958). A temperature of 680°C is commonly selected to produce high copper extraction and iron rejection efficiencies in the subsequent leaching operation. The predominance area diagrams in Figure 1 compare the Cu-O-S and Fe-O-S systems at 680°C; the selected operating point lies within the copper sulfate and hematite stability regions. The operating points denote the requisite oxygen and sulfur dioxide partial pressures for selective sulfation of the copper-bearing minerals.

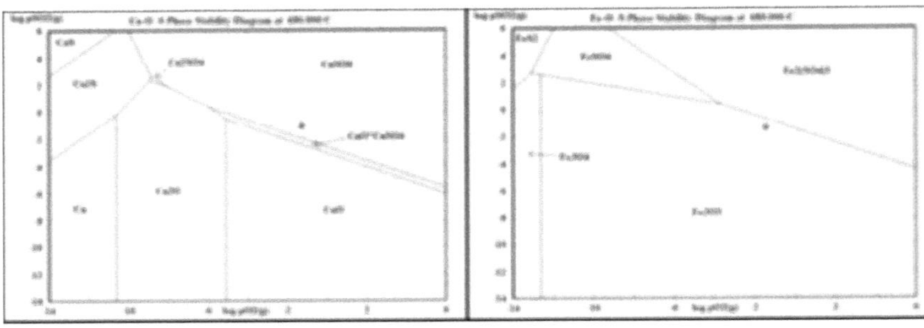

Figure 1. (Left) Predominance area diagram for the Cu-O-S system at 680°C. (Right) Predominance area diagram for the Fe-O-S system at 680°C (Diagrams produced using HSC Chemistry for Windows 7.0).

Preliminary Experiments

Sulfation roasting of the flotation concentrate was conducted in a four-inch diameter quartz tube positioned in an ATS Series 3210 three-zone tube furnace. Temperature was maintained at the desired set-point with three PID controllers. Pure silica was added to aid in maintaining a consistent bed temperature throughout the experiments. Baffles on the rotary tube ensured adequate agitation during roasting. During heat-up, the reactor was purged with argon at a rate of 600 mL/min until the desired bed temperature was reached. Thereafter, the atmosphere was controlled by adding 1000 mL/min of a premixed gas composition of 6% oxygen and 8% sulfur dioxide (balance nitrogen). A four-stage series of impinge-type scrubbers removed sulfur dioxide from the reactor off-gas. The scrubbers included two vessels filled with 500mL of 18-MΩ water followed by, two vessels filled with 500mL of 0.25M sodium hydroxide solution.

Experiments P-7 and P-8 were conducted at 680°C for 7.5 and 10 hours, respectively. In P-8, the concentrate was blended with 2% by mass sodium sulfate and brought to temperature under 500 mL/min of argon and 95 mL/min of water vapor to remove fluorine contained in the concentrate. Calcines were leached for two hours in a weak sulfuric acid medium with pH maintained below 2.0. The solid leach residue was leached for two hours in a 90°C solution of 0.14M ferric chloride and 2M hydrochloric acid. The resulting copper and silver extraction efficiencies are summarized in Table 1. The data show that the FCL method produced substantially higher silver extractions and effectively scavenged residual copper from the WAL residue samples. However, the sodium sulfate addition in experiment P-8 had little effect on cumulative copper extraction, as shown in Table 2. A potential effect on silver extraction was observed as the cumulative extraction efficiency increased from 80.6% in P-7 to 88.9% in P-8.

Table 1. Mass loss and metal extraction efficiencies (%) for preliminary experiments 7 and 8.

Sample ID	Mass Loss, %	Cu Extraction, %	Ag Extraction, %
WAL Res. - 7	69.7%	90.0%	26.7%
WAL Res. - 8	59.7%	90.0%	11.7%
FCL Res. - 7	25.9%	91.5%	73.6%
FCL Res. - 8	22.2%	88.7%	87.4%

Table 2. Cumulative extraction efficiencies (%) for copper and silver (combining both leaching stages).

Sample	Total Cu Extraction, %	Total Ag Extraction, %
P-7	99.2%	80.6%
P-8	98.9%	88.9%

Fluidized Bed Reactor

The next series of roasting experiments was performed in a fluidized-bed reactor (FBR) that consisted of a four-inch diameter stainless steel tube that was indirectly heated by a three-zone, refractory-lined electric resistance furnace. Initial experiments were performed with a 680°C bed temperature, and elemental sulfur was mixed with flotation concentrate to compensate for the concentrate's sulfur deficiency. The feed mixture was delivered by a hopper system with a screw feed mechanism; feed dropped through a 1.5-inch diameter stainless steel pipe that extended through the furnace top to the middle of the bed.

Continuous experiments were eventually abandoned in favor of semi-batch due to persistent reactor bed agglomeration. Measures taken to mitigate sintering included lower bed temperature and fluidizing with commercially pure nitrogen during heat-up. When the bed reached 650°C, the fluidizing gas composition was adjusted to provide 6% oxygen and 8% sulfur dioxide, and the roast duration was 6 hours after the oxygen level stabilized. Calcine sample C-1 was split to feed WAL experiments 1-0 and 2-0, which were conducted according to the procedure used in the preliminary experiments. As shown in Figure 2, both experiments resulted in copper extraction slightly below 90% and negligible silver extraction.

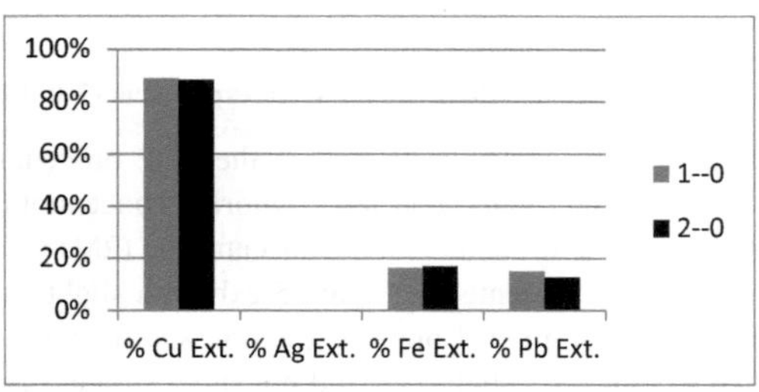

Figure 2. Metal extraction efficiencies for WAL samples 1-0 and 2-0.

Leach residue samples from experiments 1-0 and 2-0 were leached for four hours at 90°C in 2M hydrochloric acid and 0.14M ferric chloride. The variables were temperature (ambient (~23°),

60°, and 90°C) and retention time (one, two, and four hours). As shown in Figure 3, copper and silver extraction efficiencies ranged from 48% to 96% and from 17% to 52%, respectively. Experiments conducted at 90°C produced the highest extraction efficiencies; experiments performed with four hour retention time experiments consistently produced higher extraction efficiencies than those performed with two hour retention times.

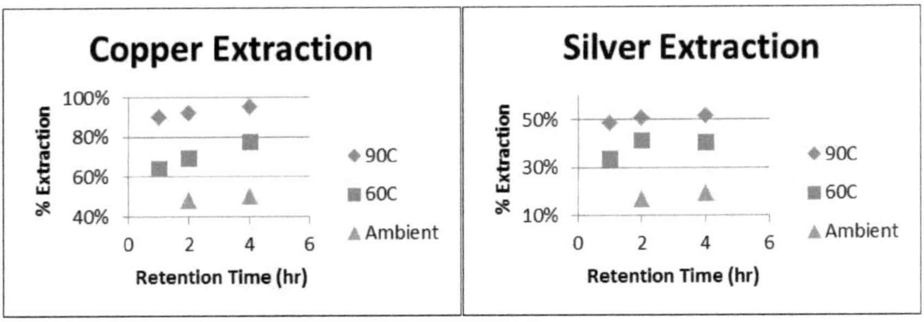

Figure 3. Copper and silver extraction efficiencies for FCL experiments 1-1 to 1-4 and 2-1 to 2-4.

In an effort to manage agglomeration in the FBR, C-3 was roasted in a 50/50 sand-flotation concentrate bed. Calcine samples used in WAL experiments 4-0 and 5-0 were produced from concentrate blended with 2 wt. % sodium sulfate prior to roasting. The calcine samples were leached under the same conditions as 1-0 and 2-0 except the retention time for experiment 5-0 increased to four hours. As indicated in Figure 4, copper extractions ranged from 65% to 75% and silver extractions were consistent at about 4%. Significant amounts of iron precipitates were observed in the leach liquors from experiments 4-0 and 5-0, which suggests that the sodium sulfate addition affected hematite formation during sulfation roasting.

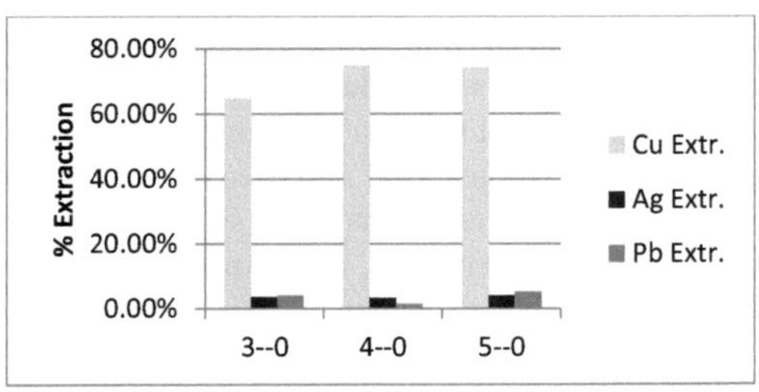

Figure 4. Metal extraction efficiencies for WAL experiments 3-0, 4-0, and 5-0.

Residue from WAL experiment 3-0 was subjected to the FCL procedure. The variables were retention time (two, four, and eight hours) and ferric chloride concentration (0.14M, 0.21M, and 0.27M); temperature (90°C) and hydrochloric acid concentration (2M) were constant. The copper and silver extraction efficiencies presented in Figure 5 exhibit a slight negative correlation with increasing ferric chloride concentration. Copper extraction significantly improved as residence time increased from two to four hours but silver did not show similar improvement. There is no advantage to extending residence time to eight hours.

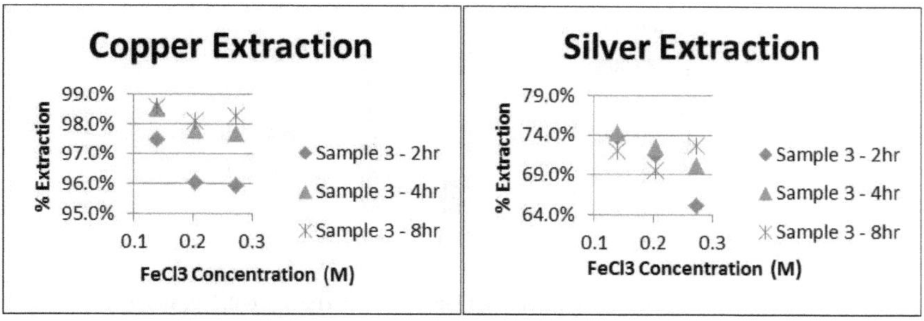

Figure 5. Copper and silver extraction efficiencies for FCL experiments performed on WAL residue 3-0.

Residue from WAL 4-0 was subjected to the FCL variables from the 3-0 FCL experiments. The only difference between the two sets of experiments is the presence of sodium sulfate during the roasting of C-4. Metal extraction efficiencies are displayed in Figure 6. Copper extraction efficiencies ranged from 94% to 97.5% with a slight negative effect associated with increasing ferric chloride concentration and a slight positive effect associated with increased retention time. Silver extraction reached a high at 79% with no evident correlation between time and ferric chloride concentration.

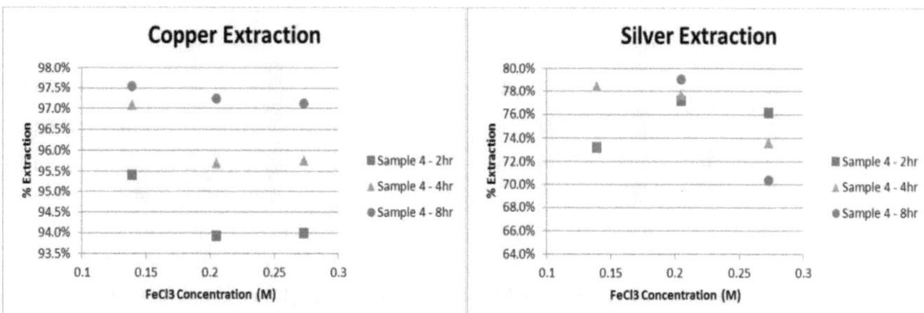

Figure 6. Copper and silver extraction efficiencies for FCL experiments performed on WAL residue 4-0.

Static Bed Muffle Furnace

Static bed roasting experiments were performed at 400°, 550°, 680°, and 900°C and times ranging from 18 to 24 hours in a Lindberg Blue M furnace. Calcine samples produced at 400°, 550°, and 680°C were subjected to the WAL process conditions in experiments 6-0, 7-0 and 9-0. As presented in Figure 7, copper extractions were consistent at 94% at the three roasting temperatures. Silver extraction exhibited the largest change, increasing from 1.8% at 400°C to 87.4% at 680°C.

A peculiar inverse correlation was observed between WAL extraction efficiency and the degree of agitation during the roast. Calcine samples produced in the FBR calcines exhibited less than 90% copper and negligible silver extraction; calcine produced in the rotary tube furnace yielded 90% copper and 26.7% and silver extraction, and the calcine produced in the static bed muffle furnace yielded 94% copper and 97.4% silver extraction.

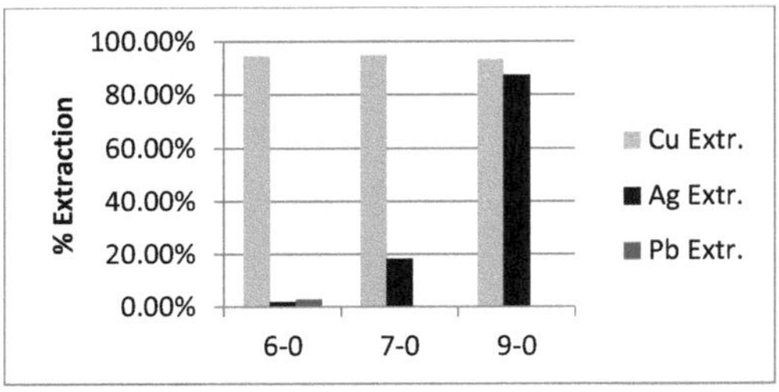

Figure 7. Metal extraction efficiencies for WAL experiments 6-0, 7-0, and 9-0.

In experiments 6-1, 7-1, and 9-1, calcine samples (from roast experiments 6-0, 7-0, and 9-0) were directly leached in the FCL process and experiments 6-2, 7-2, and 9-2 were performed on WAL residue samples. The experiments were conducted at 90°C with a four hour retention time; the lixiviant was 1.5M ferric chloride and 2M hydrochloric acid. The corresponding metal extraction efficiencies are compared in Figure 8. In each experiment performed on calcine samples, copper extraction exceeded 98% and lead extraction exceeded 95%. Silver extraction efficiency of 90% was achieved with the calcine roasted at 680°C but was as low as 30% with calcine roasted at 550°C.

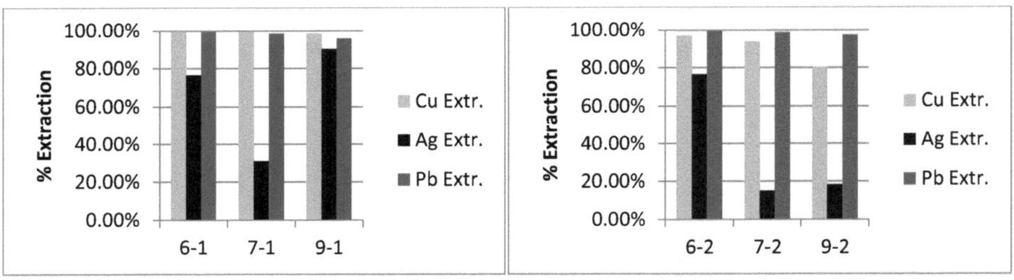

Figure 8. (Left) Metal extraction efficiencies for FCL experiments 6-1, 7-1, and 9-1. (Right) Metal extraction efficiencies for FCL experiments 6-2, 7-2, and 9-2.

The results of the WAL residue leaching experiments evidenced a negative correlation to increasing roasting temperature. Leaching WAL residue from calcine roasted at 400°C resulted in 96% copper extraction but only 80% extraction efficiency was obtained from calcine produced at 680°C. Similarly with silver, the 76% extraction was achieved by leaching WAL residue from the 400°C roast but substantially lower extraction efficiencies were achieved with WAL residue samples derived from 550° and 680°C calcine sample.

Table 3 shows the cumulative (WAL + FCL) copper and silver extraction efficiencies. When compared with extraction efficiencies obtained by direct leaching of calcine, copper and silver extractions were almost identical in the two- and single-stage leaching processes.

Table 3. Cumulative extraction efficiencies for copper and silver (combining both leaching stages).

Sample Series	Cu Extraction, %	Ag Extraction, %
6	99.8%	77.0%
7	99.7%	30.5%
9	98.7%	89.7%

Sintering Characterization

Samples of sintered calcine/concentrate were characterized to investigate the cause for the low temperature sintering phenomena. GC-MS analysis confirmed a significant amount of organic material in the flotation concentrate. Scanning electron microscopy (SEM) and energy dispersive analysis of x-rays (EDAX) was then employed to analyze sintered calcine samples. Spot analysis was conducted on samples from C-4 and C-5 using SEM-EDAX to determine the composition of the liquid phase formed during sulfation roasting. An SEM micrograph displaying the spot location and EDAX spectrum for sample C-4 is shown in Figure 9.

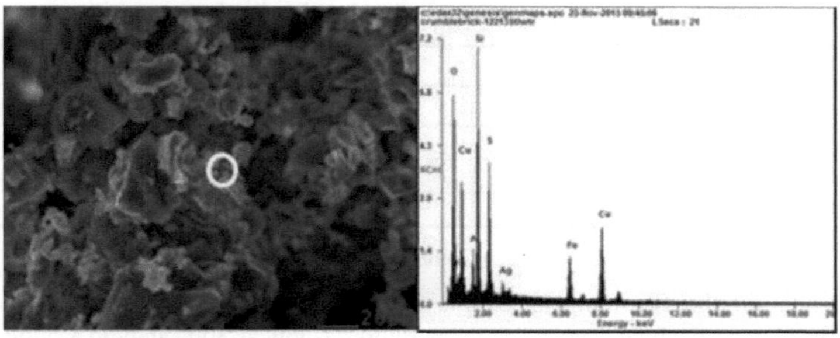

Figure 9. (Left) SEM micrograph depicting location of spot analyis for EDAX. (Right) EDAX spectrum of spot analysis of calcine produced at 400°C.

Spot analysis results for C-4 revealed the particles were connected by a high silicon content liquid phase that also contained varying amounts of copper, iron, and silver. The consistently strong silicon peak suggests a siliceous liquid phase, and the C-5 spot analysis provided confirmatory evidence. In each spot, strong sulfur peaks were detected in conjunction with the silicon, so it is considered probable that sulfate compounds produced during the roasting process may contribute to the sintering.

Confirmatory Experiments

A bulk sample of calcine produced at 400°C in a muffle furnace was leached via the WAL procedure, and the WAL residue was split into five samples for further evaluation of the FCL procedure The FCL experiments were conducted at 90°C and a four hour retention time. The ferric chloride concentration was varied (0.3M, 1.5M, and 3M) with 2M hydrochloric acid constant. As shown in Table 4, copper extraction increased from 94.4% to 99.1% as the ferric chloride concentration increased from 0.3M to 3.0M. However, silver and lead extraction efficiencies were consistent at 71% and 99%, respectively and did not respond to changes in ferric chloride concentration.

Table 4. Metal extraction efficiencies (%) based on ferric chloride concentration.

Leachate	Extraction, %		
$FeCl_3$ (M)	Cu	Ag	Pb
0.3	94.4	69.9	98.9
1.5	97.1	71.2	99.3
3.0	99.1	70.9	99.6

The fourth WAL residue split was leached in 0.3M ferrous chloride and 2M hydrochloric acid, and the fifth in 5M hydrochloric acid without iron chloride. Results of the lixiviant change are displayed in Table 5. Lead extraction remained consistent at 99%. Iron extraction from 56% to 93% by the use of ferrous chloride in the same concentration as ferric chloride. The absence of iron chloride had a minimal effect in the 5M hydrochloric acid experiment. Copper extraction shows minimal increases in the ferrous chloride and 5M hydrochloric acid experiments to 97.6% and 96.3%, respectively. Silver extractions exhibit significant decreases of 15% with ferrous chloride and 43% with 5M hydrochloric acid.

Table 5. Metal extraction efficiencies (%) based on lixiviant composition.

Lixiviant Composition	Extraction, %		
	Cu	Ag	Pb
0.3M $FeCl_3$ and 2M HCl	94.4	69.9	98.9
0.3M $FeCl_2$ and 2M HCl	97.6	55.4	99.4
5M HCl and no $FeCl_x$	96.3	27.1	98.8

Conclusions

Organic matter in the flotation concentrate is the suspected cause of the low-temperature sintering phenomenon, which occurred at temperatures as low as 300°, in both a static bed muffle furnace and a FBR. Phases involved are suspected to be components of the organic ash, specifically the silicon-containing organics, and a combination of the calcine products.

Copper extraction efficiency varied due to the inconsistent calcine quality with 90% extraction achieved by conducting the WAL for two hours at ambient temperature and pressure with pH maintained below 2.0. Silver extraction in the WAL process was generally low.

Eighty percent silver extraction was attained by leaching WAL residue for four hours in a 90°C solution of 2M hydrochloric acid and 0.14M ferric chloride. Copper and lead extractions exceeded 93% and 98%, respectively, in all FCL experiments. Scavenging copper and removing lead with high efficiency minimizes greatly the loss of copper to the process tails and avoids concentrating lead in the process tails eliminating environmental hazards.

Direct FCL of calcine produced the best overall extraction efficiencies. When calcine was leached for four hours in a 90°C solution of 2M hydrochloric acid and 1.5M ferric chloride, extraction efficiencies were 98% for copper, 90% for silver, and 96% for lead.

References

Harry Mudgett, D. J. (2011). *Process Development Study for Roast-Leach of Copper-Silver Flotation Concentrate.* Golden, CO.

R.Coussement, F. T. (1964). Fluid-bed roasting reactions of copper and cobalt sulfide concentrates. *Journal of Metals*, 831-834.

R.S. Boorman, R. S. (1984, September). The RPC sulphation roast process. *Mining Magazine*, pp. 225-235.

Sulfate roasting Copper-Cobalt sulfide concentrates. (1958). *Journal of Metals*, 134-136.

W.A. Griffith, H. D. (n.d.). TMS Paper Selection. *Development of the Roast-Leach-Electrowin process for Lakeshore.* New York.

Drying, Roasting, and Calcining of Minerals
Edited by: Thomas P. Battle, Jerome P. Downey, Lawrence D. May, Boyd Davis, Neale R. Neelameggham, Sergio Sanchez-Segado, and P. Chris Pistorius
TMS (The Minerals, Metals & Materials Society), 2015

CHLORINATION ROASTING OF RARE EARTH ELEMENT OXIDES

Daniel W. Gaede, Bryce D. Ruffier, Jerome P. Downey, Jannette L. Chorney, Larry G. Twidwell, Ryan J. Foy, Katelyn M. Lyons

Montana Tech of the University of Montana
1300 West Park Street, Butte, MT, 59701 USA

Keywords: Rare Earth Elements; Roasting; Thermodynamics, Chlorination

Abstract

The Army Research Laboratory (ARL) has an interest in the recovery and processing of rare earth elements. At Montana Tech, elevated temperature chlorination of rare earth oxides is presently being investigated as a potential methods leading to the selective separation and recovery of rare earth elements from mineral ores and concentrates. After preliminary evaluation, ytterbium oxide was selected as a representative surrogate material for more extensive chlorination studies. Methods used to determine the conversion and recovery include X-ray diffraction (XRD) and inductively coupled plasma (ICP) analysis. These data have been used to establish and confirm process conditions that consistently yield high conversion efficiencies. Based on the promising results, further experimentation is underway to ascertain the metallurgical response of other rare earth oxides and carbonates to the roasting process. The end goal is to create a robust process for industrial application that has greater efficiency and reduced costs as compared to current practice.

Introduction

Rare earth elements (REE) are critical to many defense applications as well as being essential components of electronic devices such as cellphones, medical devices such as MRI's and PET scanners, and batteries for hybrid vehicles. The United States may face supply uncertainty because worldwide production is currently dominated by China[1]. Demand for REE has sharply increased in recent years and, due to continued advances in materials development, it is expected to continue to increase.

Rare earth elements frequently occur in the host ores as small concentrations of REE oxides or carbonates, and the similarities of their physical and chemical properties make separation of individual REE very challenging. Current commercial separation methods are often capital intensive and produce large amounts of waste, including contaminated wastewater. A process that could effectively and economically separate the REE elements while diminishing waste generation would greatly benefit rare earth resource development.

Conversion of rare earth oxides to rare earth chlorides is often the initial step in the pyrometallurgical and hydrometallurgical processing of REE. Rare earth chlorination has been studied by several researchers with varying success [2,3,4,5,6]. Chlorination can be effected by roasting rare earth oxides (or carbonates) in the presence of a chloridizing reactant such as ammonium chloride; the conversion proceeds according to the general net chemical reaction shown in equation 1.

$$RE_2O_{3(s)} + 6NH_4Cl_{(s)} \rightarrow 2RECl_{3(s)} + 6NH_{3(g)} + H_2O_{(g)} \quad (1)$$

The effects of varying the NH_4Cl:REO mole ratio and temperature on the conversion efficiency were first examined using the free energy minimization algorithm in HSC Chemistry 7.0 (Outotec)[7], which can be used to predict stable phases that form under various process conditions. Based on the results of preliminary thermodynamic assessments, four reagent grade rare earth oxides were selected for the scoping experiments: holmium oxide (Ho_2O_3), neodymium oxide (Nd_2O_3), samarium oxide (Sm_2O_3), and ytterbium oxide (Yb_2O_3).

The equilibrium concentrations of the four studied elements as a function of temperature is presented in Figure 1. The temperatures predicted for the onset of chlorination are summarized in Table 1.This information was useful in establishing ranges of parameters (i.e. temperatures and NH_4Cl:RE_2O_3 mole ratios) examined in subsequent scoping experiments.

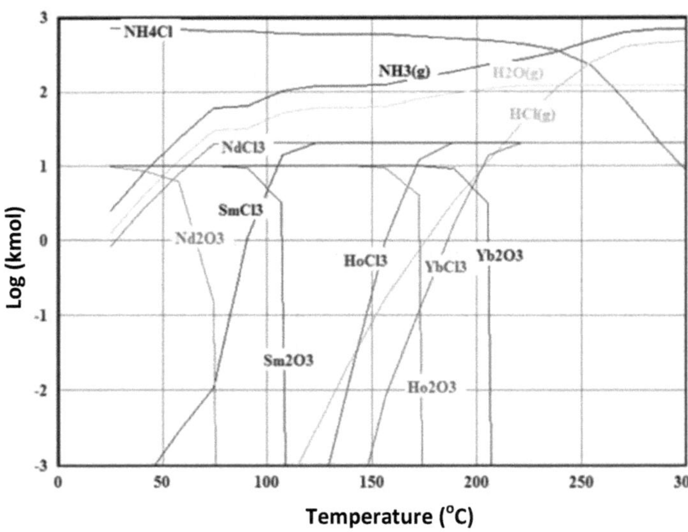

Figure 1: Thermodynamic model of REO to RECl conversion as function of temperature for Ho, Nd, Sm, and Yb.

Experimental

In general, the experiments involved roasting a mixed charge of ammonium chloride and one of four selected rare earth oxides in a tube furnace. The variables included roasting temperature, time at temperature, and the mole ratio of chloridizing agent to rare earth oxide in the furnace charge. Based on the thermodynamic modeling results and other considerations, the four rare earth oxides chosen for the scoping experiments were: Nd_2O_3, Ho_2O_3, Yb_2O_3, and Sm_2O_3. Individual rare earth oxide chemicals were purchased from commercial suppliers at 99.9% minimum purity as was the ammonium chloride at 99.5% purity. Appropriate masses of rare earth oxides and ammonium chloride were blended at either an 18:1 or 24:1 mole ratios (NH_4Cl:REO); the mole ratios were partly based on the results of a prior study[8]. Excess ammonium chloride is necessary to minimize the formation of rare earth oxychlorides during the roasting process.

The blended chemicals were placed in a mullite boat, which was centered within an alumina containment tube that passed through the heated zone of an Advanced Test Systems (ATS) Series 3210 tube furnace. The tube furnace was held at the predetermined temperature throughout each experiment. High purity argon was flowed through the containment tube at 0.250 liters/minute. Prior to entering the containment tube, the argon passed through an oxygen filter to remove trace amounts of oxygen, which could cause oxychloride formation. Rare earth oxychlorides (REOCl) are an undesirable reaction product that can form if oxygen (O_2) enters the furnace atmosphere during the roasting process. Roaster exhaust gases were scrubbed through a series of bubblers to prevent oxygen backflow and to remove any residual chloride. Samples were weighed before and after each furnace run to document mass changes incurred during roasting. When any residue was found inside the tubes, it was removed and weighed after the experiment.

Following each experiment, the solid matter was retrieved from the mullite boat and prepared for XRD analysis. XRD data were collected using a Rigaku diffractometer and scintillation detector with a 5-90° 2θscanning range, 0.5 seconds hold time, and 0.02°step. The receiving slit width was 0.3 mm with power at 40Kv and 20 mA. PDXL version 1.8.1.0 was used to analyze the patterns, to determine phases and conversion efficiencies. Following XRD, samples were leached for a minimum of two hours in 18 mega ohm water at a nominal pH 6 and then pressure filtered through Whatman size 1 filter papers. The leach solutions were sent to an outside lab for confirmatory ICP analyses (of the possible reaction products, only rare earth chlorides are soluble in cold water).

Results and Discussion

The initial scoping parameters were based upon a previous chlorination roasting study performed on neodymium and samarium oxides by Bell[8]. In the present study, the scoping experiment results revealed that neodymium, samarium, and holmium each exhibited exceptionally high oxide to chloride conversion efficiencies. However, ytterbium was less responsive. Therefore, it was decided that conversion parameters should be determined based on the most recalcitrant oxide, Yb_2O_3.

Design Expert Version 9.01 (Stat Ease Corporation)[9] was used to create a three variable Response Surface experimental matrix to examine the effects of varying roasting time and temperature and the NH_4Cl:Yb_2O_3mole ratio on chlorination efficiency. The experimental design matrix (including results) is presented in Table 2.The "Factors" are the three experimental variables and the "Responses" are the percentages of four species (three ytterbium compounds and residual ammonium chloride) that were detected in the experimental products. The roasting experiments were conducted in concert with the previously described method and equipment configuration. As evident in Table 1, experiments 8, 11, 15 and 17 yielded the highest chloride conversion efficiencies. The data also indicate that high (85 to 96%) conversion efficiency is attainable and that oxychloride formation is avoidable.

Table 1. Three Variable Response Surface Matrix of Chlorination Roast Experiments.

Std. Expt. No.	Factors			Responses				% Conversion to RECl
	Temp., °C	Time, Hrs	NH_4Cl:oxide, mole ratio	Experimental Product Composition (%)				
				Yb_2O_3	NH_4Cl	YbOCl	$YbCl_3$:$6H_2O$	
1	150	1	6	73.50	26.50	0	0	0
2	400	1	6	27.40	0	46.90	25.70	11.77
3	150	4	6	72.90	27.10	0	0	0
4	400	4	6	38.45	19.92	41.63	0	0
5	150	1	24	28.50	71.50	0	0	0
6	400	1	24	4.00	0	8.60	87.40	72.38
7	150	4	24	28.80	71.20	0	0	0
8	400	4	24	1.80	7.00	0	91.03	96.37
9	150	2.5	15	57.6	42.40	0	0	0
10$^{\Omega}$	400	2.5	15	1.10	7.00	30.70	45.60	41.77
11	275	1	15	6.10	23.20	0	70.70	85.70
12	275	4	15	7.40	9.20	8.50	74.90	65.40
13	275	2.5	6	55.10	0	0	44.90	29.65
14	275	2.5	24	0	29.00	0	71.00	100
15	275	2.5	15	7.00	13.70	0	79.30	85.42
16	275	2.5	15	8.20	19.70	0	72.10	81.97
17	275	2.5	15	5.01	17.62	0	77.38	88.87
18	275	2.5	15	12.51	21.22	0	66.27	73.26
19	275	2.5	15	8.20	21.80	0	70.00	81.53
20	275	2.5	15	10.00	18.20	0	71.80	78.78

Ω indicates results obtained through a leach and mass balance.

Design space models were calculated for the four specific responses: Yb_2O_3, NH_4Cl, $YbCl_3$:$6H_2O$ and (oxide-to-chloride) Conversion Efficiency. Statistically valid models were successfully determined for all of the response data. The Conversion Efficiency model is considered the most important response because it can be used to identify combinations of variables required to maximize conversion of ytterbium oxide to ytterbium chloride.

Figure 2 presents a series of four contour visualizations of the experimental results. The two diagrams at the top of the figure illustrate the combined effects of varying roasting temperature and the NH_4Cl:Yb_2O_3 mole ratio at constant time (1 and 4 hours in the top diagrams at left and right, respectively). The ring-shaped contours identify regions where the model predicts a specific oxide-to-chloride conversion efficiency (1.0 = 100% conversion, 0.8 = 80% conversion, etc.). Similarly, the two diagrams at the bottom of Figure 2 illustrate the combined effects of varying roasting temperature and time at constant NH_4Cl:Yb_2O_3 mole ratio. The ready availability of these diagrams, which are proven to be statistically sound, facilitates identification of practical windows of operating parameters. The relatively large areas that correspond to 90 to 100 percent conversion efficiency suggest that a very robust process can be designed through proper selection of operating parameters.

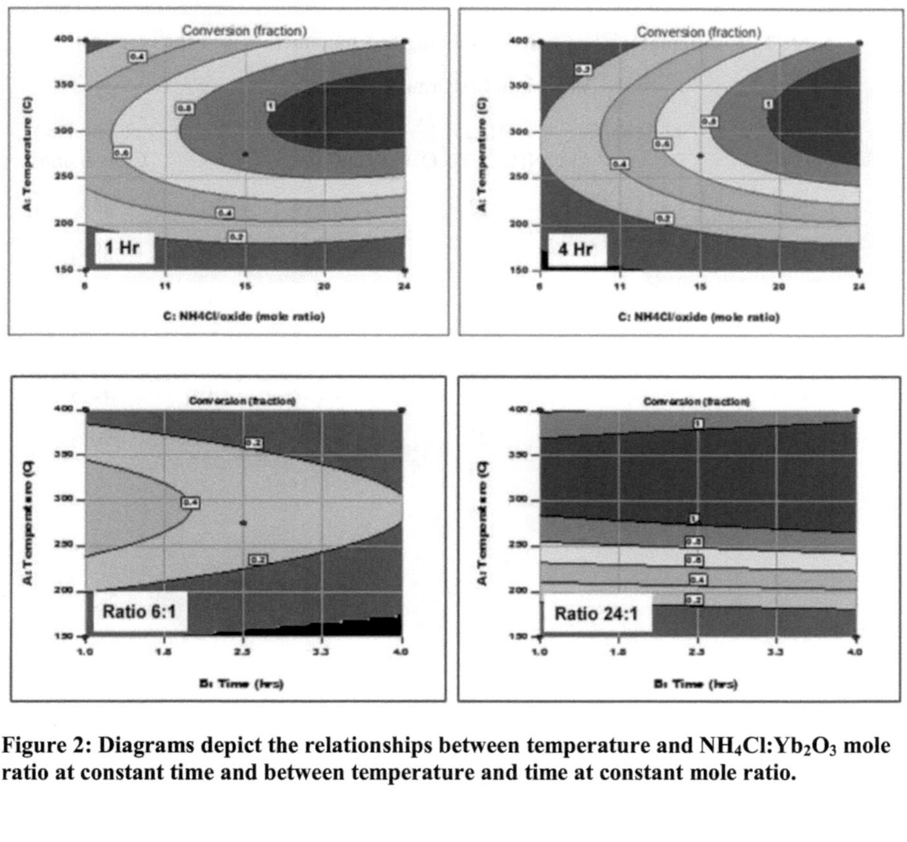

Figure 2: Diagrams depict the relationships between temperature and NH_4Cl:Yb_2O_3 mole ratio at constant time and between temperature and time at constant mole ratio.

A series of six confirmatory experiments was designed to demonstrate the validity of the Conversion Efficiency model and to illustrate its usefulness in selecting operation parameters. In this exercise, the variables were temperature (290 °C and 330°C) and the $NH_4Cl:Yb_2O_3$ mole ratio (12, 18 and 21). Time at temperature was held constant at 1 hour in each of the experiments. The roasting experiments were performed according to the same general procedure as the previous scoping and design matrix experiments.

The results of the confirmatory experiments are summarized in Table 2 and graphically depicted on the wireframe plot in Figure 3, which expresses conversion efficiency as a function of temperature and $NH_4Cl:Yb_2O_3$ mole ratio. The six dots superimposed on the contour surface correspond to the conversion efficiencies obtained in the six confirmatory experiments. The confirmatory data precisely conform to the model predictions (i.e. no outliers). Consequently, the model can be employed for predictive purposes with a high degree of confidence.

Table 2. Confirmatory test matrix showing test parameters and the response.

Ytterbium Confirmation Tests					
T °C	Time (hr)	Ratio ($NH_4Cl:Yb_2O_3$)	NH_4Cl (grams)	Yb_2O_3	Response Conversion
290	1	12	0.814	0.5	86%
290	1	18	1.222	0.5	88%
290	1	18	1.222	0.5	81%
290	1	21	1.425	0.5	85%
330	1	18	1.222	0.5	92%
330	1	21	1.425	0.5	90%

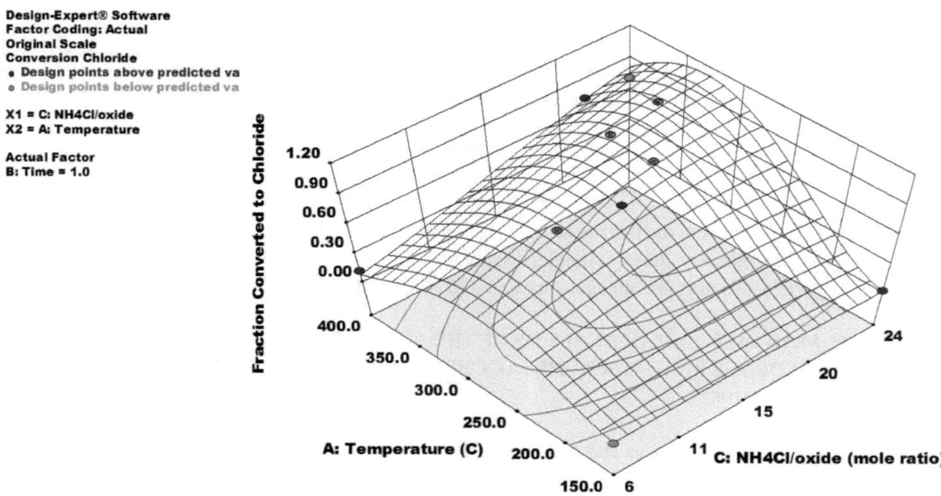

Figure 3: 3-D wireframe plot showing the response of the oxide-to-chloride Conversion Efficiency matrix. The six dots superimposed on the 3-D response surface correspond to the conversion efficiencies listed in Table 3.

Conclusions

Roasting parameters have been determined to optimize the conversion yield of Yb_2O_3 to $YbCl_3$. A time of one hour, a ratio of 21:1, and a temperature of 330 °C resulted in a conversion efficiency of approximately 90%. The confirmation tests fit the response model well, as shown in Figure 3 providing confidence in the experimental design. Refined roasting parameters for chlorination did not produce any REOCl. This was identified as a non-desirable product and the goal was to prevent and/or reduce its formation. These parameters will be applied to additional REE to determine their conversion efficiencies.

Refined parameters will be applied to other REO, to gangue constituents, to REE synthetic ore that will be created in the laboratory, and to REE ore. These parameters will be expanded and their effects will be monitored for the following REE's; Gd, Eu, Tb, Sm, Nd, and Dy. The effects of gangue minerals that naturally occur within the ore itself will be evaluated. Refinement of chloride conversion efficiency may allow the possibility of more efficient separation of REE using a simpler technique than is used today. Results of this experimentation are being applied to the other REO, gangue constituents (Al_2O_3, CaO, SiO_2, and Fe_2O_3, all reagent grade), to a synthetic ore, and finally to ore and concentrate samples from U.S. deposits; to create a viable industrial process.

Acknowledgements

Funding for this project was provided by the Army National Laboratory through the University of Alabama at Birmingham. The authors gratefully acknowledge ANL and UAB.

References

[1] U.S. Department of the Interior and U.S. Geological Survey, "Mineral Commodity Summaries 2014," 28 February 2014. [Online]. Available: http://minerals.usgs.gov/minerals/pubs/mcs/2014/mcs2014.pdf. [Accessed 23 September 2014].

[2] G. Meyer and P. Ax, "AN ANALYSIS OF THE AMMONIUM CHLORIDE ROUTE TO ANHYDROUS RARE-EARTH METAL CHLORIDES," *Materials Research Bulletin,* vol. 17, pp. 1447-1455, 1982.

[3] F. J. Pomiro, G. G. Fouga and A. E. Bohe, "Kinetic Study of Europium Oxide Chloriniation," *The Minerals, Metals & Materials Society and ASM International,* vol. 44B, pp. 1509-1519, 2013.

[4] E. A. Brocchi, R. Navarro and F. Moura, "A chemical thermodyanamics review applied to V2O5 Chlorination," *Thermochimica Acta,* vol. 559, pp. 1-16, 2013.

[5] M. V. Bosco, G. G. Fouga and A. E. Bohe, "Kinetic study of neodymium oxide chlorination with gaseous chlorine," *Thermochimica Acta,* vol. 540, pp. 98-106, 2012.

[6] J. P. Gaviria, L. G. Navarro and A. E. Bohe, "Chlorination of Lanthanum Oxide," *The Physical Journal of Chemistry,* vol. 116, pp. 2062-2070, 2012.

[7] Outotec Research, "HSC 7.0," Outotec Research, Antti Roine, 2014.

[8] D. Bell, Vapor Phase Removal of Rare Earth Oxides for the Use with Spent LWR Fuel Recycling, M.S. Thesis, Montana Tech, 2012.

[9] Stat-Ease, Inc., "Design Expert 9.0," Stat-Ease, Inc., Minneapolis, MN, 2014.

[10] M. R. Stefania Massari, "Rare earthelementsascriticalrawmaterials:Focusoninternationalmarkets," *Resource Policy ,* no. 38, pp. 36-43, 9 August 2013.

[11] J. H. a. L. Niinisto, "Thermoanalytical Study on the Reactions of Selected Rare Earth Oxides with Ammonium Halides," *Thermochimica Acta,* no. 37, pp. 155-160, 4 September 1980.

Drying, Roasting, and Calcining of Minerals
Edited by: Thomas P. Battle, Jerome P. Downey, Lawrence D. May, Boyd Davis, Neale R. Neelameggham, Sergio Sanchez-Segado, and P. Chris Pistorius
TMS (The Minerals, Metals & Materials Society), 2015

BROMINATION ROASTING OF RARE EARTH OXIDES

Bryce D. Ruffier, Daniel W. Gaede, Jerome P. Downey, Larry G. Twidwell, Jannette L. Chorney, Ryan J. Foy, Katelyn M. Lyons

Montana Tech of the University of Montana;
1300 West Park St.; Butte, MT 59701, USA

Keywords: Rare Earth, Bromination, Roasting

Abstract

The Metallurgical and Materials Engineering Department at Montana Tech is investigating various methods of extracting and refining rare earth elements from mineral ores and concentrates. As part of this research, an elevated temperature "roasting" process has been evaluated as a means of converting the rare earth elements contained in various matrices to bromides as a pretreatment step in preparation for downstream rare earth element extraction and recovery operations. Laboratory and bench-scale experiments have been performed to assess the effects of varying temperature (150 to 400° C), time (1 to 4 h), and the ammonium bromide to rare earth oxide molar ratio (6 to 24) in the roaster charge. The results show that nearly complete bromination of the rare earth oxide is achievable when the roast is performed under optimum conditions.

Introduction

China accounted for more than ninety percent of the world's rare earth production in 2012 and 2013. China recently downsized their exports of rare earths, citing domestic needs and environmental concerns. In addition, Chinese export taxes increased causing a substantial increase in rare earth metal prices [1]. Projected shortages of rare earth elements provided impetus for natural resource development and renewed interest in the development of more efficient and economical techniques of rare earth metal extraction and refining. Economic recovery of rare earth elements is challenged by the scarcity of deposits and complexity of extraction and separation operations that results from the similar physical and chemical properties of rare earth elements [2]. The European Union, U.S. Department of Energy, and General Electric Company have declared rare earth elements a high risk raw material [3].

Conversion of critical rare earth element oxides to their respective bromides could facilitate selective separation of rare earth elements. Bromination has been examined in a few recent studies. For example, Jorma Holsa and Lauri Niinis demonstrated positive results in converting rare earth oxides to rare earth oxide bromides by reaction with ammonium bromides [4]. The graph in Figure 1 was produced with HSC Chemistry 7.0 for Windows to illustrate the possible reaction products of a roasting four rare earth oxides (holmium oxide, samarium oxide, neodymium oxide, and ytterbium oxide) in the presence of ammonium bromide. The results indicate that, from the standpoint of thermodynamics, converting the oxides to their respective bromides is feasible by roasting the rare earth oxides in the presence of ammonium bromide at relatively modest temperatures.

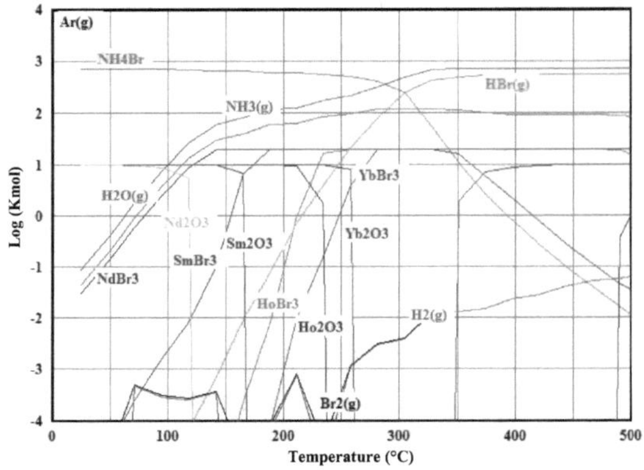

Figure 1: HSC 7.0 model shows the possible reaction products of roasting holmium oxide, samarium oxide, neodymium oxide, and ytterbium oxide with ammonium bromide.

Preliminary calculations identified a range of process conditions under which it is thermodynamically feasible to convert rare earth element oxides to their respective bromides according to Equation 1 (the symbol RE is used as a generic representation of various rare earth elements).

$$RE_2O_{3\,(s)} + 6\,NH_4Br_{(s)} \rightarrow 2\,REBr_{3\,(s)} + 6NH_{3\,(g)} + 3H_2O_{(g)} \quad [1]$$

A series of scoping experiments was performed to evaluate the metallurgical response of the same four reagent grade rare earth element oxides. The results revealed ytterbium oxide is the most recalcitrant of the four oxides. Further experiments were performed with reagent grade oxides to more precisely identify the roasting temperature, time (at temperature), and the molar ratio of bromidizing agent to ytterbium oxide that result in maximum conversion efficiency. Ammonium bromide was used exclusively as the bromidizing agent in this work. The objectives of the present study are to establish reliable process conditions and to demonstrate the effectiveness of bromination roasting with synthetic materials.

Experimental

Scoping experiments were executed in the Fisher Isotemp model 189 tube furnace depicted in Figure 1. The rare earth oxide and ammonium bromide were manually mixed in a mullite boat to provide the desired bromidizing agent to rare earth oxide molar ratio. The mullite boat was centered inside an alumina containment tube that was positioned within the tube furnace. A Type K thermocouple was positioned proximate to the sample inside the alumina containment tube with the thermocouple tip situated directly above the powder. Argon purge gas, purified by flowing through an oxygen filter, continually flowed through the containment tube at 250 mL/min throughout each experiment.

Figure 2. Scoping experiments were performed in this tube furnace and alumina containment tube. The Type K thermocouple union is visible at the left.

Each of the four rare earth oxides was evaluated at 400 °C with residence times of 1 and 2 hours at brominating agent to oxide molar ratios ($NH_4Br:RE_2O_3$) of 18:1, and again at 2 and 4 hours at a molar ratio of 24:1. Experimental products were analyzed by x-ray diffraction spectroscopy (XRD) to characterize the material that remained after the experiment. Experimental parameters and corresponding XRD results are summarized in Table 1.

Table 1: Scoping experiment conversion efficiencies

	RE Oxide to Bromide Conversion Efficiency			
Mole Ratio	$NH_4Br:Re_2O_3$ = 18:1		$NH_4Br:Re_2O_3$ = 24:1	
Time, h	1	2	2	4
Precursor:				
Nd_2O_3	100%	100%	100%	100%
Sm_2O_3	100%	100%	100%	100%
Ho_2O_3	100%	100%	100%	100%
Yb_2O_3	41%	58%	NA	100%

After XRD analyses revealed ytterbium oxide is least responsive of the four rare earth oxides evaluated in the scoping experiments, ytterbium oxide was selected for more extensive experimentation to establish conditions for maximum conversion to ytterbium bromide. Design Expert 9.0 statistical analysis software was employed to develop the response surface matrix of experiments outlined in Table 2. The variables were temperature (150 °C, 275 °C, and 400 °C), ammonium bromide to ytterbium oxide molar ratios (6:1, 15:1, 24:1), and residence times at the target temperature (1, 2.5, and 4 hours).

Table 2. Experimental matrix used to optimize roasting parameters for conversion of ytterbium oxide to ytterbium bromide.

Expt. No.	T, °C	Time, h	NH_4Br:REO Ratio	NH_4Br, g	Yb_2O_3, g
1	150	1	6	0.7456	0.5000
2	400	1	6	0.7456	0.5000
3	150	4	6	0.7456	0.5000
4	400	4	6	0.7456	0.5000
5	150	1	24	2.9825	0.5000
6	400	1	24	2.9825	0.5000
7	150	4	24	2.9825	0.5000
8	400	4	24	2.9825	0.5000
9	150	2.5	15	1.8640	0.5000
10	400	2.5	15	1.8640	0.5000
11	275	1	15	1.8640	0.5000
12	275	4	15	1.8640	0.5000
13	275	2.5	6	0.7456	0.5000
14	275	2.5	24	2.9825	0.5000
15	275	2.5	15	1.8640	0.5000
16	275	2.5	15	1.8640	0.5000
17	275	2.5	15	1.8640	0.5000
18	275	2.5	15	1.8640	0.5000
19	275	2.5	15	1.8640	0.5000
20	275	2.5	15	1.8640	0.5000

The optimization experiments were performed in an ATS Series 3210 tube furnace and in a GSL-1100X tube furnace. Photographs of the furnaces are presented in Figures 3 and 4, respectively.

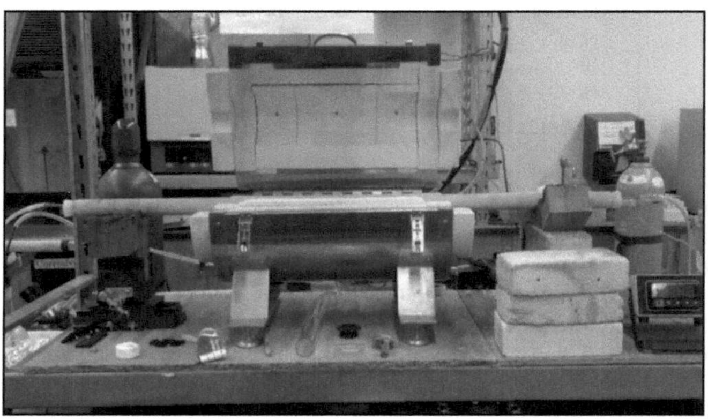

Figure 2: ATS Series 3210 tube furnace used to perform bromination roast experiments.

Figure 3: GSL-1100X tube furnace used to perform bromination roasting experiments.

In preparation for each experiment, masses of reagent grade ytterbium oxide and ammonium bromide (in the desired molar ratio) were placed in the center of a mullite boat, and the mass of the loaded boat was measured and recorded. Ammonium bromide was placed in the boat first so that, upon decomposition of the ammonium bromide, nascent hydrogen bromide gas would pass upward through the ytterbium oxide layer. The alumina containment tube was constantly purged with 250 mL/min argon purge throughout experiments and as the solid products were allowed to furnace cool. Optimization experiment products were analyzed by XRD according to standard parameters of 5° to 90° scan range, 0.5 second hold time, a scintillation detector, and 0.3 mm slit opening. The XRD patterns were evaluated using PDXL version 1.8.1.0.

Results and Discussion

Optimization experiment XRD results are summarized in Table 3, and the experiment labels correspond to those in Table 2. Table 3 presents the mass percentages of the reaction (i.e. Yb_2O_3, NH_4Br, and $YbBr_3 * 6H_2O$) identified by the XRD analysis, and the normalized percentages of the significant compounds. Conversion efficiencies were entered into Design Expert 9.0 to generate a statistically valid model. Models generated by Design Expert 9.0 are replicated in Figure 4 and Figure 5.

After the model was constructed, four points were chosen for confirmatory experiments and to verify experimental reproducibility. One additional point was selected from outside the 90% conversion zone. Again, the solid products were analyzed by XRD. The product samples were also leached in 50 mL of water. Water leaching provides an excellent means of verifying conversion efficiency because the desired product, ytterbium bromide, is highly soluble whereas ytterbium oxide and other possible intermediate reaction products are practically insoluble. The leach solutions were submitted for ICP analysis.

Table 3. Optimization experiment product compositions. The normalized values reflect the percentages of compounds that contain rare earth elements.

Expt. No.	Yb_2O_3 Wt. %	NH_4Br Wt. %	REE $Br_3 \cdot 6H_2O$ Wt. %	Normalized Yb_2O_3 Wt. %	Normalized REE $Br_3 \cdot 6H_2O$ Wt. %
1	57.3	42.7	0	100	0
2	100	0	0	100	0
3	61.2	38.8	0	100	0
4	100	0	0	100	0
5	71.7	28.3	0	100	0
6	--	--	68.8	--	--
7	20.3	79.7	0	100	0
8	2.1	0	97.9	2.1	97.9
9	30.1	69.9	0	100	0
10	3.5	30.9	65.6	5.0	94.9
11	16.3	51.9	31.9	33.8	66.2
12	6.9	34.7	58.4	10.6	89.4
13	29.9	24.3	45.8	39.5	60.5
14	5.1	38.6	56.3	8.3	91.7
15	8.2	33.2	58.6	12.3	87.7
16	7.0	29.3	63.7	9.9	90.1
17	8.3	49.7	42.0	16.5	83.5
18	5.5	43.7	50.8	9.8	90.2
19	6.8	30.3	62.9	9.8	90.2
20	7.6	33.8	58.6	11.5	88.5

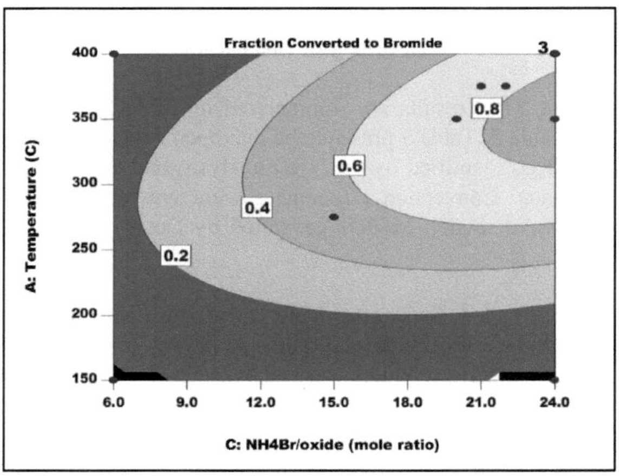

Figure 4: Design Expert 9.0 response surface contour illustrating bromination conversion efficiency as a function of temperature and NH_4Br: Yb_2O_3 molar ratio (1 hour time at temperature). Five superimposed points represented conditions selected for confirmatory experiments[6].

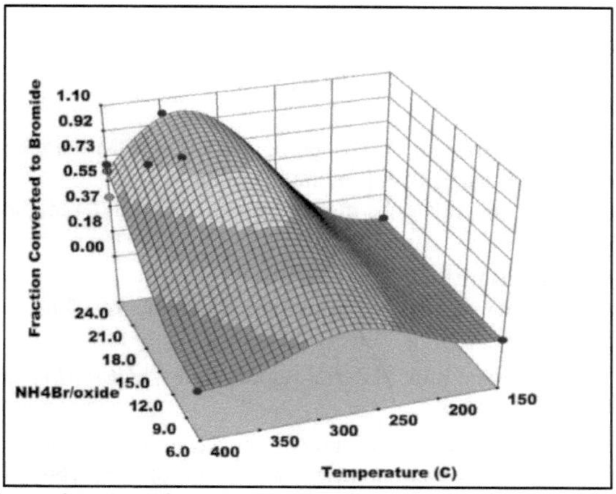

Figure 5. This 3-dimensional surface model illustrates bromination conversion efficiency as a function of temperature and NH_4Br:Yb_2O_3 molar ratio (1 hour time at temperature). The five data points superimposed in the high conversion region represent the results of the confirmatory experiments[6].

Based on the consolidated results of the statistical analysis and confirmatory experiments, there is a very high degree of confidence that roasting ytterbium oxide in the presence of ammonium bromide (at an initial NH_4Br to Yb_2O_3 molar ratio of 24:1) for 4 hours at 350 °C will result in greater than 90% conversion of ytterbium oxide to ytterbium bromide. An additional series of experiments was performed to assess the effectiveness of the stated apparatus. The procedures were identical to those employed during the optimization and confirmatory experiment series.

Conclusion

Experimental data analyzed with Design Expert 9.0 show that the bromination roasting process has promise as a method to convert rare earth oxides to their respective bromides. Of the four rare earth oxides evaluated in this study, ytterbium oxide provide least responsive to the roasting process in scoping experiments. However, the results of a subsequent series of optimization experiments reveal that greater than ninety percent conversion of reagent grade ytterbium oxide can be obtained by roasting for 4 h at 350 °C in the presence of ammonium bromide at a NH_4Br:Yb_2O_3 molar ratio of 24:1. Bromination roasting of rare earth oxides may open new opportunities to selectively separate rare earth elements. Ongoing research is focused on evaluating the metallurgical response of other rare earth oxides to the bromination roasting process, including confirmatory studies on synthetic and actual ore samples.

Acknowledgements

Funding for this project was provided by the Army National Laboratory through the University of Alabama at Birmingham. The authors gratefully acknowledge ANL and UAB.

References

[1] U.S. Department of the Interior and U.S. Geological Survey, "Mineral Commodity Summaries 2014," 28 February 2014. [Online]. Available: http://minerals.usgs.gov/minerals/pubs/mcs/2014/mcs2014.pdf. [Accessed 23 September 2014].

[2] M. R. Stefania Massari, "Rare earthelementsascriticalrawmaterials:Focusoninternationalmarkets," *Resource Policy* , no. 38, pp. 36-43, 9 August 2013.

[3] A. Y. K. a. S. Hung, "Manage Raw Material Supply Risks," *CEP Magazine,* pp. 28-35, September 2014.

[4] J. H. a. L. Niinisto, "Thermoanalytical Study on the Reactions of Selected Rare Earth Oxides with Ammonium Halides," *Thermochimica Acta,* no. 37, pp. 155-160, 4 September 1980.

[5] Outotec Research, "HSC 7.0," Outotec Research, Antti Roine, 2014.

[6] Stat-Ease, Inc., "Design Expert 9.0," Stat-Ease, Inc., Minneapolis, MN, 2014 (www.statease.com).

Drying, Roasting, and Calcining of Minerals
Edited by: Thomas P. Battle, Jerome P. Downey, Lawrence D. May, Boyd Davis, Neale R. Neelameggham, Sergio Sanchez-Segado, and P. Chris Pistorius
TMS (The Minerals, Metals & Materials Society), 2015

THE ADVANTAGES OF THERMAL ANALYSIS PRIOR TO BENCH-SCALE ROASTING

Tyler Salisbury[1], Jesse White[1]

[1]Hazen Research, Inc.; 4601 Indiana Street; Golden, Colorado 80403, USA

Keywords: Thermal Analysis, TGA, Bench-Scale Roasting, Arsenopyrite

Abstract

Research projects contain many stages of development from the conception of an idea to a full-scale commercial operation. The transition from laboratory-scale research to bench-scale roasting can represent a significant increase in complexity and cost, depending on how many variables are being evaluated. Thermogravimetric analysis (TGA) can simulate roasting conditions and produce a small sample for chemical analysis or materials characterization. A simultaneous thermogravimetric and differential scanning calorimetry (TG–DSC) can identify endothermic and exothermic reaction temperatures. Using thermal analysis, some of the process variables can be tested and eliminated such as ore and reagent blends, retention times, operating temperatures, and some atmospheres.

Introduction

As a project progresses beyond the initial laboratory-scale tests and into the bench-scale work, the complexity of the testing matrix and the cost can increase. Thermal analyses techniques have improved to better understand the factors that drive a reaction. One analysis technique that can be used as a stepping stone between laboratory- and bench-scale work is thermogravimetric (TG) and differential scanning calorimetry (DSC), both independently and simultaneously. This analysis has the ability to evaluate reagent blends, retention times, processing temperatures, and some atmospheres. As an example, this paper will explore the different processing methods for arsenopyrite (FeAsS) ores.

As the use and demand for gold continues, clean or high-grade ore sources are being consumed. Arsenopyrite presents a processing challenge because leachable arsenic is a characteristic waste and is therefore regulated under the Resource Conservation and Recovery Act (RCRA). In addition, arsenic can inhibit gold extraction by standard cyanidation leach. To improve the gold extraction by leaching, the arsenopyrite needs to be oxidized, decomposed, or removed from the sample. A literature search located a number of different roasting methods to accomplish this task. A summary of selected conditions is shown in Table I.

Table I. Summary of Arsenopyrite Roasting Conditions

Temperature, °C	Atmosphere	Additive	Notes	Reference
>800	100% of stoichiometric O_2			[1]
500–750	Air or O_2 deficient	Hydrated lime		[2]
850–900	Inert/reducing		Stage 1	[3]
650	Oxidizing		Stage 2	
450–850	Mildly oxidizing		Stage 1	[4]
	Strong oxidizing		Stage 2	
750–<850	Reducing	Sulfur 3:1		[4]
530–600	Inert/reducing		Long retention times	[5]

Using this information, a testing matrix was developed; the samples were roasted and analyzed to determine the effect on the arsenic.

Experimental Plan

The objective of the roasting plan is to remove the arsenic and oxidize the sulfide content so that it will not interfere with a standard cyanidation leach. An arsenopyrite ore was selected as a common feed for this work. A summary of the head analysis is provided in Table II.

Table II. Summary of Feed Assay

	Analysis, wt%					
	As	Fe	S^{tot}	SO_4^{2-}	S^{2-}	C^{tot}
Sample 1	0.47	13.3	9.29	0.06	9.27	7.25

Note: Sulfide content was calculated from the total sulfur and sulfate assay.

Three main gas atmospheres were selected to represent the typical conditions found in the literature search. The use of air was discussed in a number of papers as a means to volatilize the arsenic as As_2O_3 and convert the sulfide to sulfate and SO_2 [2, 4, 6]. Roasting the ore under an inert atmosphere with a long retention time was found to be an alternate means of removing arsenic [3, 5]. However, long retention times lead to increased operating costs. As a potentially more cost-effective alternative, a two-stage roasting program was selected; the first stage would use air followed by a second stage that would use pure O_2 [4].

Using these selected atmospheres, a matrix of nine roasting conditions was selected to bracket selected conditions found in the literature. A Netzsch F3 Jupiter Simultaneous Thermal Analyzer (STA) was used to conduct the evaluation. The TG method uses a 5 mL Al_2O_3 sample cup, which allows for a larger sample size as compared with the TG–DSC method. The larger sample size results in a more representative sample and the opportunity for additional analysis of the calcine. The retention time, heating rate, and sample size were fixed and the temperature and atmosphere were varied for each sample. Retention time was set at 60 min, sample heating rate was set to 10°C/min, and the sample size was set at 1,500 mg ± 200 mg for the TG samples. The testing matrix is summarized in Table III. Three experiments were added after reviewing the result; a TG analysis was added to evaluate combining an inert and air roast. One TG–DSC analysis was added to observe the endothermic and exothermic nature of the reaction for conditions that affected the arsenic.

Table III. Testing Matrix

Sample ID	Temperature, °C	Atmosphere	Retention Time, min	Method	Objective
MTG-1	500	Air	60	TG	Dead roast, volatilize As and convert S to SO_4^{2-}
MTG-2	600	Air	60	TG	Dead roast, volatilize As and convert S to SO_4^{2-}
MTG-3	750	Air	60	TG	Dead roast, volatilize As and convert S to SO_4^{2-}
MTG-4	500	Inert	60	TG	Try to decompose FeAsS to volatilize As and retain FeS
MTG-5	600	Inert	60	TG	Try to decompose FeAsS to volatilize As and retain FeS
MTG-6	750	Inert	60	TG	Try to decompose FeAsS to volatilize As and retain FeS
MTG-7	525/575	Air/O_2	30/30	TG	Two-stage roasting conditions
MTG-8	500/550	Air/O_2	30/30	TG	Two-stage roasting conditions
MTG-9	550/600	Air/O_2	30/30	TG	Two-stage roasting conditions
MTG-10	600/600	Inert/air	30/30	TG	Two-stage inert/air (remove As and convert fully to SO_4^{2-})
TG-DSC	600	Inert/air	30/30	TG–DSC	Observe the endo and exothermic response

The roasted products were analyzed to determine the arsenic, total sulfur, and sulfate (SO_4^{2-}) concentrations. Arsenic was measured using a Perkin Elmer Optima 7300 DV inductively coupled plasma–optical emission spectroscopy (ICP–OES). Total sulfur was measured using a LECO Corporation sulfur analyzer, and sulfate is measured by gravimetric precipitation using barium chloride. The sulfide content is calculated as the difference between total sulfur and sulfate concentrations.

Results and Discussion

After each experiment was completed, the calcine material was analyzed to determine the resulting chemical compositions; a summary of the results is shown in Table IV.

Table IV. Summary of Analytical Results

Sample ID	Temperature, °C	Atmosphere	Residual Mass, wt%	As	S^{tot}	SO_4^{2-}	S^{2-}
Head				0.47	9.29	0.06	9.27
MTG-1	500	Air	92.8	0.43	3.29	8.47	0.47
MTG-2	600	Air	89.5	0.43	4.19	10.1	0.82
MTG-3	750	Air	87.5	0.44	2.33	4.53	0.82
MTG-4	500	Argon	97.2	0.38	9.09	0.00	9.09
MTG-5	600	Argon	91.2	0.29	8.86	0.42	8.72
MTG-6	750	Argon	89.9	0.31	9.05	0.00	9.05
MTG-7	525/575	Air/O_2	89.3	0.51	2.67	8.12	-0.04
MTG-8	500/550	Air/O_2	89.0	0.46	2.80	8.51	-0.04
MTG-9	550/600	Air/O_2	88.7	0.46	2.78	8.47	-0.04
MTG-10	600/600	Argon/air	91.7	0.32	5.32	18.3	-0.79

Note: Sulfide content was calculated from the total sulfur and sulfate assay. Negative values represent zero sulfide content and are a result of analytical variations of the other analyses.

The analytical results show that the air roasting was sufficient to convert the sulfide to sulfate. However, there was minimal arsenic removal. Calcine material from the inert atmosphere roasting showed an improvement of the arsenic removal in the final product. Of the conditions tested, the samples averaged 30% As removal from the calcine. The largest arsenic removal was observed when roasting at 600°C, which resulted in a 37.7% removal. Similar to the air roasting, the two-stage roasting showed minimal arsenic removal and had a higher conversion of sulfide to sulfate. Upon reviewing the results from the first nine roasting experiments, an additional TG experiment (MTG-10) was outlined to increase arsenic removal while maintaining the conversion of sulfide to sulfate. This experiment was able to achieve 32.3% removal of arsenic and fully convert any remaining sulfide as sulfate.

Up to this point, the results that have been presented are similar to results that could be obtained from an evaluation in a 4 in. diameter rotary batch kiln system at Hazen using a larger sample size. By analyzing these conditions in the STA as TG samples, we were able to obtain additional information. Using the data analysis software Proteus®, a curve of the mass loss as a function of temperature can be generated. The derivatives of the mass loss can be calculated to identify the temperature at which certain reactions occur. The diagram in Figure 1 is a combined graph of the experiments in air (MTG-1 through MTG-3) with the peaks of the derivatives identified.

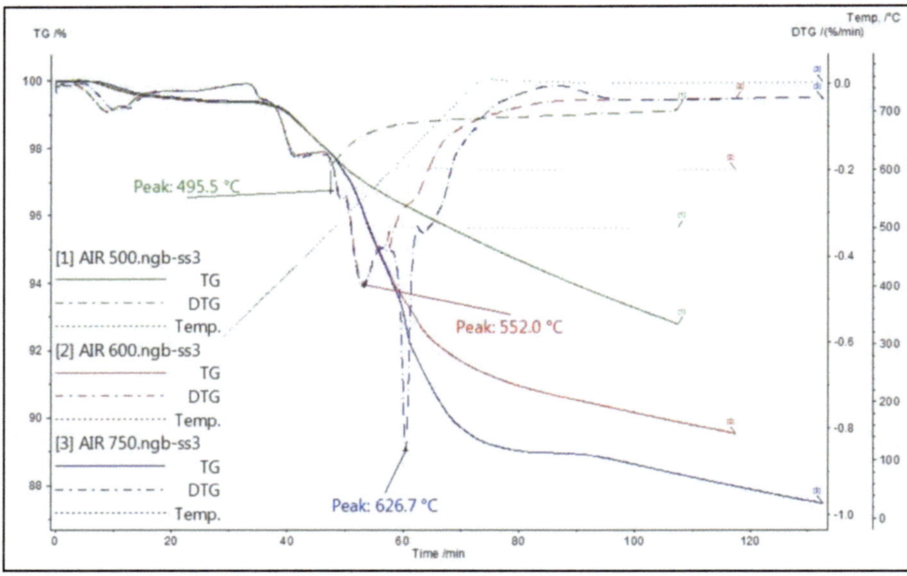

Figure 1. Thermogram of Air Experiments

Additionally, the diagram in Figure 2 is of the inert gas experiments (MTG-4 through MTG-6) with the peaks of the derivatives identified.

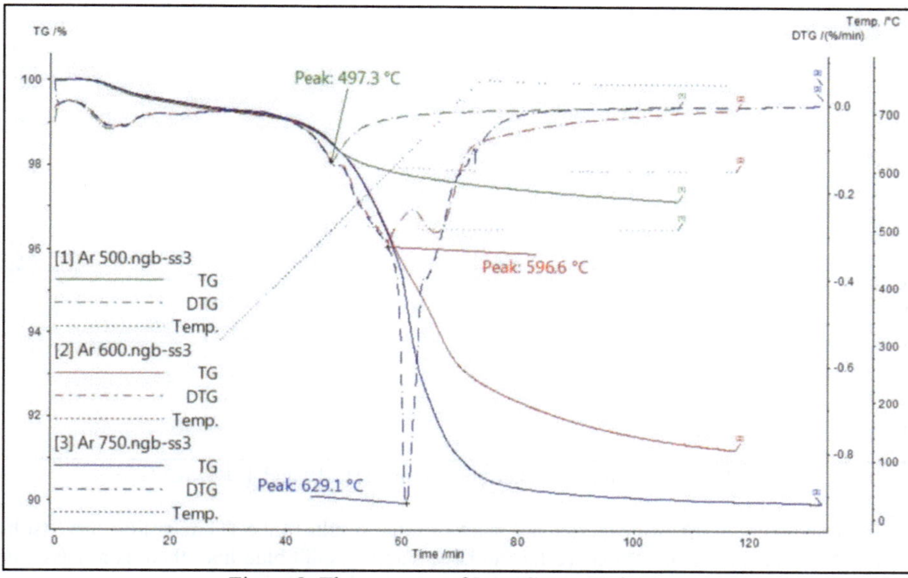

Figure 2. Thermogram of Inert Gas Experiments

Both graphs display similar points at which reactions are occurring: at 496, 552–596, and 628°C. Using these points, additional experiments can be designed to heat and hold isothermally at the temperatures of interest.

Additionally, these graphs show the trend of mass loss as the system is held isothermally. In the air experiments, there is a steady decreasing trend that can be correlated to the removal of the organic carbon content and/or sulfide as SO_2. The inert experiments showed a slow rate of mass loss during the isothermal stages; however, this scenario supports the idea of long retention times being necessary, as was found in the literature.

The additional TG experiment (MTG-10), which targeted arsenic removal (inert stage) and conversion of sulfide to sulfate (oxidizing), was run twice. The first experiment was conducted only as a TG experiment to generate material for analytical evaluation and secondly as a simultaneous TG–DSC analysis to determine the heat of reaction and mass loss. The diagram in Figure 3 is the combined thermogram of the TG and TG–DSC experiments. Each sample was run under the same condition; however, the sample size for the TG–DSC was 61 mg as a result of sample crucible size limitation (85 μL) of the TG–DSC instrument.

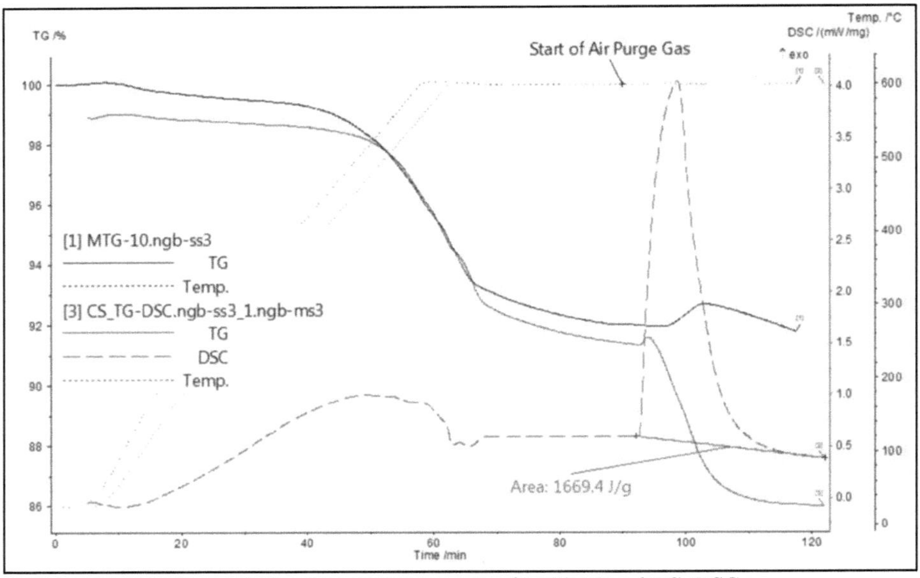

Figure 3. Combined Thermogram of MTG-10 and TG–DSC

This thermogram, of the two experiments, shows similar results in the TG mass loss curve (solid lines). Near the end of the TG curve for the large sample (solid blue line) there is a mass gain when the oxygen is turned on. This is likely due to sample oxidation and not having sufficient gas-to-solid interaction to volatilize the organic carbon content. By calibrating the instrument signal, the DSC curve (green dashed line) can be used to determine the heat requirement at each stage. The exothermic release corresponds with the introduction of air and is identified by the large DSC peak and rapid mass loss. This energy release of oxidation reaction was measured to be 1,669.4 J/g. This analysis provides useful information about the amount of heat that the feed material can provide to a reactor in a larger-scale operation.

Conclusions

As a project progresses through the development stages, the complexity will increase along with a corresponding increase in cost. Thermal analysis can provide valuable information in a cost-effective manner with minimal sample use. Some of the advantages of using TG and TG–DSC analysis are:

- Each experiment requires 1–3 g of material and is able to generate a sample for chemical analysis.
- The heating rate, retention time, gas atmosphere, and feed composition are customizable, allowing for the evaluation of a wide range of conditions.

- Using the analysis software, a graph of the mass loss as a function of temperature can be generated. These graphs, along with the derivative of the curves, can be used to identify reaction temperatures for future test work.

- During an isothermal stage, the mass loss is recorded as a function of time, which can be interpreted to determine when a reaction is complete.

- Analyzing a sample using TG–DSC provides energy data for different reactions. The exothermic or endothermic requirements can be quantified and used to help in the design of larger-scale operations.

While TG and TGA–DSC evaluations can offer valuable insights in the process, there are limitations to their use.

- The instrument is measuring milligram mass changes in a sample and must be held stationary. This approach minimizes the mixing or fluidizing of a process sample, thereby limiting the gas-to-solid interaction.

- The thermal analyzer is limited on a maximum sample size of 5 mL in the largest sample cup. For the example case of arsenopyrite, there was insufficient material to be subjected to a cyanidation leach.

Thermal analysis can provide a valuable tool for the development of a process. It can help to explore a wide range of operating conditions and reduce the number of bench-scale tests that may be required. If the example work with arsenopyrite went to a bench scale program, the results would suggest running a two stage experiment at 600°C, with a first stage under inert condition followed by a second stage with an air atmosphere, for a total of 60 min. The larger bench-scale systems are still needed to generate sufficient sample to observe the effect of mechanical action in the system and to evaluate downstream processes. However, with the minimal sample use and the information that can be obtained, thermal analysis work can offer focus of the bench-scale work to further evaluate a process.

References

1. Angel Vian, Conrado Iriarte, and Angel Romero, "Fluidized Roasting of Arsenopyrites, A Theory of the Mechanism of the Dearsenification Process," *Industrial & Engineering Chemistry Process Design and Development,* 2 (3) (1963), 214-223.

2. Ann M. Hagni, Richard D. Hagni, Patrick R. Taylor, "Process Mineralogy of Roasted Pyrite and Arsenopyrite," *JOM,* 44 (4) (1992), 36-38.

3. Thomas D. Heath and Norbert G.T. Holle, Method for Removing Arsenic from Arsenopyrite Ores. US Patent 2,867,526, filed March 14, 1957, and issued January 6, 1959.

4. Jean-Marc Lalancette, Precious Metal Recovery from Refractory Sulphidic Ores. WO 1998006878 A1, filed August 12, 1996, and issued February 19, 1998.

5. N. Chakraborti and D.C. Lynch, "Thermodynamics of Roasting Arsenopyrite," *Metallurgical Transactions B*, 14B (2) (1983), 239-251.

6. Eric J. Grimsey and Mark G. Aylmore, "Roasting of Arsenopyrite," in *Gold '90: Proceedings of the Gold '90 Symposium, Salt Lake City, Utah, February 26 to March 1, 1990* (Littleton, CO: Society for Mining, Metallurgy, and Exploration, Inc.), 397-409.

7. P.M. Swash and P. Ellis, "The Roasting of Arsenical Gold Ores: A Mineralogical Perspective" in *GOLD 100, Proceedings of the International Conference on Gold*, ed. C.E. Fivaz and R.P. King (Johannesburg, South Africa: The South African Institute of Mining and Metallurgy, 1986), 235-257.

8. M.W. von Bernewitz, "Roasting Gold-Silver Sulphide Ores and Concentrates," *Canadian Mining Journal,* 61 (1940), Part I, April, 221-224; Part I conclusion, May, 298-299; Part II: "Roasting Gold-Silver Telluride Ores and Concentrates," June, 365-370; Part III: "Roaster Operations and Plant Control," July, 438-441; Part IV: "Roaster Operations and Plant Control," August, 519-524; Part IV: "Testing and Experimental," October, 651-656.

Drying, Roasting, and Calcining of Minerals
Edited by: Thomas P. Battle, Jerome P. Downey, Lawrence D. May, Boyd Davis, Neale R. Neelameggham, Sergio Sanchez-Segado, and P. Chris Pistorius
TMS (The Minerals, Metals & Materials Society), 2015

EXTRACTION OF INDIUM FROM ZINC OXIDE FLUE DUST BY MICROWAVE SULFATION ROASTING AND WATER LEACHING

Jun Chang [1,2] Jin-hui Peng [1,2], Li-bo Zhang [1,2], Jing Chen [3*]

[1]Key Laboratory of Intensification Metallurgy in Yunnan Province, Kunming University of Science and Technology, Kunming 650093, China
[2]Faculty of Metallurgical and Energy Engineering, Kunming University of Science and Technology, Kunming 650093, China
[3]School of Chemical Science and Technology, Yunnan University, Kunming 650091, China

Keywords: Indium; Microwave sulfation roasting; Water leaching

Abstract

The leaching of indium from refractory zinc oxide flue dust(ZOFD) is attempted to utilize microwave sulfating roasting process, followed by water leaching. The ZOFD and roasted product were characterized by XRD and SEM/EDS analysis. The effects of various roasting parameters, such as roasting temperature/duration, the proportion of sulfuric acid, as well as the leaching parameters such as the leaching duration and liquid/solid ratio, were assessed. The results indicate that nearly 97% of indium could be leached under the following conditions: mass ratio of concentrated sulfuric acid to ZOFD to be 0.5, roasting temperature of 150 ℃ and duration of 1.5 hours; while the optimal leaching conditions hold at temperature of 25 ℃ for 1 hour, with 4:1 of liquid/solid ratio. The indium leaching rate by the microwave roasting increased by 8% as compared with the conventional roasting process under identical conditions.

Introduction

Indium is an important metal by virtue of its strong light permeability and conductivity which are being used in a variety of high-tech fields, especially in liquid crystal displays as indium tin oxide film[1,2]. However, Indium resources are very limited, which does not exist in nature independently. It is widely distributed in sphalerite with indium content below 0.1%.

Recently, indium-bearing secondary resource, such as copper-smelting ash[3], zinc oxide flue dust[4], spent catalysts[5], blast furnace sludge[6], indium-tin oxide (ITO) scrap[7], alloy wire scrap[8], have been used for recovering indium. In virtue of the above, zinc oxide flue dust (ZOFD), which contains 0.1%~0.3% (mass fraction) indium[4], is a potential resources for

Foundation item: National Key Basic Research Program of China(No. 2014CB643404) ; Yunnan Province Young Academic Technology Leader Reserve Talents (No. 2012HB008).
Corresponding author: Jing Chen; Tel.: +86 871 65138997; E-mail address: JingChen_ynu@163.com

extracting indium. However, there are no indium independent mineral phases in ZOFD, and indium in the form of isomorphism is enwrapped by zinc ferrite ($ZnFe_2O_4$) or other solid forms. Only 60–70% indium can be extracted by conventional leaching process.

Microwave roasting in metallurgy has been a subject of interest during these years [9,10]. It is being increasingly applied to improve the yield of extracted metal and reduce processing duration, especially with the increasing demand for more environmental friendly processes. It is proved beyond doubt that recovery of metals from industrial residues can be increased significantly with the application of microwave energy compared to conventional heating technologies[11-13], which can be attributed to an effective mineral breakage and liberation[14]. Considering the effectiveness of microwave pretreatment, the present work attempts to investigate a novel hydrometallurgical process to improve the extraction rate of indium from ZOFD through application of microwave roasting, followed by leaching with water at room temperature. A proposed process flow sheet is illustrated in Figure 1.

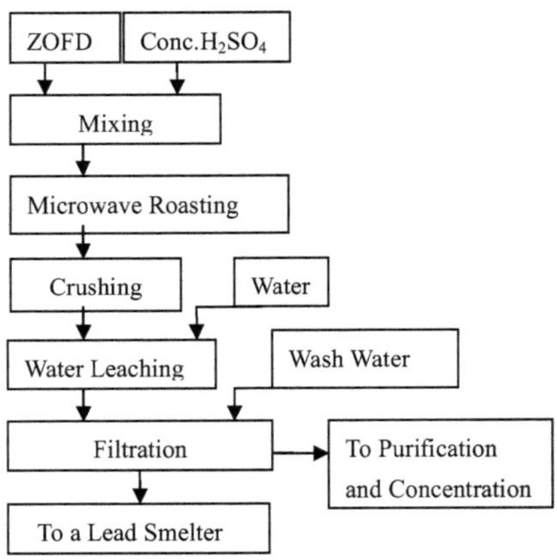

Figure 1. Proposed flow sheet for indium extraction from ZOFD

Experimental

Materials

The ZOFD samples were received from a zinc production plant in Mengzi, Yunnan, China, by roasting the hydrometallurgical zinc residue in the rotary kiln at high temperature. The samples were characterized using X-Ray Diffractometry (XRD, SmartLab, Rigaku) and Scanning Electron Microscopy/Energy Dispersive Spectrometer (SEM/DES, XL30ESEM-TMP, Philips and EDAX), the results are shown in Figure 7a and Figure 2 respectively. As Figure 7a indicate, the composition of ZOFD is very complex, which includes zincite (ZnO), lead oxide (PbO), franklinite ($ZnFe_2O_4$), magnesium silicate (Mg_2SiO_4), and so on. Different morphology and aggregation were observed in Figure 2. The SEM is supplemented by EDS analysis, which demonstrated the large amount of Zn, Pb and Fe in the ZOFD.

The chemical composition is presented in Table I. All the reagents used in the present work were of high laboratory grade, while water utilized was double distilled.

Table I. Main chemical composition of zinc oxide flue dust (mass fraction, %)

Components	Zn	Pb	Fe	S	Ca	As	Sn	In
% wt	34.22	17.70	11.78	2.88	2.03	1.51	1.40	0.15

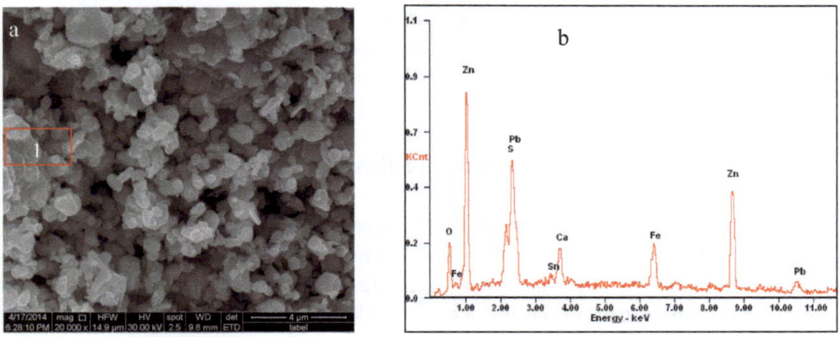

Figure 2. SEM image (a) of zinc oxide flue dust and EDS (b) of region 1

<u>Microwave Reactor</u>

The roasting was carried out in self-made microwave heating equipment described in Figure 3 which has a temperature control system along with vacuum pump to drive away the gases formed during roasting, the working power is 0.5 kilowatt.

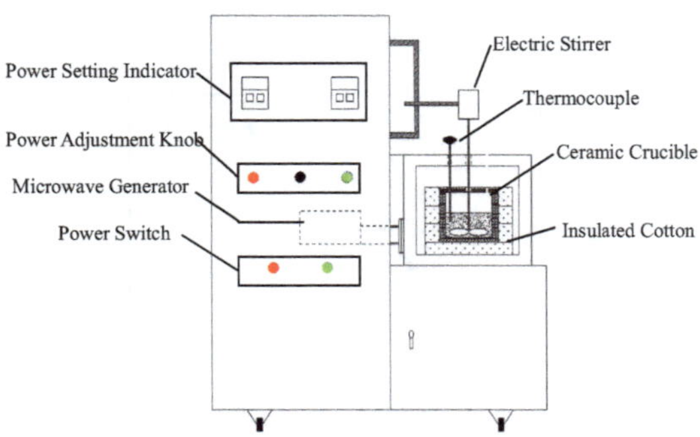

Figure 3. Sketch map of the microwave roasting device

Experimental procedure

ZOFD, 100 g, was added to concentrated sulphuric acid in a ceramic crucible and placed in the microwave heating device for roasting at a certain temperature. After cooling down the sinter was weighed, ground, and immersed in a sealed glass bottle for leaching with water at room temperature, the solid-liquid mixture was magnetically stirred at a rotational speed of 200 ~ 250 rpm. After solid/liquid separation and the washing of leach residue, both the solids and the filtrate were retained and the content of the indium was determined by atomic absorption spectrometry (Z2000 HITACHI). The reactions of this process are as follows.

$$In_2O_3+3H_2SO_4 \rightarrow In_2(SO_4)_3+3H_2O \quad (1)$$

$$2ZnFe_2O_4+8H_2SO_4 \rightarrow 2Fe_2(SO_4)_3 + 2ZnSO_4 + 8H_2O \quad (2)$$

$$ZnO+H_2SO_4 \rightarrow ZnSO_4+H_2O \quad (3)$$

$$PbO+H_2SO_4 \rightarrow PbSO_4+H_2O \quad (4)$$

Results and discussion

The experiments were conducted by varying the roasting temperature, duration of roasting and sulfuric acid dosage. The roasted products were leached with water to estimate the leaching rate of indium. The leaching was conducted under the different solid-liquid ratio and leaching duration. The experimental results were detailed as following.

Effect of sulfuric acid dosage on roasting

The effect of initial sulfuric acid dosage on the roasting was assessed by varying the ratio of acid to the ZOFD sample in the range of 0.2 to 0.6, while keeping the roasting temperature and duration to be at 150 ℃ and 2 hours respectively. The effect of sulfuric acid dosage on the indium leaching rate is presented in Figure 4. The results indicate, the leaching rate of indium increased along with the mass ratio of sulfuric acid to ZOFD reaching the maximum leaching rate of about 97% at a ratio of 0.5 . Any further increase in the acid to solid ratio was not found to increase the leaching rate and reduce the producing cost.

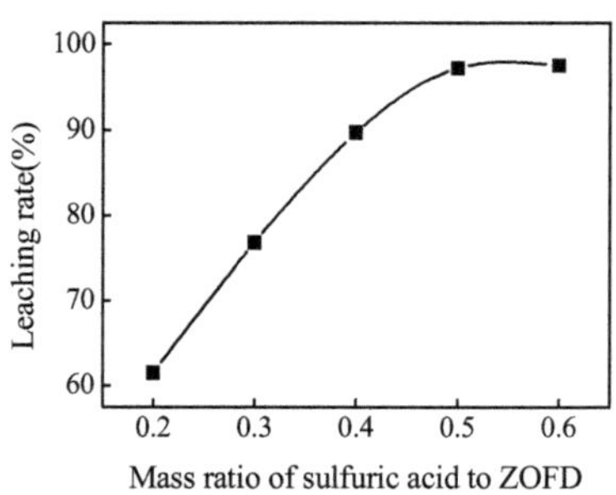

Figure 4. Effect of the dosage of sulfuric acid on indium leaching rate

Effects of temperature and duration of roasting

Both roasting temperature and duration have been reported to have significant effect on the leaching rate[15,16]. Figure 5a reports the effect of roasting temperature on the leaching rate, with the maximum being around 97%. The samples were held at constant temperature for 2 hours upon reaching the set temperature. As can be observed from the figure that the temperature is significant below 150 ℃. It can be stated that the concentrated H_2SO_4 accompanied microwave irradiation is more effective for sulphation up to 150 ℃, and the faster formation of zinc sulfate from the franklinite matrix leading to a higher leaching rate of indium. However, at temperatures beyond 150 ℃, no increase in the leaching rate could be observed, which could be attributed to H_2SO_4 partially decomposed, producing SO_3 or SO_2, resulting in reduction the degree of sulphation, thus the recovery of indium decreases.

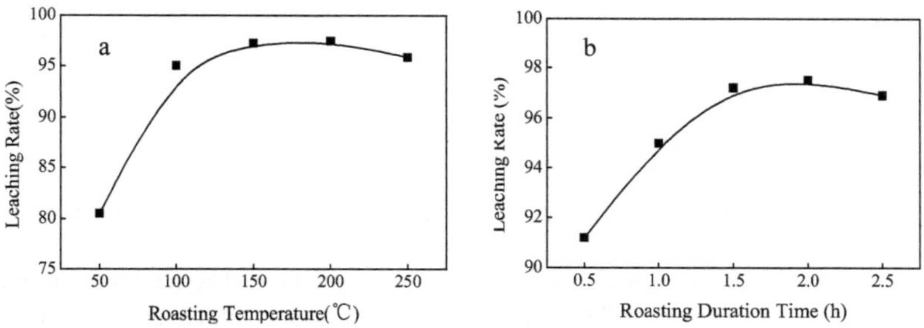

Figure 5. Effect of roasting temperature (a) and duration time (b) on indium leaching rate

The effect of the duration of roasting at the temperature of 150℃ is shown in Figure 5b. It is observed that the leaching rate of indium increases with roasting time up to 1.5 hours. There is little difference between duration time of 1.5 hours and 2 hours, and the leaching rate of indium was near 97%. After 2 hours, the slightly decreases in the leaching rate may be attributed to the promotion of secondary reactions and the formation of some insoluble compounds covered on the surface of soluble salts at prolonged duration[17], which is expressed below:

$$ZnO + ZnSO_4 \rightarrow ZnO \cdot ZnSO_4 \qquad (5)$$

where diffluent zinc sulfate was transformed into insoluble complexes. Therefore, the optimum roasting reaction duration is 1.5 h.

Effects of liquid/solid ratio and leaching duration

The roasting is followed by leaching to estimate the leaching rate of the indium. The effect of liquid-solid ratio and leaching duration are shown in Figure 6. Several experiments were performed at the optimal roasting conditions (mass ratio of sulfuric acid to ZOFD at 0.5, microwave roasting temperature at 150℃ for 1.5 hours, leaching temperature at 25 ℃).

Figure 6a shows that the leaching of indium increases significantly with the increase of liquid/solid ratio while it reaches an asymptote at higher solid-liquid ratio. At low liquid-solid ratio the mixture was visibly viscous and contributed to significant mass transfer resistance, eventually resulting in low leaching rate. However, with the increase in liquid-solid ratio beyond 3, the proportion of liquid was visibly found to be in excess resulting in better mixing and an overall increase in the leaching rate. Any further increase in the liquid-solid ratio beyond 4 was not found to contribute to increase in the leaching rate, and high liquid/solid ratio can result in large liquid circulation.

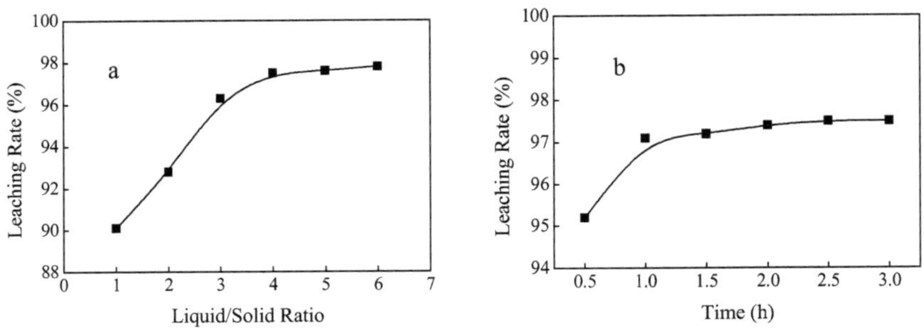

Figure 6. Effect of liquid/solid ratio(a) and leaching time(b) on the indium leaching rate

Figure 6b shows the effect of treatment duration on indium leaching. The leaching rate was observed to reach the maximum of 97% within 1 hour, hence any further increases in the leaching duration was not contributing to further improve in the leaching rate. This is consistent with high solubility of indium sulfate in the water. Thereafter, indium dissolution hardly increases after the rapid dissolution. In this study, indium leaching of 1hour is recommended undoubtedly.

Phase transformation in sulfate roasting

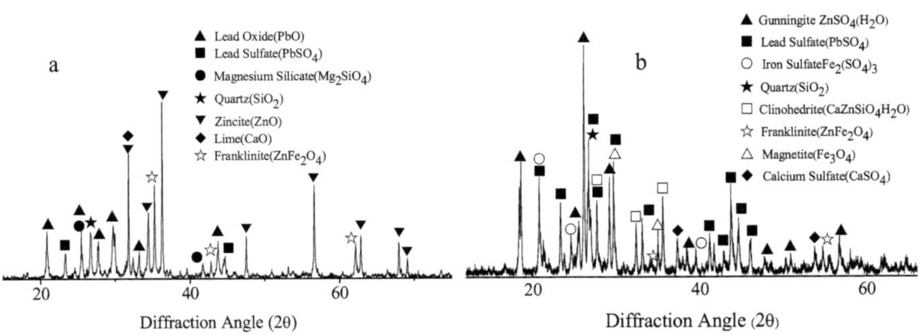

Figure 7. XRD patterns of zinc oxide flue dust (a) and calcine (b)

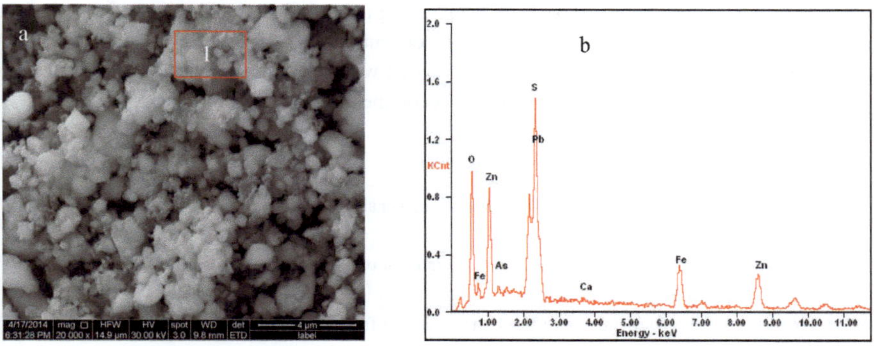

Figure 8. SEM image(a) of ZOFD calcine and EDS (b) of region 1

Figure 7b is the X-ray diffraction patterns of the calcined resultant obtained under optimal roasting conditions. It can be seen that the phases in the calcined resultant changed obviously as compared to the non-roasted ZOFD. This is attributed to that the indium-bearing mineral is an excellent microwave absorber. Rapidly heating of ore minerals in a microwave transparent matrix generated thermal stress of sufficient magnitude to create microcracks along mineral boundaries. So that the mineral was converted into sulfate after reaction. It was found from the SEM image in Figure 8a that the inclusions in ZOFD became loose and increased contact area between reactants, thus the indium leaching rate was greatly improved. The Figure 8a is supplemented by EDS analysis, which is shown in Figure 8b. Compared with Figure 2b, the elements S and O were markedly increased because of the addition of sulfuric acid, the other elements content remain unchanged.

Comparison of two pretreatment technologies

In order to study the difference between conventional roasting and microwave roasting, the experiments were carried out utilizing conventional roasting under the identical experimental conditions. The results show that the leaching rate with conventional roasting was 89.2%, while one with microwave roasting was 97.2%. This indicated that microwave pretreatment could play an important role in the improvement of indium liberation.

Conclusions

1) An efficient method of indium extraction from ZOFD is proposed in the present work. During sulfating roasting, indium recovery is affected by sulfuric acid addition amount, roasting temperature, and time. The leaching rate increases with the increase of amount of sulfuric acid and the reaction time. During the water leaching, reaction duration and liquid/solid ratio have no significant effect on indium extraction.

2) The highest indium recovery occurs at the mass ratio of acid to the ZOFD sample of 0.5, duration time of 1.5 hours, and temperature of 150 ℃ during roasting; and at room temperature, a liquid/solid ratio of 4 and a duration time of 1 hour during water leaching.

3) The microwave roasting could break the containing indium phase inclusion and increased the indium leaching rate by 8% compared with the conventional roasting at the same conditions. A significant increase in the leaching rate as compared with the conventional process render this process to be economical cost effective, and be recommended for further design and scale up.

References

1. C.X. Li et al., "Kinetics of indium dissolution from sphalerite concentrate in pressure acid leaching," *Hydrometallurgy*, 105 (1-2) (2010), 172-175.
2. S. Virolainen et al., "Recovery of indium from indium tin oxide by solvent extraction," *Hydrometallurgy*, 107 (1-2) (2011), 56-61.
3. G.C. Zhu et al., "Recovering indium with sulfating roasting from copper-smelting-ash," *Rare Metals*, 26 (5) (2007), 488-491.
4. X.H. Li et al., "Indium recovery from zinc oxide flue dust by oxidative pressure leaching," *Transaction of Nonferrous Metals Society of China*, 20 (S1) (2010), 141-145.
5. R.S. Marinho et al., "Recovery of platinum,tin and indium from spent catalysts in chloride medium using strong basic anion exchange resins," *Journal of Hazardous Materials*, 192 (3) (2011), 1155-1160.
6. X.M. Shen et al., "Ultrasonic-Assisted Acid Leaching of Indium from Blast Furnace Sludge," *Metallurgical and Materials Transactions B*, 44 (6) (2013), 1324-1328.
7. Y.H. Li et al., "Recovery of indium from used indium–tin oxide (ITO) targets,"*Hydrometallurgy*, 105 (3-4) (2011), 207-212.
8. M.A. Barakat, "Recovery of lead, tin and indium from alloy wire scrap," *Hydrometallurgy*, 49 (1-2) (1998), 63-73.
9. M. Al-Harahsheh and S.W. Kingman, "Microwave-assisted leaching—a review," *Hydrometallurgy*, 73 (3-4) (2004), 189-203.
10. B. Nanthakumar, C.A. Pickles and S. Kelebek, "Microwave pretreatment of a double refractory gold ore," *Minerals Engineering*, 20 (11) (2007), 1109-1119.
11. H.K. Krishnan, D.B. Mohanty and K.D. Sharma, "The effect of microwave irradiations on the leaching of zinc from bulk sulphide concentrates produced from Rampura–Agucha tailings," *Hydrometallurgy*, 89 (3-4) (2007), 332-336.
12. R.K. Amankwah and C.A. Pickles, "Microwave roasting of a carbonaceous sulphidic gold concentrate," *Minerals Engineering* , 22 (13) (2009), 1095-1101.
13. M. Al-Harahsheh and S.W. Kingman, "The influence of microwaves on the leaching of sphalerite in ferric chloride," *Chemical Engineering and Processing: Process Intensification*, 46 (10) (2007), 883-888.
14. Z.B. Zhang et al., "Effects of microwave pretreatment on zinc extraction from spent catalyst saturated with zinc acetate," *Transaction of Nonferrous Metals Society of China*, 20 (S1) (2010),182-186.
15. Q.X. Yan et al., "Extraction of lithium from lepidolite using chlorination roasting-water leaching process," *Transaction of Nonferrous Metals Society of China*, 22 (7) (2012), 1753-1759.
16. Y.L. Zhang, X.J. Yu and X.B. Li, "Zinc recovery from franklinite by sulphation roasting," *Hydrometallurgy*, 109 (3-4) (2011), 211-214.
17. M.D. Turan, H.S. Altundoğan and F. Tümen, "Recovery of zinc and lead from zinc plant residue," *Hydrometallurgy*, 75 (1-4) (2004),169-176.

Drying, Roasting, and Calcining of Minerals
Edited by: Thomas P. Battle, Jerome P. Downey, Lawrence D. May, Boyd Davis, Neale R. Neelameggham, Sergio Sanchez-Segado, and P. Chris Pistorius
TMS (The Minerals, Metals & Materials Society), 2015

BEHAVIOR OF ARSENIC, ANTIMONY AND BISMUTH AT ROASTING TEMPERATURES

R. Padilla, M. C. Ruiz

Department of Metallurgical Engineering, University of Concepcion
Edmundo Larenas 285, Concepcion, 4070371, Chile

Keywords: Arsenic, Antimony, oxidation, enargite, volatilization

Abstract

Arsenic, antimony and bismuth are commonly found in copper concentrates usually as the minerals enargite, stibnite and bismuthinite, respectively. Concentrates having high levels of arsenic and antimony must be pretreated before smelting. Roasting is the preferred method for removing these toxic impurities from concentrates; However, the removal of As, Sb and Bi in oxidizing roasting by volatilization may not be complete due to the slow reaction rates at 500-700 °C. Oxidizing/volatilization data were obtained from pure mineral mixtures and from arsenic rich copper concentrate. The reaction mechanism for the volatilization of arsenic and antimony were discussed. Near 100% As and over 95% Sb volatilized in the oxidation of a mixture of pure minerals at 700 °C, 1% oxygen, while over 95% As, and 35% Sb volatilized at 600 °C in 1% oxygen atmosphere when using a copper concentrate containing 3.51% As, 0.14% Sb, and 0.01% Bi. Bismuth did not volatilize.

Introduction

Arsenic, antimony, and bismuth are major deleterious impurities in copper sulfide concentrates, where these elements occur mainly as sulfide minerals. Common minerals of these elements are enargite (Cu_3AsS_4), tennantite ($Cu_{12}As_4S_{13}$), tetrahedrite ($Cu_{12}Sb_4S_{13}$), stibnite (Sb_2S_3), and bismuthinite (Bi_2S_3). In the smelting of copper concentrates containing these minerals each of the deleterious elements is distributed in the slag, mate and gas phases [1] depending on the temperature and partial pressure of oxygen prevailing in the smelting process. When the copper concentrates contain high amounts of these impurities, especially arsenic, the concentrates must be pretreated to eliminate them before smelting. The common pretreatment process is roasting where the unwanted elements should be removed through the gas phase. However, in industrial practice the removal of these elements is far from complete because at the low roasting temperatures (500- 700 °C), kinetics is a preponderant factor over thermodynamics and thus the volatilization rate in each case will determine whether the elimination through the formation of volatile species is feasible or not.

The vaporization kinetics of arsenic from enargite in oxidizing atmosphere has been studied recently by Padilla et al. [2], who proposed a reaction mechanism which includes sequential reactions for enargite during heating in N_2-O_2 atmospheres. The reactions are as follows:

First stage: $$4Cu_3AsS_4 + 13\ O_2(g) = 6Cu_2S(s) + As_4O_6(g) + 10SO_2(g) \quad (1)$$

Second stage: $$6Cu_2S\ (s) + 9O_2(g) = 6Cu_2O + 6SO_2(g) \quad (2)$$

Third stage: $$6Cu_2O(s) + 3O_2(g) = 12CuO(s) \quad (3)$$

The overall reaction is the sum of reactions (1), (2), and (3):

$$4Cu_3AsS_4(s) + 25O_2(g) = 12\ CuO(s) + As_4O_6(g) + 16SO_2(g) \quad (4)$$

According to this mechanism, arsenic from enargite is removed in the first stage reaction as $As_4O_6(g)$ species. These investigators showed that the rate of elimination of arsenic was strongly and positively influenced by the partial pressure of oxygen in the atmosphere.

Regarding stibnite (Sb_2S_3) in oxidizing atmospheres, Zivkovic et al. [3], and Padilla et al. [4] studied the mechanism of the oxidation/volatilization. The latter investigators proposed the following mechanism for the behavior of stibnite in oxidizing atmospheres for the temperature range of 700 -1150 °C:

$$Sb_2S_3(s) + 4.5O_2(g) = Sb_2O_3(s,l) + 3SO_2(g) \quad (5)$$

$$Sb_2O_3(s,l) = Sb_2O_3(g) \quad (6)$$

$$Sb_2O_3(s,l) + 0.5O_2(g) = 2SbO_2(s) \quad (7)$$

These investigators reported also that at temperatures below 700 °C the volatilization of antimony was negligible; therefore, reaction (5) was probably the only reaction to occur at low temperature. Above 700 °C, the antimony starts to volatilize; however, the fraction of antimony volatilized at a given condition depends on the relative rates of reactions (6) and (7). Therefore, the use of high oxygen concentrations in the gas phase may be counterproductive because SbO_2 is a non-volatile solid at these temperatures.

In the case of the behavior of bismuthinite in oxygen containing atmospheres, Johnson et al. [5] reported that Bi_2S_3 was stable up to 400 °C and it oxidized above this temperature. Padilla et al. [6] in their study on the kinetics of oxidation of Bi_2S_3 with N_2-O_2 mixtures proposed the following decomposition and oxidation consecutive reactions for the temperature range 400 - 1000°C.

$$Bi_2S_3(s,l) + O_2(g) = 2Bi(l) + 3SO_2(g) \quad (8)$$

$$2Bi(l) + 1.5O_2(g) = Bi_2O_3(s,l). \quad (9)$$

According to these authors bismuth would not volatilize from bismuthinite at roasting temperatures of 400-700 °C.

The objective of the present paper was to compare the behavior of the oxidation/volatilization of arsenic, antimony and bismuth from pure and mixed minerals and from an industrial copper concentrate containing arsenic and antimony at roasting temperatures in N_2-O_2 atmospheres.

Experimental Work

Materials and Procedure

The study on the volatilization of arsenic, antimony and bismuth at 500 -700 °C was carried out in two experimental setups: a vertical furnace TGA apparatus and a horizontal tube furnace setup. Standard experimental procedures were followed when using both setups [2, 3], where 50 mg and 2 g samples were used, respectively. All experiments were carried out under a constant flow of nitrogen-oxygen mixtures. The materials used were: i) high grade enargite mineral sample, pure Sb_2S_3, and pure Bi_2S_3, ii) a mixture of a chalcocite concentrate with the pure minerals (MO), and iii) an industrial copper concentrate rich in enargite (MH) from CODELCO, Chile. The chemical analysis of the enargite sample is shown in Table I. Tables II and III show the mineral composition and chemical analysis of the mixture of pure minerals (MO), respectively, while Table IV shows the chemical analysis of the copper concentrate (MH). The mineralogical analysis of the samples were also done using X-Ray Diffraction (XRD) Spectroscopy, and the XRD spectrums of the enargite sample and the industrial copper concentrates are shown in Figures 1 and 2, respectively.

Table I. Chemical Composition of the Enargite Mineral

Size Fraction	%Sb	%As
200/270 #	1.78	16.3

Table II. Mineral Composition of the Sample MO

Enargite, %	Stibnite, %	Bismuthinite, %	Chalcocite, %
15.76	1.39	1.23	81.60

Table III. Chemical Composition of the Sample MO

As, %	Sb, %	Bi, %	Cu, %
2.54	1.28	1.00	72.2

Table IV. Chemical Composition of the Copper Concentrate MH

%Cu	%Fe	%As	%Sb	%Bi	%Zn	% Pb
27.78	13.58	3.51	0.14	0.01	0.70	0.08

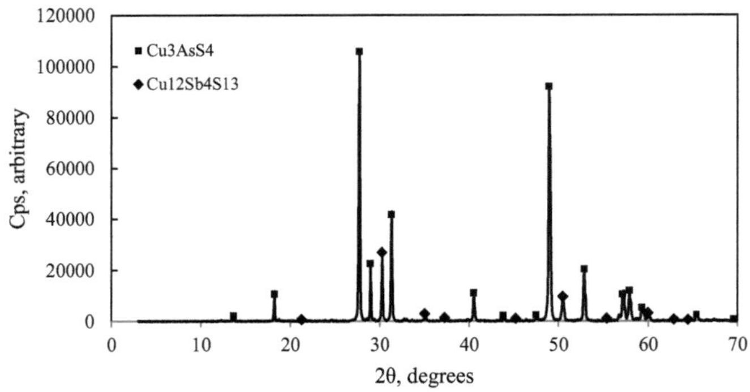

Figure 1. XRD spectrum of the enargite mineral sample.

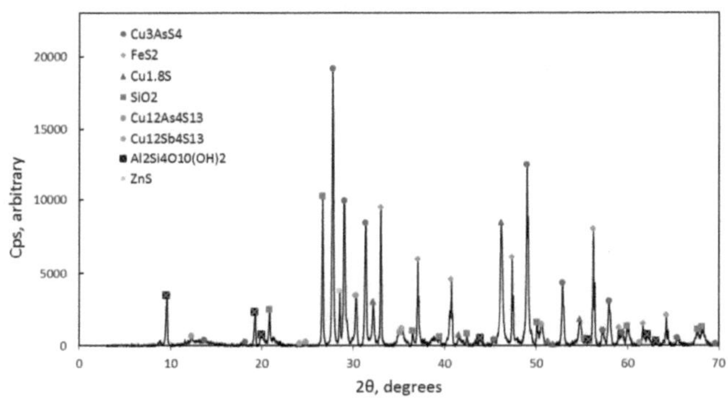

Figure 2. XRD spectrum of the copper concentrate with high content of arsenic.

Results

Results of TGA Experiments

Figures 3 and 4 show the results obtained for the oxidation of pure minerals, i.e., enargite mineral, pure stibnite and pure bismuthinite reacted at the same conditions. In these figures the data are shown as fraction of weight loss as a function of reaction time. The fraction of weight loss was defined as $(W_o-W_t)/W_o$, where W_o is the initial weight and W_t is the weight at time t.

The effect of the partial pressure of oxygen on the rate of weight loss for the enargite and stibnite samples heated at 500 °C are shown in Figure 3 for 5% O_2 and 21% O_2 in the gas phase. In Figure 4, we can see a comparison of the behavior of enargite, stibnite, and bismuthinite at 700 °C in a gas phase with 21% O_2. From the results presented in both figures, it becomes clear that the weight loss curve for enargite is composed of three consecutive reaction stages corresponding to reactions (1) to (3), which are unmistakably defined by the sudden change in the slopes of the curve. In the first stage, the fraction of weight loss of enargite increases quickly up to about 0.39, which agrees well with the theoretical weight loss for complete removal of arsenic from enargite according to the reaction (1) [2]:

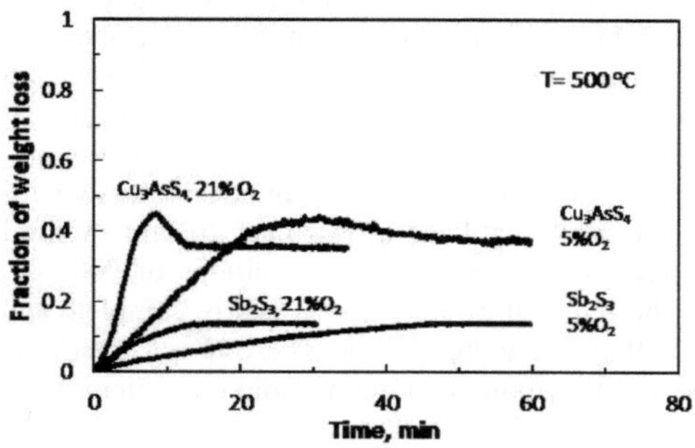

Figure 3. Oxidation of enargite and stibnite pure samples at 500 °C .

The second and third stages correspond to the oxidation of the copper sulfide product of reaction (1). It can also be observed that an increment in the oxygen partial pressure of oxygen from 5 to 21% increases significantly the rate of arsenic volatilization at 600 °C.

In the case of stibnite, it can be observed that the rate of weight loss is slower as compared to the enargite sample, and at 500 °C it increases only up to a maximum value of about 0.14 for both oxygen partial pressure of 5 and 21 %, which agrees well with the theoretical weight loss for the oxidation of Sb_2S_3 to valentinite (Sb_2O_3) according to reaction (5). Thus during roasting at 500 °C, stibnite will only oxidize to valentinite and will not volatilize appreciably. At 700 °C the maximum weight loss of stibnite reached about 0.18 indicating a small amount of volatilization.

Regarding the bismuthinite, as seen in Figure 4, in 21% oxygen atmosphere at 700 °C, the maximum weight loss is about 0.18. This value also agrees very well with the theoretical weight loss of the bismuthinite's sulfur according to the decomposition reaction (8). Thus, in roasting at 700 °C, bismuth will remain in the calcine as liquid bismuth since appreciable volatilization of liquid bismuth occurs only at temperatures higher than 900 °C [6].

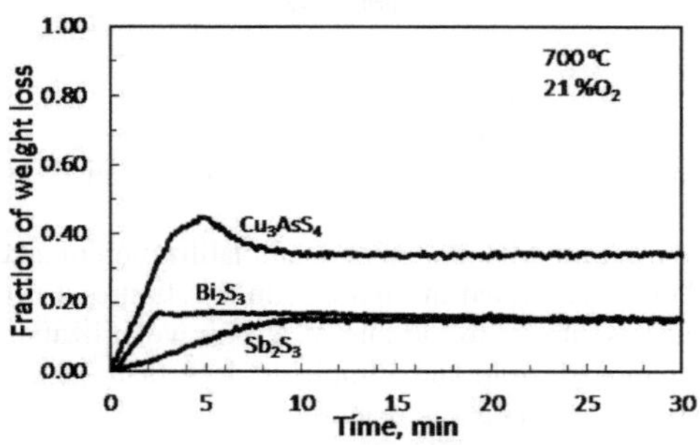

Figure 4. Oxidation of enargite, stibnite and bismuthinite at 700°C in 21 % oxygen.

Results of Batch Oxidation Experiments

The oxidation/volatilization of the prepared mixture of pure minerals MO (mixture of enargite, stibnite, bismuthinite and chalcocite), and the industrial copper concentrate MH were studied by using the horizontal tube reactor setup.

The volatilization behaviors of antimony and arsenic from the MO sample at 600 °C in 5% O_2 atmosphere are shown in Figure 5. In this figure, the percent of arsenic and antimony volatilized are shown as a function of time. Bismuth is not shown because under these low temperatures and partial pressures of oxygen it did not volatilize from the MO sample. However, it is interesting to note in this figure that a substantial volatilization of both arsenic and antimony occurs from the mixture sample MO. As seen in the figure, more than 95 % of arsenic was volatilized in less than 15 min and more than 85 % of Sb was volatilized in less than 20 min. A similar volatilization behavior of arsenic and antimony was observed at the conditions of 700 °C and 1 % O_2, as shown in Figure 6. At 700 °C, the removal of arsenic was nearly complete and the removal of antimony increased to more than 95 % in about 20 min. These results clearly are very different from the results obtained for the volatilization of antimony from pure stibnite.

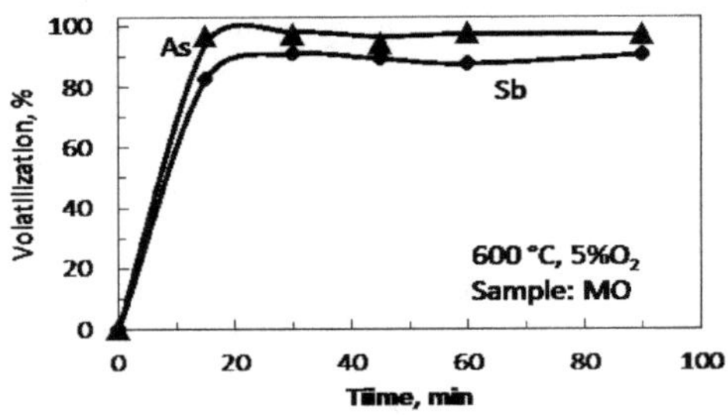

Figure 5. As and Sb volatilization from sample MO at 600 °C and 5% oxygen.

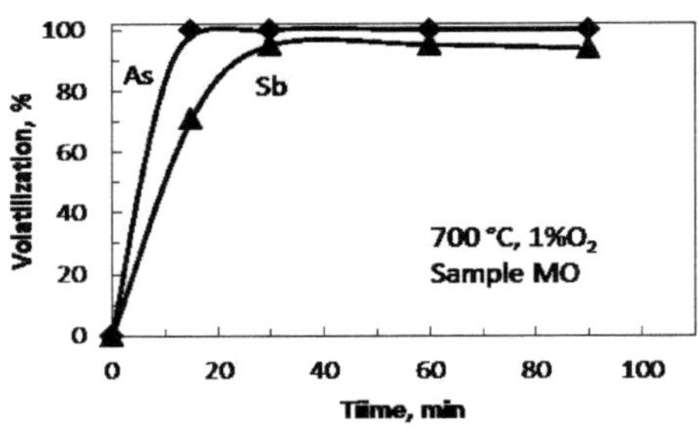

Figure 6. As and Sb volatilization from sample MO at 700 °C and 1% oxygen.

In the case of the industrial concentrate MH, the oxidation/volatilization of arsenic and antimony at 700 C and 1% O_2 atmosphere is presented in Figure 7. In this figure, it can be observed that the rate of arsenic volatilization is similar to the rate of arsenic volatilization from the sample MO. About 98 % As was removed from the sample in less than 10 min. The antimony volatilization from MH was about 45% in 45 min reaction time.

Figure 8 compares the data for the volatilization of antimony from the prepared sample MO to the volatilization from the copper concentrate MH. These data were obtained at 600 with 1% oxygen. Similar data were obtained at 700 °C and 5% oxygen. In Figure 8 we can observe a pronounced difference in antimony volatilization from MO and MH samples. The enhancement of the antimony volatilization particularly in the MO sample can only be attributed to interactions between the gaseous oxides formed in the oxidation process.

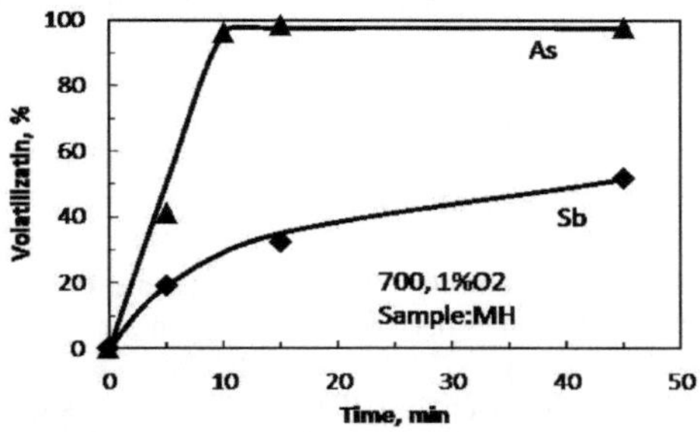

Figure 7. Arsenic and antimony volatilization from the copper concentrate MH at 600 °C

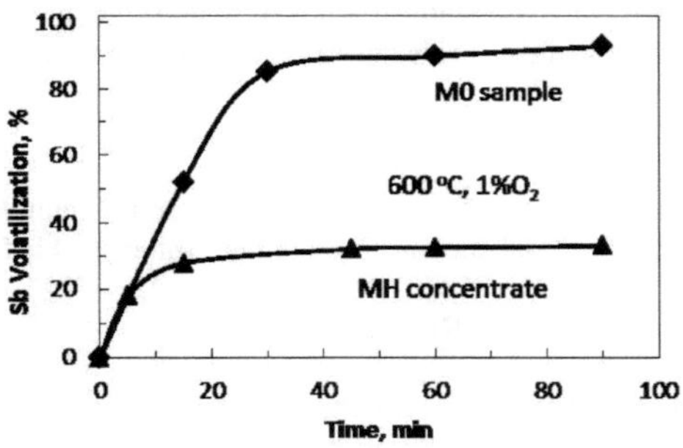

Figure 8. Volatilization of arsenic and antimony from the prepared mixtures MO and the copper concentrate MH.

According to the available thermodynamic data, the volatilization of antimony oxide is technically feasible only over the melting point of Sb_2O_3 at 655 °C [7] because below this temperature the partial pressure of Sb_4O_6 (g) is very low. Thus, at the roasting temperatures studied here (500 – 700 °C) the partial pressure of the Sb_4O_6(g) varies from 7.4×10^{-4} to 0.08 atm at 500 °C and 700 °C, respectively. Thus a condensed antimony oxide is stable at lower than 700 °C, as was verified by the results shown in Figure 4, where a negligible weight loss is shown after the oxidation of the stibnite. At the same roasting temperatures in the presence of oxygen, arsenic is present as As_4O_6(g), irrespective of the partial pressure of oxygen prevailing in the roasting atmosphere. However, As_4O_6(g) can react with Sb_2O_3 to form complex gaseous species as reported by Drowart et al. [8], and Li and Hager [9]. Thus, the volatilization of antimony could occur also through the formation of these complex gaseous species as follows:

$$1.5As_4O_6(g) + Sb_2O_3 = 2As_3SbO_6(g) \quad (10)$$

$$As_4O_6(g) + 2Sb_2O_3 = 2As_2Sb_2O_6(g) \quad (11)$$

$$As_4O_6(g) + 6Sb_2O_3 = 4AsSb_3O_6(g) \quad (12)$$

These reactions (10) to (12) would explain the enhanced vaporization of antimony in the presence of enargite as shown in Figures 5 and 6. However, the lower volatilization shown for the case of the copper concentrate MH may be due to the presence of other gaseous and solid oxides present, formed from the complex copper concentrates MH that hinders the reactions (10) through (12).

Conclusions

The following can be concluded from the experimental data:

- Arsenic can be eliminated from concentrates containing enargite through the gas phase at roasting temperatures of 500 -700 °C, irrespective of the oxygen pressure in the gas phase.
- Antimony does not volatilize in appreciable amounts from pure stibnite at roasting temperatures of 500 – 700 °C, but the extent of volatilization increases when stibnite is mixed with arsenic minerals such as enargite. This can be attributed to the formation of complex volatile oxides of arsenic and antimony.
- Bismuth does not volatilize at roasting temperatures of 500-700 °C from bismuthinite or from mixed arsenic and bismuth minerals. It remains as liquid bismuth in the calcine.

Acknowledgements

The authors acknowledge the National Fund for Scientific and Technological Development (FONDECYT) of Chile for the financial support through the Project # 1110590.

References

1. W. G. Davenport, et al. *Flash Smelting, Control and Optimization*, (Warrendale, PA: The Minerals Metals and Materials Society, 2001).
2. R. Padilla, A. Aracena, and M.C. Ruiz, "Reaction Mechanism and Kinetics of Enargite Oxidation at Roasting Temperatures," *Metallurgical and Materials Transactions B*, 43B(2012), 1119-1126.
3. Z. Zivkovic, et al., "Kinetics and Mechanism of Sb2O3 Oxidation Process," *Thermochimica Acta*, 383(2002), 137-143.
4. R. Padilla, G. Ramirez, M.C. Ruiz, "High-Temperature Volatilization Mechanism of Stibnite in Nitrogen-Oxygen Atmospheres," *Metallurgical and Materials Transactions B,* 41B(2010), 1284-1292.
5. V. S. Johnson, et al., "Structural Investigation of the High-Temperature Oxidation of the Bismuth Sulfide Using TPO-MS and in Situ X-Ray Diffraction Techniques," *Industrial Engineering Chemistry Research,* 43(2004), 3127-3132.
6. R. Padilla, R. Villa, M.C. Ruiz, "Kinetics of the Oxidation of Bismuthinite in Oxygen-Nitrogen Atmospheres," *Metallurgical an Materials Transactions B,* 42(2011), 1017-1022.
7. Roini: *HSC Chemistry 5.1*, (Outokumpu Research Oy, Pori, Finland, 2002).
8. J. Drowart, S. Smoes, A. Vanderauwera-Mahieu, "The Atomization Enthalpies of AsSb(g), As_4(g), Sb_4(g), As_3Sb(g), As_2Sb_2(g), $AsSb_3$(g), As_3SbO_6(g), $As_2Sb_2O_6$(g), and $AsSb_3O_6$(g). A Thermodynamic Study by the Mass Spectrometric Knudsen-Cell Method," *The Journal of Chemical Thermodynamics,* 10(1978), 453-464.
9. T. Li and J.P. Hager, "New Chemistry for the Recovery and Separation of Arsenic and Antimony," *The Reinhardt Schuhmann International Symposium on Innovative Technology and Reactor Design in Extractive Metallurgy,* ed. D.R. Gaskell, et al. (Warrendale, PA: The Metallurgical Society of AIME, 1986), 845-868.

Drying, Roasting, and Calcining of Minerals
Edited by: Thomas P. Battle, Jerome P. Downey, Lawrence D. May, Boyd Davis, Neale R. Neelameggham, Sergio Sanchez-Segado, and P. Chris Pistorius
TMS (The Minerals, Metals & Materials Society), 2015

CHARACTERIZATION OF PHYSICO-CHEMICAL CHANGES DURING THE ALKALI ROASTING OF NIOBIUM AND TANTALUM OXIDES

Sergio Sanchez-Segado, Ahmad Fahmi Ruzaidi, Yuan Zhang and Animesh Jha

Institute for Materials Research, University of Leeds LS2 9JT UK

Keywords: Niobium, Tantalum, Alkali roasting, Kinetics, Thermodynamics

Abstract

Niobium and tantalum are considered critical materials because the increasing demand in the energy sector which is reliant on limited abundance of these two metals in minerals. Traditionally, niobium and tantalum mineralsare digested using hydrofluoric acid at moderate temperatures, producing oxy-fluorides which are separated by solvent extraction, involving the use ofhazardous chemicals. The increasing environmental pressures are forcing the industry to search for alternative low carbon processing methods. In this paper alkali roasting based separation and beneficiation, which is not well understood in the literature, is proposed, for which the kinetics and thermodynamic equilibria for the formation of sodium tantalate and niobate with sodium hydroxide are analyzedat the temperature range of 400°C to 700°C. The apparent activation energy of formation,changes from 31.0±0.8 to 22.3±1.2 kJ mol^{-1} and from 23.8±1.2 to110.2±12.8 kJ mol^{-1}, for tantalate and niobate, respectively. The reaction mechanism is examined on the basis of the analysis of reaction products using X-ray powder diffraction (XRPD) and X-ray fluorescence (XRF).

Introduction

The chemical behaviour of niobium (Nb) and tantalum (Ta) are quite similar due to their electronic structures which not only determine their mineralogical association in nature, but also pose major challenge in separation and purification. Demand for these two metallic elements continues to soar in the field of energy, mobile communication and medical implant devices against the limited supply of mineral concentrates from deposits which are often very lean. Niobium is mainly used in steel and ferroalloys as ferro-niobium, aerospace and superalloys as alloying element, capacitors for the electronic and mobile communication devices and specialized corrosion resistance alloys for low and high temperatures [1-3]. Tantalum as an alloying element has proven itself as an exceptional material for bone implants, causing low side effects due to toxicity. These emerging need have pushed the materials price up in the commodity market due to limited supply of materials and high processing costs. Consequently within European Union has classed these two materials as "critical",since 2010 [4]. In this paper we examine the chemistry of the process which might potentially yield an environmentally better extraction route for niobium and tantalum.

Niobium and tantalum are components of 100s of different minerals, which mainly consistof complex salts of niobic and tantalic acid together with titanium, zirconium, tin, rare earths (RE), thorium, uranium and alkali-earth metals in variable ratios [5]. The most important mineral sources of niobium and tantalum may be classifiedas: tantalo-niobates which are salts of niobic and tantalic acids such as columbite and tantalite or titano(tantalo)-niobates in which titanium is the main component associated with niobium, tantalum and lanthanides. In Table 1 the chemical composition of the main tantalum and niobium minerals is presented. However, tantalite and columbite are the main sources for their industrial production. Commonly,these minerals are highly stable and only can be decomposed by fusion with alkali salts or digestion with concentrated hydrofluoric acid or its mixtures with sulphuric acid. The digestion with hydrofluoric acid has actually been the main industrial method, involving fewer steps than the fusion with alkali salts [6].

Mineral	Composition	Ta_2O_5 content %	Nb_2O_5 content %
Columbite	$(Fe,Mn)(Nb,Ta)_2O_6$	1-40	40-75
Euxenite	$(Ca,RE,U,Th)(Nb,Ta,Ti)_2O_6$	1-6	22-30
Fergusonite	$(RE)(Ba,Ta)O_4$	4-43	14-46
Microlite	$(Na,Ca)(Ta,Nb)O_6F$	60-70	5-10
Pyrochlore	$(Na,Ca)Nb_2O_6F$	2	40-65
Samarskite	$(Fe,Ca,U,Y,Ce)_2(Nb,Ta)_2O_6$	15-30	40-55
Strueverite	$(Ti,Ta,Nb,Fe)_2O_4$	7-13	9-14
Tantalite	$(Fe,Mn)(Nb,Ta)_2O_6$	42-84	2-40

Table 1 Chemical composition of the main tantalum and niobium minerals.

Since the ore quality is determined by the concentrations of oxide, the presence of gangue minerals during the decomposition of concentrates with hydrofluoric acidyields solutions with large amounts of impurities which hamper the separation of high purity niobium and tantalum compounds.The consumption of 70% HF is 1.2 to 2.5 ton per ton of concentrate depending on the composition of the concentrate, the original mineral (the Nb_2O_5:Ta_2O_5 and FeO: MnO ratios), and the processing conditions. Other problems associated with this technology are the evaporation loss of 6-7 vol% of HF during the decomposition at 250°C, the prohibitively highrecovery costsand the equipment maintenance [7-10].
Some research work has been carried out to remove hydrofluoric acid from the process via chlorination and carbo-chlorination with gaseous mixtures of N_2-Cl_2 and N_2-Cl_2-CO respectively.Also hydrothermal leaching with KOH, at temperatures up to 300°C and times of processing around 1 hour, has been tested [7-12]. In this work, we have studied the alkali roasting of high purity niobium and tantalum pentoxides at moderate temperature, with the aim of reporting fundamental physico-chemical and structural data which are actually unknown by the scientific community.

Experimental

Materials and reagents

Powders of Nb_2O_5 and $Ta_2O_5$with 99.9 wt% purity (metal basis) were purchased from Alfa Aesar. NaOH pellets (98wt %) were purchased from Fisher Scientific. The Nb_2O_5 and Ta_2O_5 powders were used asreceived whereas the NaOH pellets were crushed to make powder using a mortar and pestle before mixing with the oxide powders.

Alkali-roasting

By following the stoichiometry of reactions (1) and (2), about 3 g of oxide were mixed homogeneously with the stoichiometric amount of NaOH in a ceramic crucible. The crucible was placed inside a resistance furnace (Instron, TF105/3/1 ZF with temperature controller Eurotherm 2116, precise to ±1 °C) and preheated to a predetermined temperature ranging from 400 to 700 °C. The isothermal holding time was varied between20 and100 minutes in ambient air. After the reaction, the molten product was quickly taken out, allowed to cool at room temperature, and weighed. A portion of the sample was leached with distilled water at room temperature for 30 min to carry out X-Ray fluorescence analysis (XRF). The experiments were also repeated using two times the stoichiometric amount of NaOH.

Product characterization

The products were analysed by several methods: X-ray powder diffraction (XRPD) and X-ray Fluorescence (XRF). XRPD was carried out on a Bruker D8 Advance X-ray diffractometer with CuKα radiation. The patterns were collected from 10 to 80 ° (2θ values) with 0.05 ° intervals at a rate of 0.75 reading per second. The data were analysed using PANalyticalX'pertHighscore Plus v3.0 software.XRF was performed with an Olympus Innovex X-5000 portable XRF analyser over the washed samples.

Results

Equilibrium thermodynamic analysis

The reactions which may occur during the alkali roasting process are summarized below:

$$Nb_2O_{5(s)} + 2NaOH_{(s)} \rightarrow 2NaNbO_{3(s)} + H_2O_{(g)} \quad (1)$$

$$Ta_2O_{5(s)} + 2NaOH_{(s)} \rightarrow 2NaTaO_{3(s)} + H_2O_{(g)} \quad (2)$$

$$NaOH_{(s)} + 1/2CO_{2(g)} \rightarrow 1/2Na_2CO_{3(s)} + 1/2H_2O_{(g)} \quad (3)$$

$$Na_2CO_{3(s)}+H_2O_{(g)} \rightarrow Na_2CO_3{*}\ H_2O_{(s)} \quad (4)$$

Thermo-chemical properties for $Nb_2O_{5(s)}$, $Ta_2O_{5(s)}$,$Na_2CO_{3(s)}$, $Na_2CO_3{*}H_2O_{(s)}$, $NaOH_{(s)}$, $NaTaO_{3(s)}$, $NaNbO_{3(s)}$ and $H_2O_{(g)}$ were found inBarin and Knacke [13]. In Figure 1, the computed values of changes in the Gibbs energy for reactions (1) to (4) are plotted against the temperature.

From Figure 1 it is evident that sodium tantalate is the most stable phase followed by sodium niobate. However, the formation of sodium carbonate and sodium carbonate monohydratevia reactions(3) and (4) is expected under the cooling condition, which cannot be discarded.

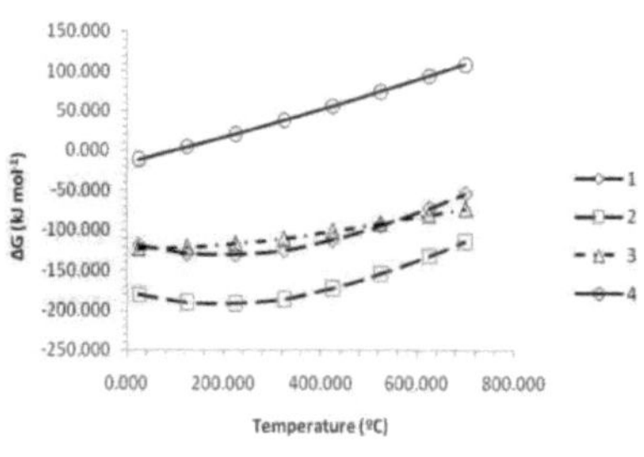

Figure 1. Temperature dependence of the Gibbs free energy values for reactions (1) to (4).

Kinetic analysis

From XRF analysis of the roasted samples, the fractional conversion of pure oxides into alkali complex was calculated using the equation (8):

$$\text{Conversion} = \frac{Mols\ of\ NaNbO_3\ or\ NaTaO_3}{Initial\ number\ of\ mols\ of\ Nb_2O_5\ or\ Ta_2O_5} \times \frac{1}{2} \quad (8)$$

The conversion values were plotted against time. Each isothermal curve was fitted to a power-form curve shown in Figure 2. As can be seen,the fraction converted increases with temperature and time. Moreover, the higher values of fractional conversion were achieved for tantalum pentoxide underidentical experimental working conditions, as for the niobium complexation reaction. This observation indirectly suggests that the sodium tantalite may be more stable than sodium niobate, as shown in Figure 1.

Various kinetics models were adopted for the analysis of solid-gas reactions,which are mechanistically different and are known as the nucleation, geometrical contraction, diffusion and reaction order models [14]. The evolution of a solid-state reaction maybe controlled either by the reaction at the interface or by the diffusion of the reactants and reaction products through the product layer. The Spencer and Topley model (SP) [15] proposes a geometrical contraction of the reactant layer via advancing reaction interface in which the rate of product layer growth is controlled by the advancing reaction interface toward the centre of the particle. Assuming a spherical geometry for the particles, the SP model equation is as follows:

$$1-(1-f)^{1/2} = kt \qquad (9)$$

Diffusion plays an important role when the reactants and products have different crystal lattices. Jander (J) [16] proposed the mechanism for diffusion controlled reactions based on the parabolic law shown in equation (10). On the other hand, Ginstling and Brounshtein (GB) [17] related the growth of the product layer with the decrease in interface area in equation (11) which is useful for reactions, controlled by gaseous diffusion through the product layer.

$$(1-(1-f)^{1/3})^2 = kt \qquad (10)$$

$$1-2/3f-(1-f)^{2/3} = kt \qquad (11)$$

Where f is the fraction converted and k is the rate of reaction. The isothermal kinetics data, shown in Figure 2, are fitted using equations (9) to (11) in Figure 3.

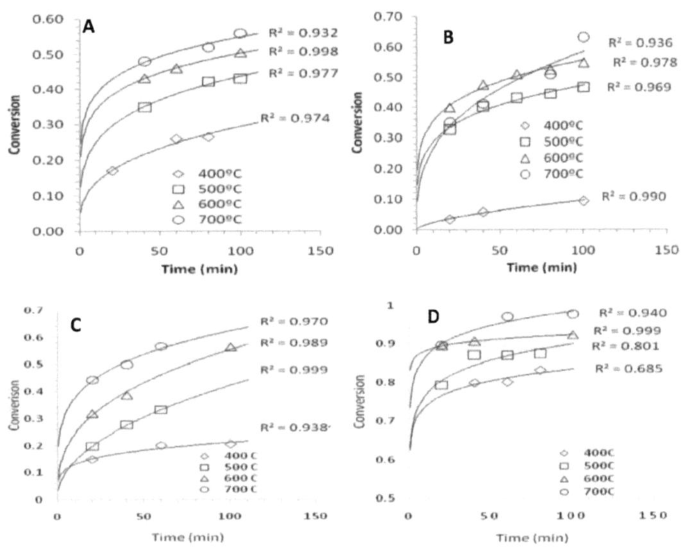

Figure 2. Kinetic curves. A. Nb_2O_5:NaOHstoichiometric, B. Nb_2O_5:NaOH double stoichiometric amount. C. Ta_2O_5:NaOHstoichiometric. D. Ta_2O_5:NaOH double stoichiometric amount.

It is apparent from Figures 3A to 3C that the Jander model best describes the measured rate data when compared with the SP or GB model. In Figure 3D the natural logarithm of the reaction rate calculated using the Jander model is plotted against the reciprocal of absolute temperature. From Figure 3D it is observed, that the rate of reaction increases with the temperature for both oxides. The derived values of the apparent activation energies for Nb_2O_5 and Ta_2O_5 were 23.8±1.2 and 31.0±0.8kJ mol^{-1}, respectively for the stoichiometric amount of NaOH. There was a dramatic increase in the derived value of activation energy from 23.8±1.2 to 110.2±12.8kJ $mol^{-1}$for Nb_2O_5, when the stoichiometric ratio of NaOH:Nb_2O_5 increased from 1:1 to 1:2. However the activation energy for Ta_2O_5 remained comparable at 22.3±1.1kJ $mol^{-1}$for NaOH:Ta_2O_5=2:1.

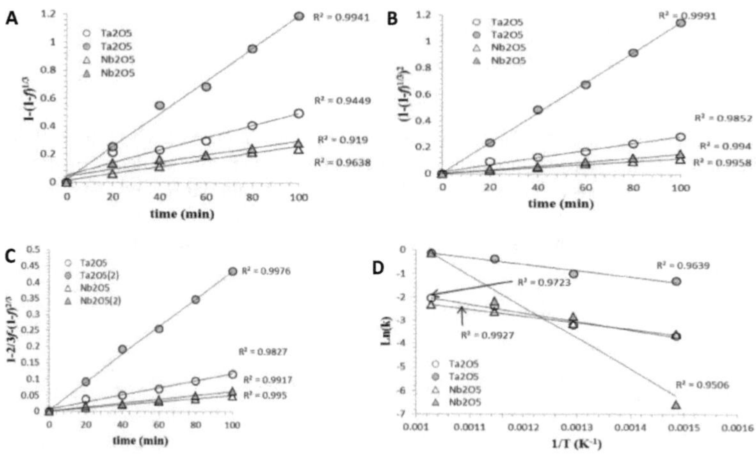

Figure 3. A. Representation of Spencer and Topley function, B. Representation of Jander function. C. Representation of Ginstling and Brounshtein function. D. Ahrrenius plot for the derivation of the activation energy according to the Jander model.*The shaded markers are referred to the double stoichiometric amount of NaOH.

Reaction products characterization

In Figure 4.A, the XRPD patterns of the roasted samples, at 100 min for the stoichiometric amount of NaOH, are compared at different temperatures. The relative intensities of Nb_2O_5 peaks decreased at the expense of the increase in the diffraction intensities of $NaNbO_3$, which formed as a result of reaction (1) between 400 and 700 °C. Although not shown, similar changes were observed for other sets of investigations with different roasting temperatures and times.For NaOH:Nb_2O_5:2:1, for example in Figures 4B and 4C, it was observed thatthe crystal structure of Nb_2O_5changed from orthorhombic to monoclinicabove 400°C. This change in crystal structure was also observed in Na_3NbO_4which transformed from a cubic to a monoclinic phase. These changes are summarized below:

$$Nb_2O_{5\,(Orthorhombic)} \rightarrow Nb_2O_{5(Monoclinic)} \quad (12)$$

$$Nb_2O_{5(s)} + 6NaOH_{(s)} \rightarrow 2Na_3NbO_{4(Cubic)} + 3H_2O_{(g)} \quad (13)$$

$$Na_3NbO_{4(Cubic)} \leftrightarrow Na_3NbO_{4(Monoclinic)} \quad (14)$$

Nafe et al. [18] in their research proposed an expression which relates the difference of the standard Gibbs energy of these two phases, from 525°C to 700°C.From equation (15) the difference in the value of Gibbs energy is -442.4 ± 8.4kJ mol^{-1} at 525 °C and -439.1 ± 9.3kJ mol^{-1} at 700 °C indicating that Na_3NbO_4 is more stable in this temperature range. However, in Figure 4C an increase inthe relative intensity of $NaNbO_3$with time is observed.

$$\Delta_f G^o_{Na3NbO4} - \Delta_f G^o_{NaNbO3} = -457.370 \pm 4.41 + (18.722 \pm 4.96) \times 10^{-3} T(K)\ [kJ\ mol^{-1}] \quad (15)$$

When $Ta_2O_5$was roasted with the stoichiometric amount of NaOH as shown in Figure 5A), two sodium tantalate phases were identified $NaTaO_3$ and $Na_8Ta_6O_{19}$, the latter does not appear to exist at 600°C, which suggests that at higher temperatures $Na_8Ta_6O_{19}$ is less stable. The following reactions may be proposed for the stoichiometric amount of NaOH:

$$Ta_2O_{5(s)} + 2NaOH_{(s)} \rightarrow NaTaO_{3(s)} + 2H_2O_{(g)} \quad (16)$$

$$3Ta_2O_{5(s)} + 8NaOH_{(s)} \rightarrow Na_8Ta_6O_{19(s)} + 4H_2O_{(g)} \quad (17)$$

$$Na_8Ta_6O_{19(s)} + H_2O_{(g)} \rightarrow 6NaTaO_{3(s)} + 2NaOH_{(s)} \quad (18)$$

For NaOH:Ta_2O_5=2:1 ratio, Na_3TaO_4 was also identified.As in case of Nb_2O_5 a crystallographic phase transition from cubic to monoclinic occurs. The intensity of the peaks of the cubic and monoclinic Na_3TaO_4 phases increase and decrease with temperature and time, as shown in Figures 5B and 5C, following a contrary sequence to the $Na_8Ta_6O_{19}$ peaks. This behaviour suggestsequilibrium between these two phases according to the scheme represented by reactions (19) to (21):

$$Ta_2O_{5(s)} + 6NaOH_{(s)} \rightarrow 2Na_3TaO_{4(s)} + 3H_2O_{(g)} \quad (19)$$

$$Na_3TaO_{4(Cubic)} \rightarrow Na_3TaO_{4(Monoclinic)} \quad (20)$$

$$6Na_3TaO_{4(Cubic/Monoclinic)} + 5\ H_2O_{(g)} \leftrightarrow Na_8Ta_6O_{19(s)} + 10NaOH_{(s)} \quad (21)$$

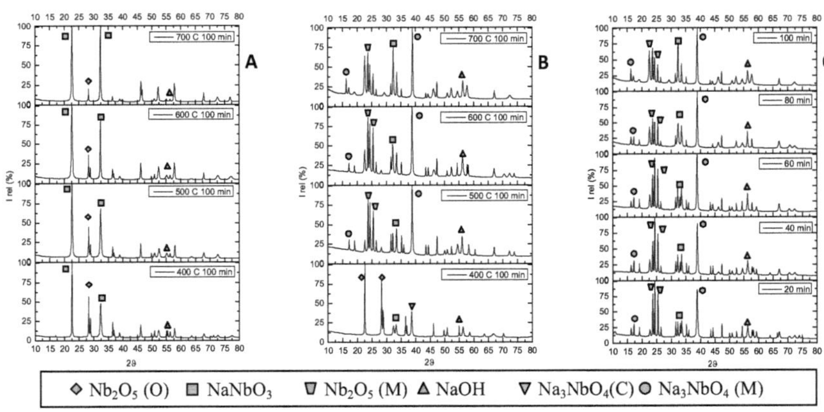

Figure 4. XRPD patterns of Nb_2O_5 roasted samples A. Stoichiometric amount of NaOH for different temperatures at 100 min. B. Double stoichiometric amount of NaOH for different temperatures at 100 min. C. Double stoichiometric amount of NaOH at 700°C for different times.*Selected peaks are label with markers.

Similar observations for the synthesis of $NaTaO_3$ with molten NaOH were reported by Kengo et al. [19]. Also it is important to highlight that no changes in Ta_2O_5 crystal structure were observed which involves that the equilibrium conditions are rapidly achieved due to the small differences in the activation energies for the roasting of Ta_2O_5 with different amounts of NaOH.

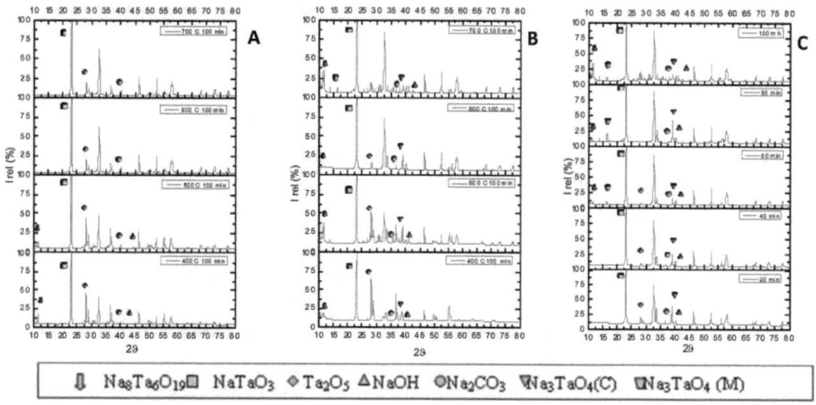

Figure 5. XRPD patterns of Ta_2O_5 roasted samples A. Stoichiometric amount of NaOH for different temperatures at 100 min. B. Double stoichiometric amount of NaOH for different temperatures at 100 min. C. Double stoichiometric amount of NaOH at 700°C for different times.*Selected peaks are label with markers.

Conclusions

The activation energies calculated for Nb_2O_5 and Ta_2O_5 roasting were 23.8±1.2, 22.3±1.1 and 110.2±12.8, 31.0±0.8 kJ mol^{-1} for NaOH:oxide ratios 1:1 and 2:1 respectively. These values point out that the roasting process is controlled by the diffusion of the reactants to the reaction interface. However, the dramatic increase observed for Nb_2O_5 at NaOH:Nb_2O_5 ratio 2:1 implies that the reaction is controlled by the crystallographic transition $Nb_2O_{5\,(Orthorhombic)} \rightarrow Nb_2O_{5(Monoclinic)}$. The crystallographic transitions indicated in reactions (14) and (20) and the reactions (17), (18),(21) are rapidly completed and consequently do not affect the rate of reaction as can be observed from the comparable activation energy values for the roasting of Ta_2O_5 at different NaOH ratios.

Acknowledgments

A. Jha acknowledged the support from the European Union's Marie Curie Fellowship grant for Dr. Sanchez-Segado.

References

[1] BGS (British Geological survey). *Niobium-Tantalum Mineral Commodity Profile* [Online 2011] [Accessed on 20 September 2014] http://www.bgs.ac.uk/mineralsuk/statistics/mineralProfiles.html

[2] R. S. Young, *Chemical Analysis in Extractive Metallurgy*. London: Charles Griffin & Company Ltd. 1971.

[3] European Commission. *Annex V to the Report of the Ad-Hoc Working Group on Defining Critical Raw Material.* [Online]. 2010. [Accessed on 28 September 2014]. http://ec.europa.eu/enterprise/policies/raw-materials/files/docs/annex-v-b_en.pdf

[4] European Commission. *Critical Raw Materials for the EU.* [Online]. 2010. [Accessed on 28 September 2014]. Available from: http://ec.europa.eu/enterprise/policies/raw-materials/files/docs/report-b_en.pdf

[5] A.N. Zelikman, O.E. Krein, G.V. Samsonov, *Metallurgy of Rare Metals.* Russia: L.V. Belyaevskaya 1964.

[6] F. Habashi, *Handbook of Extractive Metallurgy Vol.III*, Willey and sons 1997.

[7] H. Zhou, et. al. The dissolution behaviour of Nb_2O_5, Ta_2O_5 and their mixture in highly concentrated KOH solution.*Hydrometallurgy.*80 (2005), 126-131

[8] H. Zhou, et. al. Leaching of a low-grade niobium-tantalum ore by highly concentrated caustic potash solution. *Hydrometallurgy*, 80(2005), 83-89

[9] X.Wang, et al. Dissolution behaviours of Ta_2O_5, Nb_2O_5 and their mixture in KOH and H_2O system. *Transactions of Nonferrous Metals Society of China.*20 (2010), 2006-2011.

[10] X. Yang, et. al. Decomposition of Niobium Ore by Sodium Hydroxide Fusion Method. *Metallurgical and Material Transactions B.* 44(2013), 45-52

[11] J.Gonzalez, et. al. Chlorination of niobium and tantalum ore. *ThermochimicaActa* 311 (1998) 61-69.

[12] N. Kanari, et.al. An overview study of chlorination reactions applied to the primary extraction and recycling of metals and to the synthesis of new reagents.*ThermochimicaActa* 495 (2009) 42-50.

[13] I. Barin and O. Knacke. Thermochemical properties of inorganic substances Vol. 2.Springer-Verlag 1973.

[14] A. Khawam and D.R. Flangan. Solid –State kinetic models: Basics and mathematical fundamentals. *Journal of Physical Chemistry B.* 110 (2006), 17315-17328.

[15] W.D. Spencer and B. Topley. Chemical kinetics of the system $Ag_2CO_3 \leftrightarrow Ag_2O + CO_2$.*Journal of the Chemical Society.* (1929), 2633-2650.

[16] V.D. Tathavadkar et al. The soda-ash roasting of chromite minerals: Kinetics considerations. *Metallurgical and Materials Transactions B.*32 (2001), 593-602.

[17]. A.M. Ginstling and B.I. Brounshtein. Concerning the diffusion kinetics of reactions in spherical particles.*Russian Journal of Applied Chemistry.* 23 (1950), 1327-1328.

[18] H. Nafe et al. Thermodynamic characterization of the Eutectic phase mixture $NaNbO_3/Na_3NbO_4$.II: Solid state electrochemical investigation. *Journal of the American Ceramic Society* 90 (2007) 3227-3232.

[19] S. Kengo et al. Preparation of needle and plate-like $NaTaO_3$ by molten NaOH method. *Journal of the Ceramic Society of Japan.* 121 (2013) 109-112.

Drying, Roasting, and Calcining of Minerals
Edited by: Thomas P. Battle, Jerome P. Downey, Lawrence D. May, Boyd Davis, Neale R. Neelameggham, Sergio Sanchez-Segado, and P. Chris Pistorius
TMS (The Minerals, Metals & Materials Society), 2015

Mechanism of Na_2SO_4 on Refractory Gold Concentrate at Roasting Pretreatment

Qian Li, Jian-jun Hu, Yong-bin Yang, Bin Xu, Tao Jiang

School of Minerals Processing and Bioengineering, Central South University, China, Hunan, 410083

Keywords: Double refractory gold concentrate, Roasting, Sodium sulfate, Sintering

Abstract

It could significantly increase the leaching rate of refractory gold concentrate containing sulfur and arsenic by roasting pretreatment with adding certain amount of sodium sulfate, and the mechanism of that was systematically researched with DSC, XRD and SEM in this paper. Results indicated that the mechanism of Na_2SO_4 on double refractory gold concentrate at roasting pretreatment can be summed up in the following three aspects: Firstly, Na_2SO_4 can react with SO_2 and SO_3 which is produced by the oxidation of sulfides and generate $Na_2S_2O_7$(l), which plays buffer effect on sulfides exothermic oxidation, avoiding the secondary-encapsulated of gold. Secondly, the liquid phase of $Na_2S_2O_7$ can cover mineral particles' surface that possesses a sintering effect, it leads to internal cracks of the particles. Finally, mineral particles will gather with sodium sulfate particles as the center that can improve the permeability of the layer.

Introduction

Lots of researches have indicated that it can significantly improve the leaching rate of useful minerals by adding certain amount of sodium sulfate during roasting. Xue et al. whose research on gold-silver concentrate containing high sulfur reported that gold and silver leaching rates had respectively increased from 90.18% to 96.77% and 26.28% to 73.14% after adding 6% Na_2SO_4 to concentrate at roasting pre-treatment[1]. Shi et al. found that at low temperature of 300℃, Na_2SO_4 had no outstanding effect on sulfation roasting of nickel laterite, but at 700℃, the addition of sodium sulfate could significantly improve the metal sulfation rate[2]. The research on manganese oxide by bubbling SO_2 and adding Na_2SO_4 to raw mineral indicated it could significantly improve the leaching rate of manganese with a certain Na_2SO_4 dosage range from 0% to 3% and SO_2 concentrate below 1.5% , however, the roasting system (moving bed or fixed-bed) had little effect on the leaching rate of manganese[3]. The study on a multi-metal sulfide concentrate indicated that Na_2SO_4 plays a role similar to catalyst which could speed up the oxidation reaction rate from SO_2 to SO_3. What's more, the existence of sodium sulfate could accelerate the mineral particles' surface shrinkage crack[4]. Acidic alkali sulfates are also important components in the catalytic conversion of SO_2 to SO_3 in the production of sulphuric acid. In the socalled contact process, molten mixtures of alkali pyrosulphates (mainly K_2SO_4) and V_2O_5 on a Kieselguhr support act as the catalyst for oxidation[5].

In summary, there are many studies on sulfation roasting process at present, however, the research on the mechanism of that during roasting is rarely. In this paper, the mechanism of sulfation roasting in the presence of Na_2SO_4 was systematically researched with Differential Scanning Calorimetry (DSC), X-ray diffraction (XRD) and SEM.

Experimental

Raw materials

This research was conducted on a refractory gold-bearing ore coming from Yunnan province in China, representative samples, with d_{70} 0.074mm, were fine ground to -0.045mm for characterisation and experimental tests.

Table I. Chemical compositions of the gold concentrate sample (w %)

Element	Au*	Si	Ca	As	Fe	S	Sb	Al	C
Content	18.05	20.08	5.3	1.62	19.6	20.6	1.31	3.07	6.95

*The unit of gold content is g/t.

Table II. Sulfur phase analysis of gold concentrate sample (w %)

Phase type	Elemental Sulfur	Sulfide	Sulphate	Total
Content	0.35	20.10	0.15	20.60
Proportion	1.70	95.57	0.73	100.00

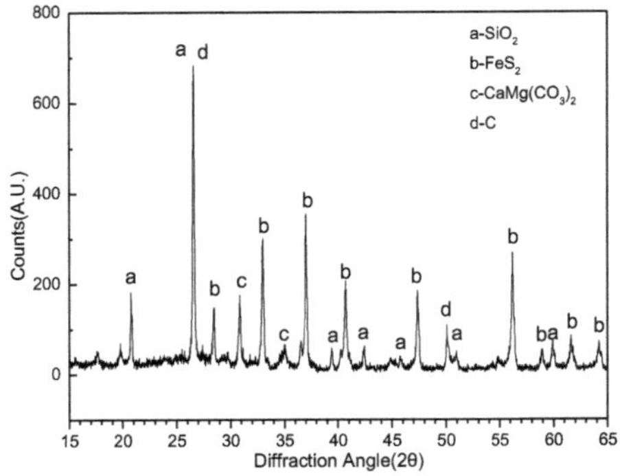

Fig 1. XRD pattern of the gold concentrate specimen

Table I shows that the gold concentrate, with 1.62%As、20.76%S、6.95%C, is a typical double refractory ore. Table II indicates that sulfur compounds rarely exist in the form of elemental sulfur and sulphate in the gold concentrate. 95.57% of sulfur is the form of sulfide which is easy to decompose at strong oxidizing atmosphere and high temperature[1].

Mineralogical characterisation by X-ray Diffraction (XRD) (see Fig 1) shows the presence of quartz and pyrite is the majority, and beyond that, carbonaceous matters also occupy a certain content.

Methods

Fig 2 shows the schematic diagram of roasting facility. The gas used in the test was O_2 (in a purity of 99.2%) and N_2(in a purity of 99.99%), the O_2 and N_2 flow was controlled by respective flow meter under the total flow rate of 1L/min, and the ratio of O_2 was 21%. The furnace temperature monitored by a thermocouple was controlled by adjusting within ±5℃.

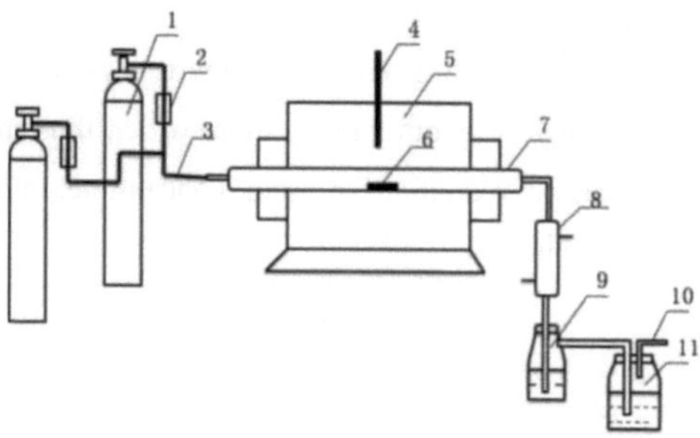

Fig 2. Schematic diagram of the roasting experimental facility

1- gas tank; 2- gas flow mete; 3-rubber hose; 4- thermocouples; 5- horizontal tube furnace; 6- raw material loader; 7- quartz tube; 8- condenser pipe; 9, 10- exhaust gas absorption equipment; 11-vent-pipe

The operation of water-immersion was that the calcine was sufficiently stirred in distilled water several times and then following with filtration and drying to reserved for the later test. The particle size was tested by ultrasonic grain size analyzer which could work without touching with water.

Results and discussion

The thermodynamic characteristics

The DSC analysis was carried out (see Figure 3) to study the influence of sodium sulfate on the exothermal.

The DSC(Differential Scanning Calorimetry) curve shows that both of DSC-0% and DSC-5% start at the same initial temperature 438℃, DSC -0% curve is a upwards parabola with the maximum at 498℃ and the termination temperature is 554℃. However, DSC-5% curve slightly decreases before reaches the maximum at 562℃ and then following with a rapid falling to 585℃. Therefore the exothermic rate of raw material is faster than the group of adding 5% Na_2SO_4 from

438℃ to 537℃, but it is quite the contrary to that beyond 537℃. As is shown in figure 3, compared with DSC-0% curve with only a intense exothermic peak, DSC-5% curve is more smoother and the temperature range of exothermic is more wider, thus the existence of Na_2SO_4 plays buffer effect on exothermic oxidation of gold concentrate, avoiding the secondary-encapsulated of gold to some extent.

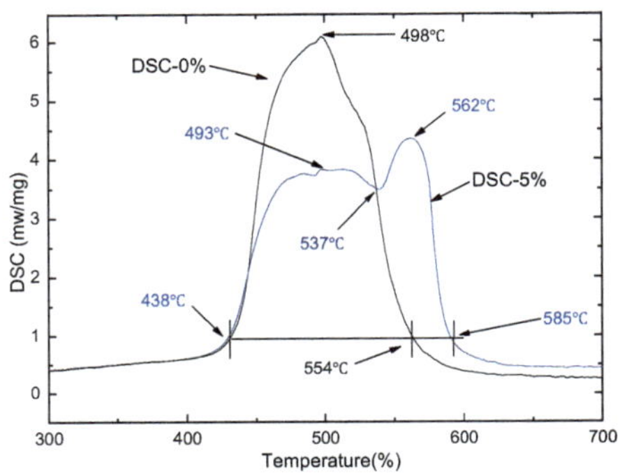

Figure 3. The comparison of DSC analysis

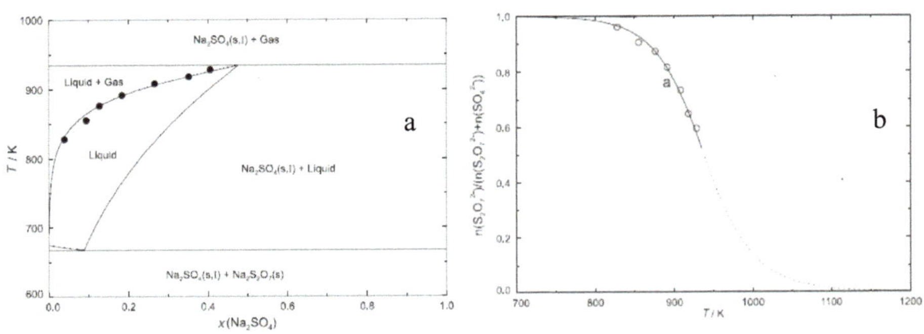

Figure 4. Thermodynamic evaluation of the Na_2SO_4 +$Na_2S_2O_7$) system

The melting point of Na_2SO_4 is 884℃, that far exceeds the roasting temperature[6], Sodium sulfate mainly plays a role in generating the intermediate product of $Na_2S_2O_7$[7,8]. Figure 4(a) is the calculated phase diagram of (Na_2SO_4+$Na_2S_2O_7$) at a total pressure of 0.1 MPa., the melting point of pure $Na_2S_2O_7$ is 401℃, and the decomposition temperature is 460℃, while Na_2SO_4 exists, the stability of $Na_2S_2O_7$ has significantly improved. Figure 4(b) is the Calculated composition of the liquid given as mole fraction of anions in the (Na_2SO_4+$Na_2S_2O_7$) systems for (liquid + gas) equilibrium as a function of temperature. The total pressure is 0.1 MPa and the composition of the gas follows the equilibrium of $SO_3 \rightleftarrows SO_2 + O_2$[9,10]. As roasting temperature rises, $n(S_2O_7^{2-})/(n(S_2O_7^{2-})+n(S_2O_7^{2-}))$ gradually declines, $Na_2S_2O_7$ is basically decomposed to Na_2SO_4 and SO_3 after heated to about 700℃.

At low temperature, the reaction of Na_2SO_4 and SO_3 generates the liquid phase of $Na_2S_2O_7$ (equations (1)-(7)), as temperature rises, on the contrary, $Na_2S_2O_7$ is decomposed to Na_2SO_4 and SO_3(equation 7-8), at this process, $Na_2S_2O_7$ as the intermediate product plays the role of "SO_3 storage"[2,4]. The liquid phase of $Na_2S_2O_7$ can cover mineral particles' surface, which protect the particles inside from violent exothermic oxidation. As temperature rises(see Figure 4(b)), the decomposition of $Na_2S_2O_7$ to Na_2SO_4 and SO_3 which is a endothermic reaction (the Standard molar enthalpy of formation is 118.68 kJ/mol[11]) accelerates, the gradual elimination of the $Na_2S_2O_7$ coverage plays buffer effect on exothermic oxidation, thus avoids the secondary-encapsulated of gold to some extent.

$$FeS_2+O_2= Fe_3O_4+SO_2 \quad (1)$$

$$FeS_2+O_2= Fe_2O_3+SO_2 \quad (2)$$

$$Fe_3O_4+O_2 = Fe_2O_3 \quad (3)$$

$$SO_2+O_2 \rightleftarrows SO_3 \quad (4)$$

$$Fe_2O_3+SO_2+O_2 = Fe_2(SO_4)_3 \quad (5)$$

$$Fe_2(SO_4)_3 = Fe_2O_3+ SO_3 \quad (6)$$

$$Na_2SO_4(s)+SO_3 \rightleftarrows Na_2S_2O_7(l) \quad (7)$$

$$Na_2S_2O_7(l)+MeO = MeSO_4+ Na_2SO_4(s) \text{（M=alkali metal）} \quad (8)$$

On the basis of the previous study, the mechanism of Na_2SO_4 on refractory gold concentrate at 700℃ was further studied. The four calcines of A,B,C and D which were operated under different conditions(see Table III) were systematically researched with particle size analysis, X-Ray Diffraction and SEM.

The comparison between raw mineral and A (see Table III) shows that particle size distribution essentially unchanged, but at the same time, surface area of A has dramatically increased by 74% because the matrix of sulfide minerals was liberated[1,12]. Whether the main chemical component or the particle size distribution, it was not found much of a difference between A, B and D, but the surface area of D take a further increase to 3738 cm^2/g. As previous study, the liquid phase of $Na_2S_2O_7(l)$ can cover mineral particles' surface, and then decomposes to the solid phase of $Na_2SO_4(s)$ and SO_3, the phase transformation leads to the solidification shrinkage on mineral particles' surface, which makes a contribution to internal cracks of the particles[4]. The roasting pre-treatment is mainly considered for liberating the ultra-fine gold from the matrix of sulfide minerals, obviously, porous skeletons of calcine can significantly improve the exposure of ultra-fine gold[12].

Table III. The comparison of particles size and surface area

Calcine Number	Material	Operation	particle size distribution /(w.%)			Surface area /(g/cm^3)
			+0.074 mm	+0.037~ -0.074mm	-0.037 mm	
-	Raw mineral	without	32.45	25.11	42.44	1473
A	Without addition	without	32.07	27.92	40.01	2565
B		Water-immersion	30.32	26.31	43.37	2769
C	With 5% Na_2SO_4	without	39.95	33.34	26.71	1656
D		Water-immersion	30.05	25.43	44.52	3738

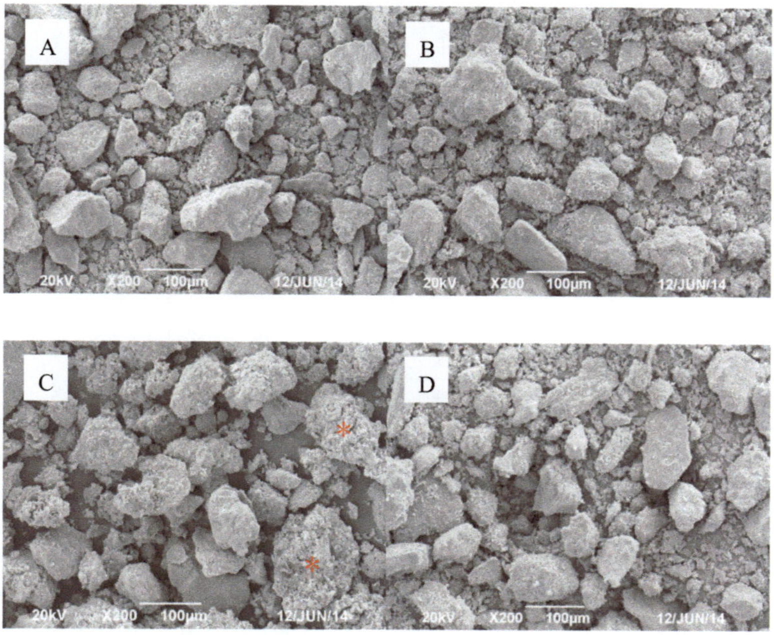

Figure 5. The comparison of SEM

As is shown in Figure 5(A), Figure 5(C) and Table III(A), Table III(C), compared with A, the particle size distribution of C has changed remarkably: the ratio of fine particles(-0.037mm) sharply decreases from 40.01% to 26.71%, at the same time, medium particles (+0.037mm - 0.074mm) and coarse particles(+0.074mm) ratios increase by 7.03% and 9.63%, respectively. What's more, we found that most of coarse particles in calcine C(see the mark * in Figure 5(C)) were hold together by fine particles with some white particles as the core. Figure 6 shows the main difference of chemical component between C and D is sodium sulfate which is the addition as well, it indicates that the existence of Na_2SO_4 play an important role in the phenomenon of the particles' adhesion. The increase of the whole particles size indicates that due to the presence of sodium sulfate, the adhesion between the particles is enhanced greatly. which leads to the ratio of fine particles presents a obvious decreasing trend by holding independent particles together. Since sodium sulfate is soluble in water, most of coarse particles in C which are adhered together by fine particles with sodium sulfate as the binder lose bonding point and spread out[13], therefore the particle size distribution of D is similar to A and B(see Table III).

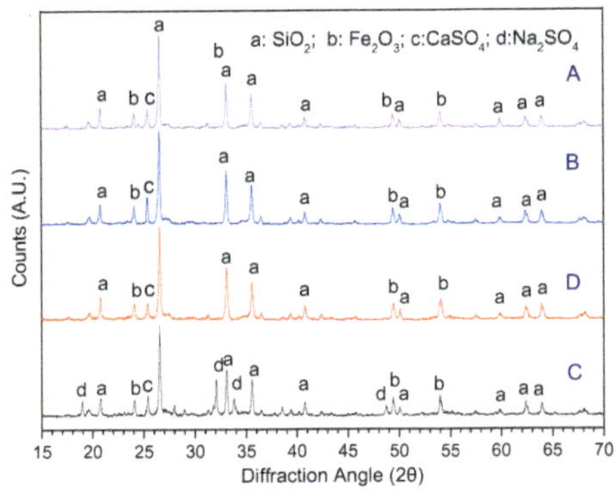

Figure 6. The comparison of XRD patterns of calcines

Conclusion

As previous study, the mechanism of Na_2SO_4 on double refractory gold concentrate at roasting pretreatment can be summed up in the following process: firstly, At low temperature, the reaction of Na_2SO_4 and SO_3 generates the liquid phase of $Na_2S_2O_7$ which can cover mineral particles' surface, as temperature rises, on the contrary, $Na_2S_2O_7$ decomposes to Na_2SO_4(s) and SO_3, the gradual elimination of the $Na_2S_2O_7$ coverage plays buffer effect on exothermic oxidation, thus avoid the secondary-encapsulated of gold to some extent. Secondly, the liquid phase of $Na_2S_2O_7$(l) decomposes to the solid phase of Na_2SO_4(s) and SO_3, the transformation of the phase leads to solidification shrinkage on the mineral particles' surface, which is advantageous to internal cracks of the particles and significantly improve the exposure of ultra-fine gold. Finally, the independent fine particles adhered together with sodium sulfate particles as the core which can improve the permeability of the layer.

Acknowledgement

The authors gratefully acknowledge the financial support from the project supported by NSFC, Grant No. 51074182.Financial supports from China Postdoctoral Science Foundation (Grant No. 2014M550422) and post-doctoral workstation of Central South University are gratefully acknowledged.

References

[1] Xue, G, Ren, W, Xu, Y, *Hydrometallurgy and analysis testing process of gold and silver* (Beijing: Science Press, 2009), 295-385.

[2] Shi Jianfeng, Wang Zhixing, Hu Qiyang, "Influence of Na_2SO_4 to sulfation roasting of nickel laterite", *Journal of Chinese society of rare earths*, 30(2012), 927-932.

[3] Yao Chenchang, Tao Daoxian, "The sulfatization of manganese ore from Dongxiangqian region", *Nonferrous Metals*, 4(1985), 112-121.

[4] Ying Bilian, Zheng Buming, "The effect of sodium sulfate on sulfating roasting of refined multi-metal sulfide ore", *Journal of Zhejiang Institute of Silk Textiles*, 8(1991), 55-59.

[5] G. Hatem, K.M. Eriksen, M. Gaune-Escard, R. Fehrmann, Top.Catal. 19 (2002) 323–331.

[6] G.E. Brodale, W.F. Giauque, *J. Phys. Chem*, 76 (1972), 737–743.

[7] L.P. Kostin, N.L. Chikishev, A.N. Ketov, V.V. Pechkovskii, Izv.Vuz. Khim. Kh. Tekh. 14 (1971) 1787–1789.

[8] L.P. Kostin, L.L. Pluzhnikov, A.N. Ketov, Zh. Fiz. Khim. 49(1975).

[9] Daniel Lindberg, Rainer Backman, "Thermodynamic evaluation and optimization of (Na_2SO_4+K_2SO_4+$Na_2S_2O_7$+$K_2S_2O_7$) system", J. Chem. *Thermodynamics*, 38(2006), 1568-1583.

[10] H. Flood, T. Fland, *Acta Chem. Scand.* 1 (1947) 781–789.

[11] M.W. Chase Jr., C.A. Davies, J.R. Downey Jr., D.J. Frurip, R.A.Mcdonald, A.N. Syverud, J. Phys. *Chem. Ref. Data Suppl*, 14(1985), 927–1856.

[12] Jiang, T, Chemistry of extractive metallurgy of gold(Changsha: Hunan science and Technology Press, 1998), 308-327.

[13] Jiang Wenwen. Study on the combustion charater of black liquor from NSSC in fluidized bed, 2008:39-49.

DRYING, ROASTING, AND CALCINING OF MINERALS

DRYING AND CALCINING

Session Chairs:
Neale R. Neelameggham
James Sever

Drying, Roasting, and Calcining of Minerals
Edited by: Thomas P. Battle, Jerome P. Downey, Lawrence D. May, Boyd Davis, Neale R. Neelameggham, Sergio Sanchez-Segado, and P. Chris Pistorius
TMS (The Minerals, Metals & Materials Society), 2015

OPTIMIZATION ON DRYING OF CuCl RESIDUE BY HOT-AIR USING RESPONSE SURFACE METHODOLOGY

Guo Zhanyong[1,2,3], Ju Shaohua[1,2,3]*, Lei Ting[4], Peng Jinhui[1,2,3], Zhang Libo[1,2,3],Jiang[1,2,3], Feng

1. National Local Joint Laboratory of Engineering Application of Microwave Energy and Equipment Technology, Kunming, Yunnan 650093, China;
2. Key Laboratory of Unconventional Metallurgy, Ministry of Education, Kunming University of Science and Technology, Kunming, Yunnan 650093, China;
3. Faculty of Metallurgical and Energy Engineering, Kunming University of Science and Technology, Kunming, Yunnan 650093, China;
4. Kunming Metallurgy College, Kunming, Yunnan 650093, China.

Keywords: CuCl Residue; Drying; Response surface methodology; Optimum conditions

Abstract

The clean utilization of the residue containing chloride, CuCl residue, produced from zinc hydrometallurgy is very important for the recycle of valuable metals. And drying is the main process in recycling. In this study, response surface methodology was used to optimize operating conditions of the drying of CuCl residue in a convection oven and desirability function used as the methodology for the optimization. Optimization factors were air temperature (80–100°C), material thickness (10–25mm) and process time (80–120min) while investigated responses was final moisture content. The optimum conditions were found to be the temperature of 95.48°C with the material thickness of 13.47mm for the process time of 89.48 min. Under optimum conditions, the predicted moisture content of CuCl residue was 4.11% while the experimental date was 4.13%. The experimental value after process optimization was found to agree satisfactory with the predicted value. After drying, XRD was used to analyze the main phases of CuCl residue. The CuCl residue is made up of CuCl,Cu_2O, ZnS and other compounds.

Introduction

Cl is a harmful element in zinc hydrometallurgical system. Thus, Cl^- contained in the solution should be removed before electrodeposits [1]. In present, one of the method removing Cl from zinc sulfate solution is adding Cu^+ to form CuCl. So the CuCl residue is one of the most important intermediate material in zinc hydrometallurgy system. In order to realize the comprehensive utilization of resources, Kunming University of Science and Technology has developed a technique of removing chlorine of CuCl residue by microwave roasting. By this technique, the copper and chlorine can be recycled [2]. Because of the initial

* **Corresponding author:** JU Shao-hua, Associate Professor, PhD; Tel: +86-871-5147924; E-mail: shj_200801@126.com

moisture content of CuCl residue is about 36%, it must be dry. To meet the dechlorination requirement, the moisture content of CuCl residue should not exceed 5% [3].

Drying is defined as a process of moisture removal due to simultaneous heat and mass transfer [4] It is also one of the most important pretreatment in comprehensive utilization of resources and is the most energy-intensive process in industry. Cost-effective is a great importance given the prevailing insecurity in comprehensive utilization of resources throughout the world.

Response surface methodology (RSM) is an effective statistical technique for optimizing complex processes. The main advantage of RSM is the reduced number of experimental trials needed to evaluate multiple parameters and their interactions. Therefore, it is less laborious and time-consuming than other approaches required to optimize a process [5]. It was widely used in many industries to optimize the process variables [6-8].

In order to obtain the desired final moisture content below of 5% in a relatively short time and low temperature the response surface method is used to optimize the process, to provide the necessary technical parameters for the CuCl residue of drying technology. And the main phases of CuCl residue was analyzed by using XRD, after drying.

1. Experiment

1.1 Materials

The raw materials is the CuCl residue after dechlorination from a zinc hydrometallurgical plant in Yunnan Province, China. The initial moisture content of CuCl residue is about 36%.

1.2 Experimental device and experiment procedures

The drying equipment (Fig.1) was modified from a electro-thermostatic convection oven with the power of 1550 W and temperature control between 10-300°C developed by Shanghai(JY-136M). Moisture loss was measured by a thermogravimetric balance, which is connected to the computer. The schematic of the experimental system is shown in Fig. 1.

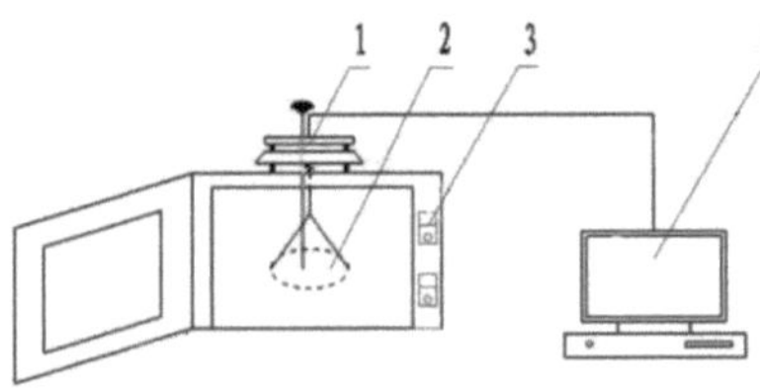

Fig.1 The schematic of microwave drying system
1-electronic balance; 2-quartz container; 3-temperature display instrument; 4-computer system;

The experiment procedures are as following: (1) an area of 50.24 mm^3 of the CuCl residue was putted into the quartz container as a thin layer and placed the disk in convection oven; (2) turned on the power of drying system and heated sample at a certain temperature with a air flow rat of 5.8 m^3/h for a certain time; (3) data collection; (4) taken out the material.

1.3 Date analysis

The moisture content of CuCl residue was calculated using the following equation:

$$M_c = \frac{m_t}{m} \times 100\% \quad (1)$$

Where M_c is the dried basis moisture content, %; m is the absolute amount of dried basis, kg; m_t is the moisture amount at the time of t, kg. The average atmospheric humidity of Kunming is about 27%, which was measured using measurer of temperature and humidity (Shanghai AZ8910).

1.4 Design of experimental

The CuCl residue belongs to high viscous materials with high moisture content. The temperature and material thickness has a deeply influence on energy transmission and steam diffusion. And drying time is the assurance of steam and energy transfer smoothly. Drying temperature (χ_1), duration (χ_2) and material thickness (χ_3) were selected as the independent variables to be optimized for the moisture content of CuCl residue.

In this work, the experimental data are analyzed using statistical software Design Expert version 7.1.5 (STAT-EASE Inc., Minneapolis, USA). The range of independent variables and their levels were presented in Table 1. The moisture content (γ) was taken as the response for the combination of the independent variables given in Table 2.

Table 1. Range of variables utilized in the experimental design

Variables	Coded variable level				
	-1.682	-1	0	1	1.682
Temperature(℃)	73.1821	80	90	100	106.818
Time(min)	66.3641	80	100	120	133.636
Thickness(mm)	5.923	10	17.5	25	26.0365

A statistical experimental design based on response surface methodology(RSM) can be an effective method of optimizing microwave drying parameters. A five-level experimental design was proposed by central composite design(CCD). Experimental data obtained from the CCD model experiments can be described in the form of the following Eq. (2):

$$\gamma = \beta_0 + \sum_{i=1}^{k} \beta_i \chi_i + \sum_{i=1}^{k} \beta_{ii} \chi_i^2 + \sum_{i=1}^{n-1} \sum_{j-i+1}^{n} \beta_{ij} \chi_i \chi_j \quad (2)$$

Where β_0 is the constant coefficient, β_i is the linear coefficient, β_{ii} is the quadratic coefficient and β_{ij} is the cross product coefficients, k is the number of factors studied and optimized in the experiment, χ_i, χ_j are the coded values of independent variables, $\chi_i\chi_j$ and χ_i^2 represent the interaction and quadratic, respectively.

2 Result and discussion

Effects of temperature, time and thickness on moisture content were studied during experimentation. The experimental conditions along with the results are shown in Table 2. The result is observed to vary in the range of 2.31% to 9.68%. The repeat runs at the center point of the variables (experiment runs 15-20) indicate the uncertainties involved in the experimental system.

Table 2 Experimental design matrix and results

Run	χ_1(°C)	χ_2(min)	χ_3(mm)	γ (%)
1	80	80	10	9.68
2	100	80	10	4.22
3	80	120	10	4.78
4	100	120	10	2.31
5	80	80	25	9.98
6	100	80	25	6.07
7	80	120	25	8.69
8	100	120	25	4.54
9	73.18	100	17.5	8.57
10	106.82	100	17.5	3.75
11	90	66.36	17.5	8.33
12	90	133.64	17.5	4.02
13	90	100	4.89	3.32
14	90	100	30.11	7.33
15	90	100	17.5	4.51
16	90	100	17.5	4.28
17	90	100	17.5	4.61
18	90	100	17.5	4.78
19	90	100	17.5	4.34
20	90	100	17.5	4.53

2.1 Development of regression model

Table 3 presents the results of various models tested with the experimental data, and the F-value is calculated for each type of model. A high F-value coupled with the low P-value indicates the suitability of the model. As software recommended, the quadratic model is the most suitable. Several indicators were used to evaluate the adequacy of the fitted model and the results are shown in Table 4. The coefficient of determination (R^2), the adjusted determination coefficient (adj.R^2) and model significance (F-value) are used to judge the adequacy of the model[9, 10].

Table 3 Statistical parameters for sequential models

Source	Sum of Squares	df	Mean Squares	F-Value	p-value	R-Squared
Linear	17.71	11	1.61	48.77	0.0002	0.8172
2FI	14.78	8	1.85	55.94	0.0002	0.8472
Quadratic	2.78	5	0.56	16.83	0.0038	0.9699
Pure Error	0.17	5	0.033			

Table 4 Analysis of variance (ANOVA) for response surface quadratic model

Source	Sum of Squares	Degree of freedom	Mean Square	F-Value	p-value Prob > F
Model	94.86	9	10.54	35.8	< 0.0001
χ_1	42.52	1	42.52	144.4	< 0.0001
χ_2	20.86	1	20.86	70.85	< 0.0001
χ_3	16.55	1	16.55	56.21	< 0.0001
$\chi_1\chi_2$	0.95	1	0.95	3.21	0.1034
$\chi_1\chi_3$	2.11E-03	1	2.11E-03	7.18E-03	0.9342
$\chi_2\chi_3$	1.99	1	1.99	6.76	0.0265
χ_1^2	6.09	1	6.09	20.69	0.0011
χ_2^2	6.19	1	6.19	21.03	0.001
χ_3^2	1.82	1	1.82	6.16	0.0324

R^2=0.9699; adj.R^2=0.9428;

For the model fitted, the coefficient of determination (R^2) was 0.9699, indicating that only 3.01% of the total variations was not explained by the model. The model F-value of 35.8 implies that model is significant. P-value less than 0.0001 shows that model terms are significant. The value of the adjusted determination coefficient (adjusted R^2 = 0.9428) also confirmed that the model was highly significant.

The regression coefficient values of Eq. (2) were listed in Table 4. Based on the F-value , χ_1 shows the largest F-value of 144.4, indicating that temperature had the most significant effect on moisture content of the CuCl residue, compared to χ_2 and χ_3. The effect of drying time on moisture content is more significant than material thickness, with F-values being 70.85 and 56.21, respectively. And the P-values also were used as a tool to check the significance of each coefficient. It can be seen from this table that the linear coefficients (χ_1, χ_2, and χ_3) and the quadratic term coefficients (χ_1^2, $\chi_2^2$and χ_3^2) were significant, with very small P-values ($P < 0.05$). The other term coefficients were not significant ($P > 0.05$).

By applying least squares method and multiple regression analysis on the experimental results, the following equation is found to relate the dependent variable after eliminating the in-significant parameters as shown in Eq. (3).

$$\gamma = 4.50 - 1.76\chi_1 - 1.24\chi_2 + 1.10\chi_3 + 0.65\chi_1^2 + 0.65\chi_2^2 + 0.35\chi_3^2 \quad (3)$$

The predicted values are calculated using the regression model.As can be seen from Fig.2, the predicted values are close to the experimental data.

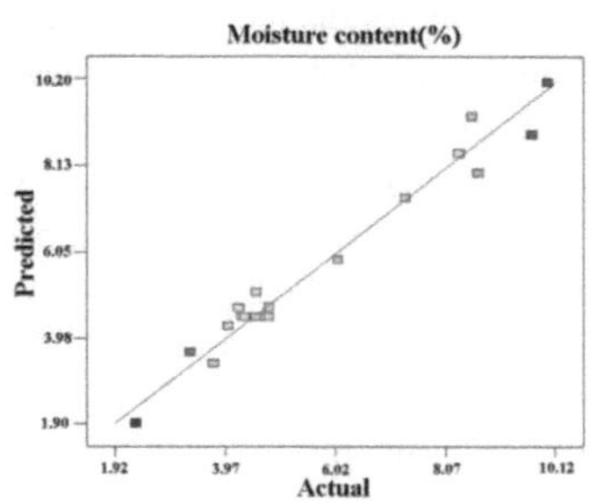

Fig.2 Predicted vs experimental moisture content of CuCl residue

2.2. Interactions of the factors

The moisture content of CuCl residue over different combinations of independent variables can be visualized through three-dimensional view of response surface plots.

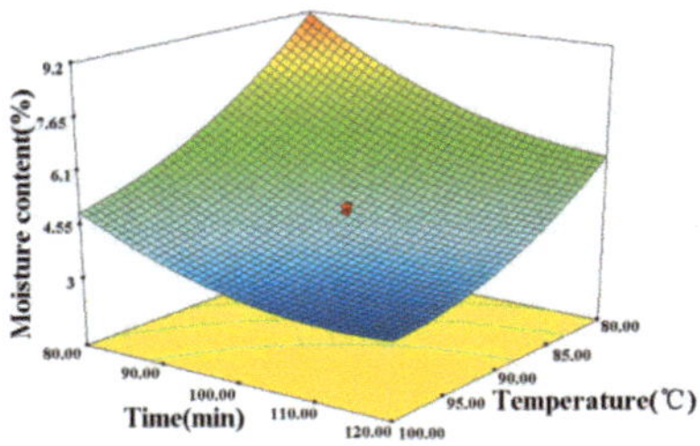

(a)The interactive effects of temperature and time

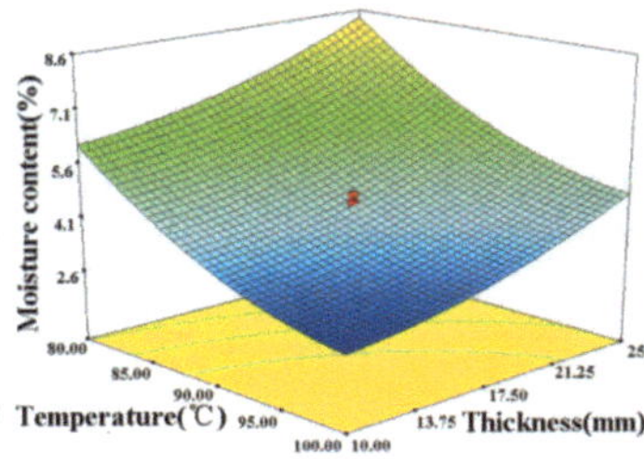

(b)The interactive effects of the temperature and thickness

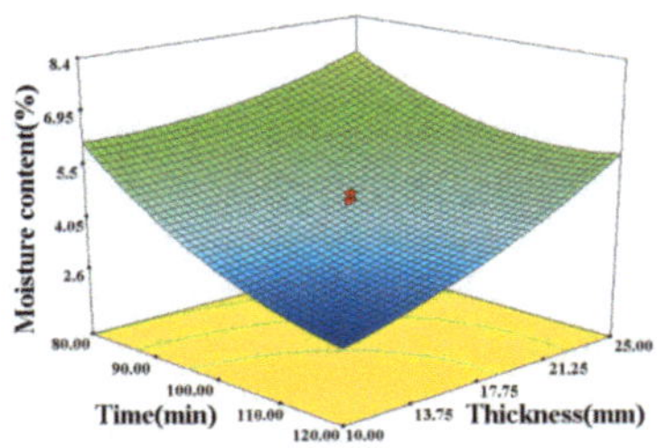

(c)The interactive effects of the time and thickness

Fig.3 The moisture content of CuCl residue over different combinations of independent variables

Fig.3(a) shows the three-dimensional response surfaces of the combined effect of drying temperature and time on moisture content of CuCl residue (thickness is fixed at 17.5mm). Fig.3(b) shows the three-dimensional response surfaces of the combined effect of drying temperature and material thickness (drying time is fixed at 100 min). Fig.3(c) showed the three-dimensional response surfaces of the combined effect of material thickness and drying time on the moisture content, while the drying temperature was fixed at 90°C. From response surface curves, it is easy and convenient to understand the interaction effects between two independent variables. As Fig.3(a) shows the moisture content decreases with an increase in the temperature and time. An increase in temperature will lead to an increase in evaporation rate. Similarly, an increase in the time can increase the extent of the diffusion. It can be seen from Fig.3(b) that the moisture content decreased evidently with decreasing of material thickness from 25mm to 17.5mm, but below 17.5mm, moisture content of CuCl residue decreased slowly. An decrease in material thickness will lead to an increase in evaporation rate. As can be seen from Fig. 3(c), the moisture content decreases with increase in drying time and decrease in material thickness. An increase in drying duration is expected to ensure internal moisture content diffusion in the material. On the contrary, an increase of thickness bring to bigger diffusion resistance.

Drying is a complicated heat and mass transfer process. The material thickness has a deeply influence on energy transmission and steam diffusion. The thicker the material thickness is, the greater the temperature gradient is. In this work, the mass transfer process will be weakened by the different direction of temperature gradient and steam diffusion, and this weakening effect will increase with the raise of material thickness.

2.3 Process optimization

In order to obtain the desired final moisture content below of 5% in a relatively short time and low temperature, the drying parameters were optimized by the Design Expert software. The experimental condition of microwave drying CuCl residue was given in Table 5. The predicted moisture content 4.11% while the experimental data was 4.13%. This experimental result agrees well with the prediction from the response surface, which indicates the success of the optimization process.

Table 5 The optimum conditions for the process

Time(min)	Thickness(mm)	Temperature(°C)	Moisture content (%)	
			Predicted	Experimental
89.49	13.47	95.48	4.11	4.13

2.4 The main Phase of CuCl residue after microwave drying

The main chemical compositions and phases of the CuCl residue after being dried are shown in Table 6 and Fig.4. It can be found from Fig. 5 (XRD-7000, Shimadzu Corporation) that the main phases of CuCl residue are CuCl, Cu_2O and ZnS.

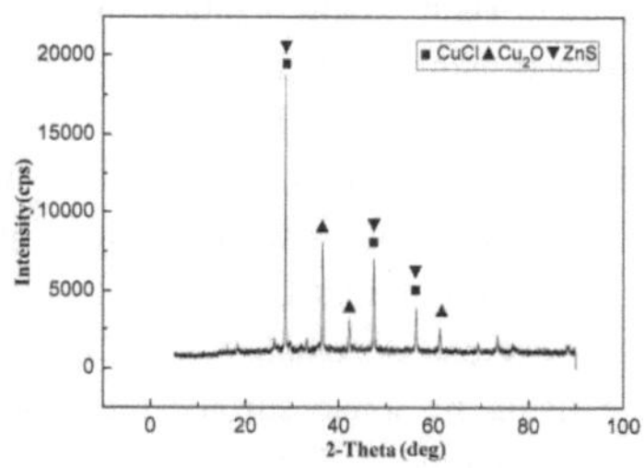

Fig. 4 X-ray diffraction pattern of microwave dried CuCl residue

Table 6 The Chemical elements of CuCl residue after drying

Composition	Zn	Cu	Cl	S
Content(%)	6.55	54.68	17.27	1.74

3 Conclusion

(1) The effects of operating parameters of drying time, material thickness and drying temperature on moisture content is investigated using analysis of variance (ANOVA), to identify the significant parameters.

(2) The optimum parameters for drying CuCl residue has been identified to be an drying time of 89 min, material thickness of 13 mm and drying temperature of 95°C. The optimum conditions result in moisture content of 4.11%, which corresponds to moisture content of 4.13% and completely meet requirement of microwave roasting CuCl residue dechlorination.

(3) The main phase of CuCl residue are CuCl, Cu_2O and ZnS.

Acknowledgements

The authors gratefully acknowledge the National Natural Science Foundation of China (NO. 51104073) for funding this work. The project was also funded as a Technology of People's Republic of China.Yunnan Provincial Science and Technology Innovation Talents scheme-Technological Leading Talent (NO. 2013HA002)

Reference:

[1]M.K. Jha, V. Kumar and R.J. Singh, "Review of hydrometallurgical recovery of zinc from industrial wastes," *Resources, Conservation and Recycling,* 33 (2001), 1–22.

[2]S. Lu, Y. Xia and C. Huang, "Removing chlorine of CuCl residue from zinc hydrometallurgy by microwave roasting," *Journal of Central South University*, 21 (2014), 1290-1295

[3]S. Lu, Y. Wei, S. Ju, J. Peng, C. Huang, G. Wu, L. Zhang, "The Contrastive Studies of Microwave and Conventional Roasting CuCl Residue from Zinc Hydrometallurgy, " *Characterization of Minerals, Metals, and Materials,* (2013), 529-540.

[4]O.V. Ekechukwu, "Review of solar-energy drying systems I: an overview of drying principles and theory," *Energy Conversion and Management*, 40 (1999), 593-613.

[5]C. Liyana-Pathirana, F. Shahidi, "Optimization of extraction of phenolic compounds from wheat using response surface methodology," *Food Chemistry,* 93 (2005), 47-56.

[6]X.L. Bai, T.L. Yue and Y.H. Yuan, "Optimization of microwave - assisted extraction of polyphenols from apple pomace using response surface methodology and HPLC analysis," *Journal of Separation Science*, 33 (2010), 3751-3758.

[7]A. Aleboyeh, N. Daneshvar and M.B. Kasiri, "Optimization of C.I. Acid Red 14 azo dye removal by electrocoagulation batch process with response surface methodology," *Chemical Engineering and Processing,* 47 (2008), 827-832.

[8]A. R. Amani-Ghadim, S. Aber and A. Olad, "Optimization of electrocoagulation process for removal of an azo dye using response surface methodology and investigation on the occurrence of destructive side reactions," *Chemical Engineering and Processing*, 64 (2013), 68-78.

[9]S. Lu, S. Sun and J. Lv, "Optimization of Microwave Roasting for Dechlorination of CuCl Residue from Zinc Hydrometallurgy," *Journal of Microwave Power and Electromagnetic Energy*, 48 (2014), 1290.

[10]G. Chen, J. Chen and C. Srinivasakannan, "Application of response surface methodology for optimization of the synthesis of synthetic rutile from titania slag," *Applied Surface Science*, 258 (2012), 3068-3073.

Drying, Roasting, and Calcining of Minerals
Edited by: Thomas P. Battle, Jerome P. Downey, Lawrence D. May, Boyd Davis, Neale R. Neelameggham, Sergio Sanchez-Segado, and P. Chris Pistorius
TMS (The Minerals, Metals & Materials Society), 2015

APPLICATION OF KUMERA STEAM DRYERS IN MINERAL PROCESSING

Carl-Gustav Berg, Shaolong Chen, Hannu Mansikkaviita

Kumera Corporation, Kumerankatu 2, 11100 Riihimäki, Finland

Keywords: Energy efficiency, emission, sustainable, maintenance

Abstract

Kumera Steam Dryer has been developed particularly for nonferrous metal concentrates with steam as heating source generated from waste heat boilers. Its unique structure enables the dryer to work at high capacity, up to 240 t/h, and high availability, over 95%. With half a century of engineering experience, and more than twenty industrial steam dryer installations worldwide during last 15 years, Kumera has demonstrated significant benefits of the steam drying technology in terms of energy efficiency and emission reduction to the end users. The dryer of this type can also be applied for other suitable mineral materials.

Introduction

Kumera supplied equipment serves large number of moist or high temperature solids, metal concentrates and fuels handled in processes like smelting, combustion and gasification. Typically low graded fuels or minerals contain 8-70 % water and 5-60 % ash. Removal of water by steam drying, and heat by cooling, is either a necessity because of subsequent process requirements, or simply very beneficial because of cost efficiency. Unnecessary moisture load and heating of water vapor, and heat losses in sub sequent process steps, can thus be eliminated, resulting in major energy savings and a substantial reduction in CO_2 and other emissions like SO_2, NO_x, metals and dust.

Energy efficiency of drying can be significantly enhanced by selection of drying technology. Modern indirect steam drying eliminates the need to use extremely hot convection gases as source of drying energy and hence securing both the product quality and eliminating the self-ignition risk. Furthermore, it can be shown that the CO_2 emission from use of fossil fuel for drying is eliminated together with SO_x and NO_x emissions. Typical energy consumption of indirect steam drying is 2800 - 3800 MJ/tH2O (for each ton of evaporated water), [1,3,4,5]. Benefits are illuminated by including applicability tests of moist oil shale and lignite. Important outcome from steam drying tests were; good product quality and elimination of self-ignition risks, when drying is done with indirect low pressure steam and low drying gas temperatures.

Kumera Steam Dryer

Energy efficiency and emissions in drying, structure and design of Kumera Steam Dryer, and applicability of technology for moist fuels like oil shale and lignite, are illuminated in this heading to the reader. One reason to select indirect steam drying is the large single line capacity, currently maximum feed of copper concentrate is 240 t/h, at 9 % moisture, (see Figure 1) [1]. The size of such a dryer is Ø 4.4 m in diameter. Rotating speed adjustable from 0.5 rpm to 2 rpm.

Figure 1. Kumera Steam Dryer, 240t/h feeding capacity for copper concentrate.

Other reasons to select indirect steam drying instead of direct heated drum dryers are lower specific energy need in drying and smaller off-gas cleaning plant.

The heating elements

The modular type steam tube arrangement (Figure 2) makes it possible to modify the heating surface area or arrangement of steam tubes without having to remove the complete tube package from the drum. The possibility to remove individual heating elements reduces the required lifting capacity of the equipment for hoisting and handling the elements. Modification of a dryer is also easily done just by changing the element design (numbers, long or short, etc.) if necessary.

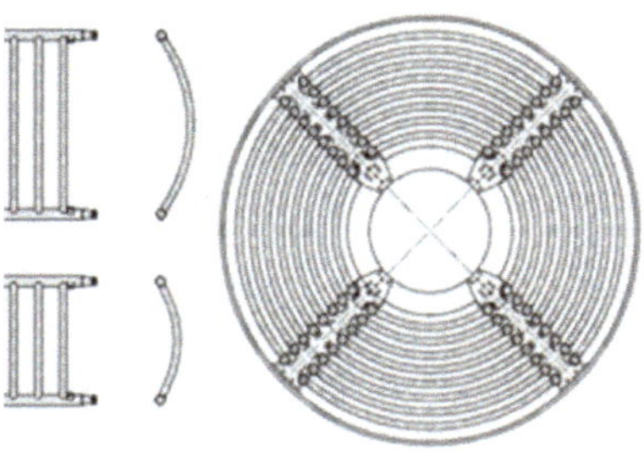

Figure 2. Heating elements in cross section of the dryer.

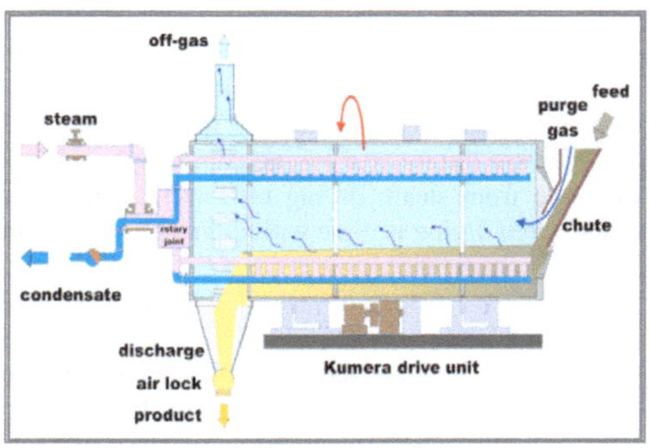

Figure 3. Indirect steam drying.

The modular type heating pipe system is therefore flexible and easily fitted to a large shell diameter, and does not limit the size of the equipment. Domestic and international patents for the design have been granted since 2000.

Indirect steam drying

Indirect Kumera Steam Dryer ensures specific needs like good material flow and gentle drying (see Figure 3). Flow of material is important and copper concentrate can be held as an ideal material to be dried in the Kumera Steam Dryer. Typical dryer performance values for copper concentrate are shown in Table 1.

Table 1. Kumera Steam Dryer performance, copper concentrate, at oil price 100 USD/barrel [1].

Performance values	8500 h/y
• Concentrate feed	240 t_{feed}/h
• Moisture in	9 %
• Moisture out	0.2..0.3 %
• Water removed by evap.	21 t/h
• Heat consumption	337 MJ/t_{feed}
• Plant wide influence, at 240 t_{feed}/h is:	
• CO_2 reduction	≈ 70 000 t/y
• SO_2 reduction	≈ 700 t/y
• NO_x reduction	≈ 1000 t/y
• Net* energy saving, at least	> 70 000 MW_th/y
⇒ Net* money saving, at least	> 3 M€/y
* Since, a conventional oil heated direct dryer uses at least 115 MJ/t_{feed} more heat than a steam dryer.	

Pilot tests

Pilot tests have been performed to copper concentrates, mining residues, apatite, oil shale and lignite as shown in Figure 4.

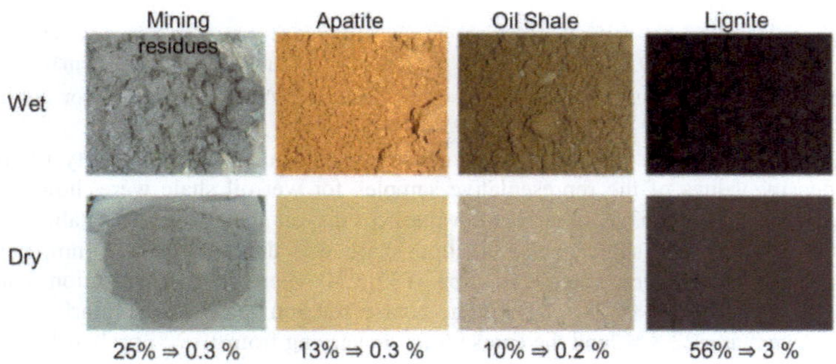

Figure 4. Tested materials and moisture change ranges.

Pilot tests were done with a transportable Kumera Steam Dryer system that includes heat recovery drying (Figure 5).

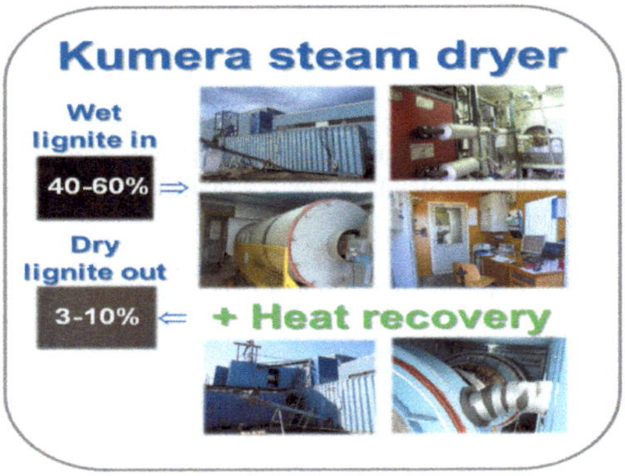

Figure 5. Transportable pilot system setup.

Lignite and oil shale properties

Hall flow meter was used as a standardized measuring device for apparent densities (Figures 6 and 7) and flowability. The meter has a conical crucible of volume 50 cm^3 with either 1/10" orifice or 2/10" orifice. The 2/10" orifice was used for dry product.

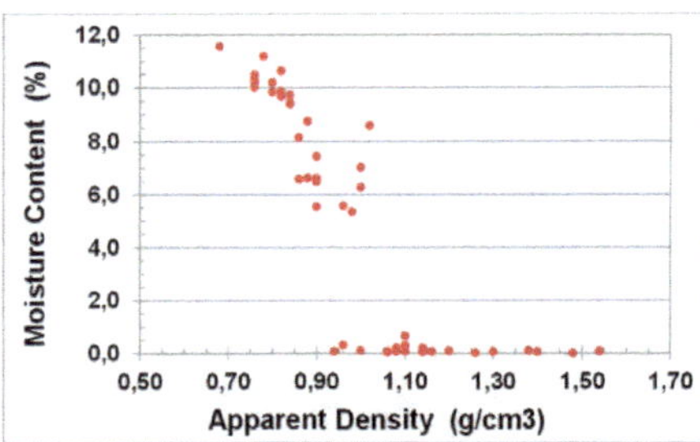

Figure 6. Apparent density for oil shale.

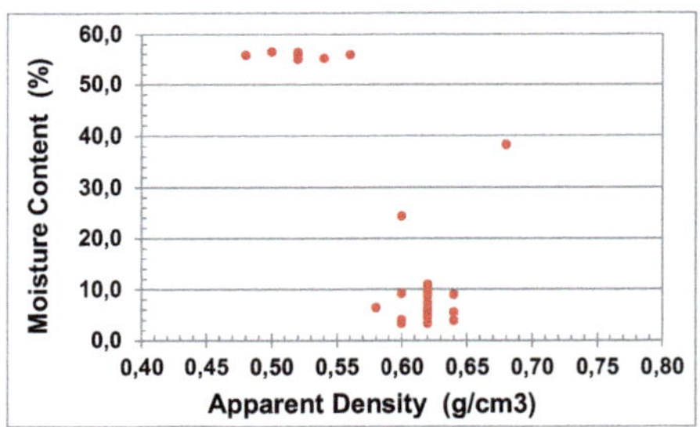

Figure 7. Apparent density for lignite.

There was lots of variation in the apparent density, again due to heterogeneity of the feed. Apparent density values of the representative samples for wet oil shale were, however, quite constant at 0.75 – 0.85 g/cm3. Similarly, the values for dry oil shale were reasonably consistent at 1.05 – 1.20 g/cm^3. Oil shale coarse fraction share was defined for > 2 mm, performed measurements gave then average coarse fraction of 37.5 %, where the coarse fraction share could vary from 23.6 % to 60 % (Figure 8). Lignite coarse fraction share was similarly defined for > 2 mm, average being 22.8 %, and the coarse fraction varying from 13.2 % to 29.6 %, Figure 9.

Flowability measurements with the Hall meter were done only for the dry product and sieved to < 2 mm fraction before measuring the time for flowing.

Figure 8. Oil shale fine and coarse fractions.

Figure 9. Lignite fine and coarse fractions.

Pilot tests for lignite and oil shale

Most important operational parameters measured during pilot tests are presented in Table 2. Oil shale dries up easily and reasonably quickly to below 1% (Figure 10). Residence time for oil shale drying from 10% moisture to below 1% is less than 30 min.

Table 2. Parameters to be measured and monitored together with rpm.

Parameter	Time of Measurement
Moisture measurement	30 min – wet and dry
Apparent density	30 min – wet and dry
Flowability	30 min – wet and dry
Feed temperature	Before drying
Discharge temperature	Continuous
Steam temperature	Continuous
Off-gas temperature	Continuous
Feed flow rate	Continuous
Steam flow rate	Continuous
Off-gas flow rate	Continuous
Dust measurement	No
CO measurement	Occasional

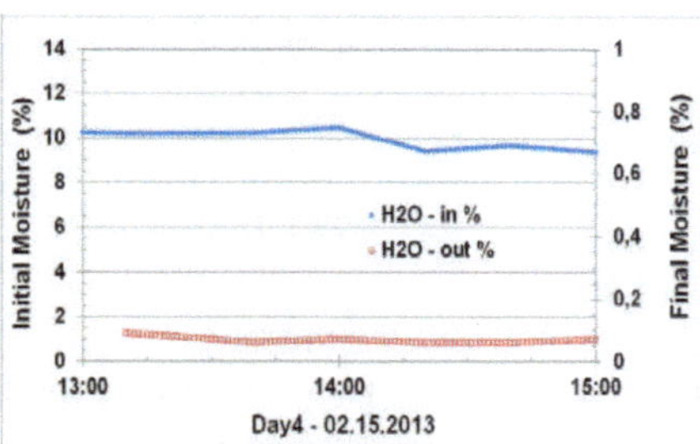

Figure 10. Oil shale moisture in-out.

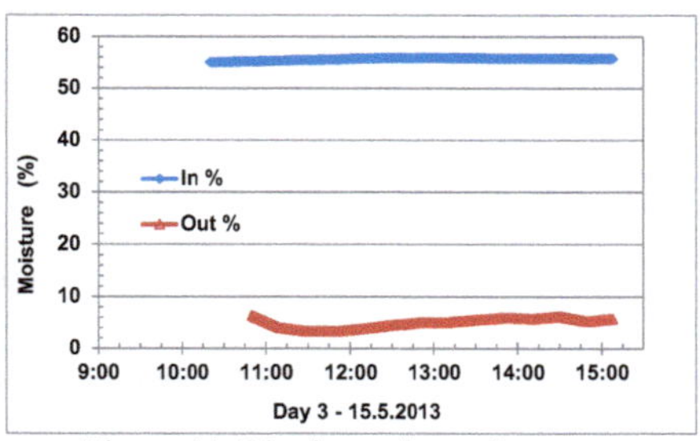

Figure 11. Lignite moisture in-out.

Lignite dries up easily down to 5 % H2O (Figure 11). The residence time is then however, clearly longer than for oil shale. A final moisture content of 2 % to 5 % H2O can only be achieved with a clearly prolonged residence time. Pilot test process temperature for Figures 10 and 11 are shown in Figures 12 and 13.

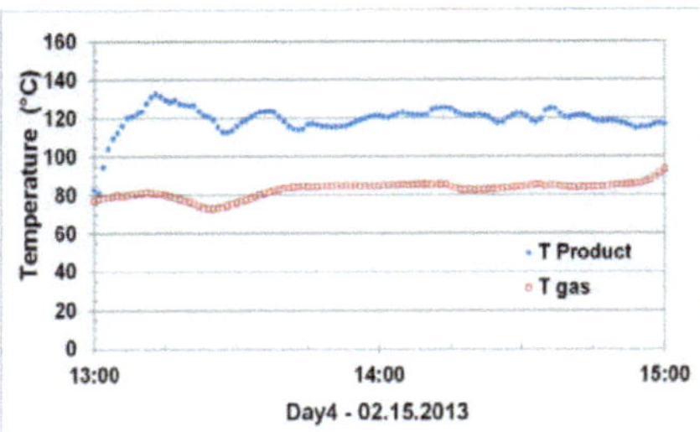

Figure 12. Oil shale temperatures.

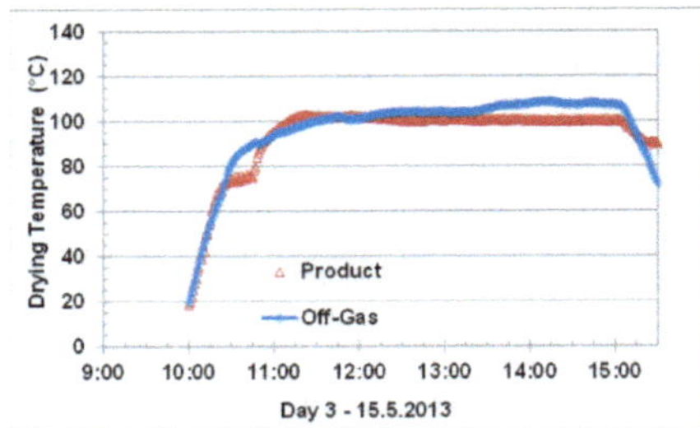

Figure 13. Lignite temperatures.

Pilot tests were conducted with 4-10 bar (g). Stable drying with no or little product degradation was proven by measuring no or very low, CO, carbon monoxide formation figures.

Results and impact of Kumera Steam Dryer

Wet material handling was done from one ton sacks emptied to hopper and fed by belt conveyors into the Kumera Steam Dryer pilot. About 6 ton of wet oil shale was dried and 5 ton wet lignite. Dried products were shipped back to customers shortly after completion of tests. Test results were analyzed in detail with theory from drying literature [2,3,4,5], and calculated into the form of Figure 14.

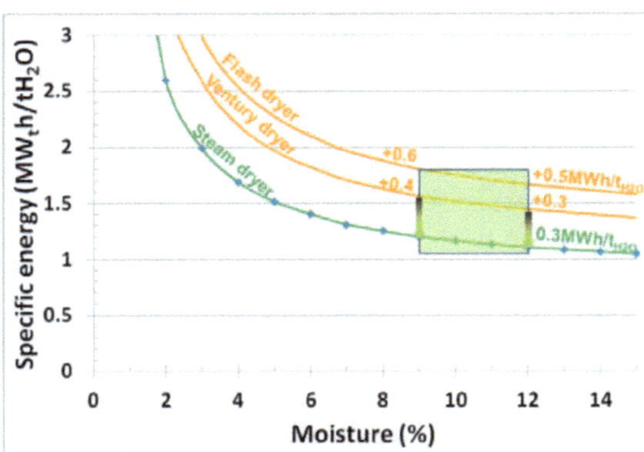

Figure 14. Specific energy for drying of oil shale as a function of the inlet moisture content [1,2,3,4,5].

Separate drying is favorable due to a number of benefits: savings in operational expenses and significantly reduced environmental footprint. Kumera Steam Dryer is currently a technology on the dryer market that significantly decreases the environmental footprint related to non ferrous concentrate drying.

Amongst the available drying technologies, steam drying is recognized as the best practice today. It consumes significantly less energy for water removal from the concentrates, while gas emissions can be almost entirely eliminated. Simultaneously, the dust emission can be substantially reduced, while at the same time also enhancing the product recovery for better profit. There has been a clear trend in a demand for increasing the feed capacity over the years. Currently, the largest steam dryers have wet copper concentrate feed capacity of 240 t/h. Kumera has also supplied large steam dryers, to smelters drying nonferrous concentrates, like zinc and lead. Consequently, due to both the very high capacity and very high availability, smelters treating over 1 million tons per year of concentrate can manage such quantity with just one Kumera Steam Dryer. Kumera Steam Dryer single line product feeding capacity for oil shale and lignite are clearly lower because of lower heat transfer, and apparent densities, and higher moisture content in the dryer feed flow. Single line capacities designed so far are in the range, from 30 t/h to 45 t/h evaporation, which yields a 4 line drying plant capacity of 120 t/h evaporation, when the drying plant is using low steam pressure. Specific energy need to dry oil shale with various drying methods and without heat recovery is shown in Figure 14.

Utilization of heat recovery in drying

The heat recovery example included here was made for oil shale. It includes a steam dryer system where waste heat from the steam dryer is utilized in a pre-dryer. The moisture content in oil shale feed 12 % and out 0.4 %. The change in primary energy usage are shown in Figures 15 and 16.

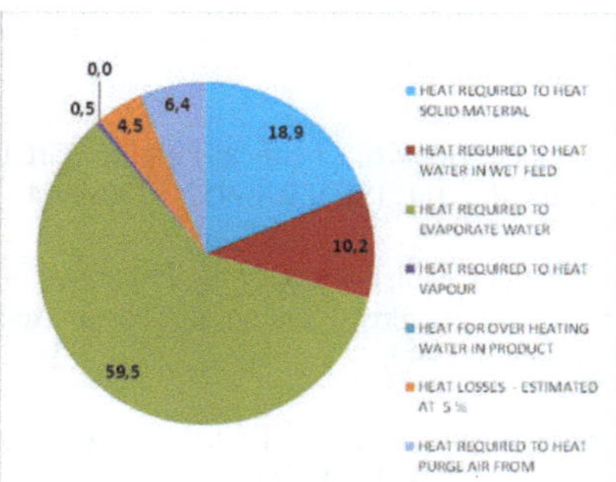

Figure 15. Energy distribution in % for oil shale drying without heat recovery.

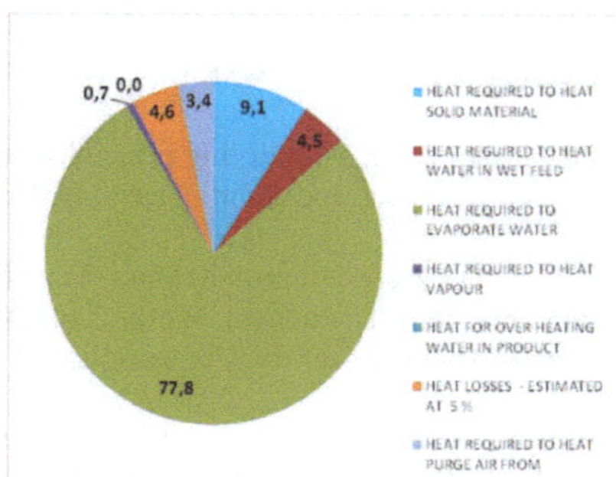

Figure 16. Energy distribution in % for oil shale drying with heat recovery to material feed.

The preliminary tests for the concept of utilizing the waste heat from drying (wet air with latent heat of water vapor) was working well for both oil shale and lignite. Pilot experiments showed that Kumera Steam Dryer system is capable of energy efficient drying of both moist oil shale and lignite. While oil shale is easily dried below 1%, lignite is dried easily to about 8%. Material is not sticky and thus applicable for rotating drying. The material heats up during pre-heating and

water content can simultaneously be reduced by 15 % to 20 %, and thus, improving already good energy efficiency by about 25 %.
Kumera Steam Dryer system, with heat recovery will hence be very energy efficient and can be cost efficiently designed to upgrade moist mineral feed or moist power plant fuel.

Conclusions

Kumera Technology Center has piloted the Kumera Steam Dryer for a number of nonferrous concentrates and minerals including abrasive oil shale and less abrasive lignite. The Kumera approach was to make an energy efficient drying system that is based on combining the best properties of the existing steam dryer and advanced heat recovery solutions. Special attention was paid to overcome problems of excessive wear and clogging of steam tubes and to reduce maintenance operations.
Kumera Steam Dryer with heat recovery drying is now a new cost effective concept for drying of moist nonferrous concentrates, and minerals or moist fuel fired power plants. Thereby increasing profit margins by reducing energy need, which cuts emission costs or increases the fixed reserve electricity production potential. All above and the possibility to use waste heat boiler steam returns extensive reductions in: carbon dioxide, sulfur dioxide, nitrogen oxides, ash, dust and heavy metals. Kumera Steam Dryer is hence a key when creating cost efficiency without compromising sustainability.

References

1. C-G. Berg et al. Optimum drying solution for copper concentrate. Proceedings of Copper 2013, Santiago, Chile, December 4-8, 2013. Volume 3, 49–60.

2. Zdzislaw Pakowski, Adamski R. and Kwapisz S., Effective diffusivity of moisture in low rank coal during superheated steam drying at atmospheric pressure. Chemical and Process Engineering 2011, 33 (1), 43-51. DOI: 10.2478/v10176-012-0004-3.

3. Carl-Gustav Berg, Pulp drying principles and applications. Book 6, Chemical Pulping Part 1, Papermaking Science and techn.. Fapet Oy, Helsinki, Finland, 2011. ISBN 978-952-5216-00-4.

4. Carl-Gustav Berg, Heat and Mass Transfer in Turbulent Moist Air Drying Processes — Experimental and Theoretical Work. Åbo Akademi University Press. Doctoral Thesis. Acta Academiae Aboensis, ser. B. Monograph. 153 p. Åbo, Finland, 1999.

5. Arun S. Mujumdar, Handbook of Industrial Drying, Third Edition, Edited by Arun S. Mujumdar, CRC Press 2007, ISBN: 978-1-57444-668-5, New York, USA, 2007.

Drying, Roasting, and Calcining of Minerals
Edited by: Thomas P. Battle, Jerome P. Downey, Lawrence D. May, Boyd Davis, Neale R. Neelameggham, Sergio Sanchez-Segado, and P. Chris Pistorius
TMS (The Minerals, Metals & Materials Society), 2015

DIELECTRIC PROPERTIES AND MICROWAVE DRYING CHARACTERISTICS OF CuCl RESIDUE

Guo Zhanyong[1,2,3], Ju Shaohua[1,2,3]*, Lei Ting[4], Peng Jinhui[1,2,3], Zhang Libo[1,2,3], Jiang Feng[1,2,3]

1. National Local Joint Laboratory of Engineering Application of Microwave Energy and Equipment Technology, Kunming, Yunnan 650093, China;
2. Key Laboratory of Unconventional Metallurgy, Ministry of Education, Kunming University of Science and Technology, Kunming, Yunnan 650093, China;
3. Faculty of Metallurgical and Energy Engineering, Kunming University of Science and Technology, Kunming, Yunnan 650093, China;
4. Kunming Metallurgy College, Kunming, Yunnan 650093, China.

Keywords: CuCl residue; Microwave drying; Dielectric properties; Heating curve

Abstract

The basics of microwave drying process of CuCl residue were studied by comparing the dielectric property of the residue with different moisture contents. The dielectric properties were measured by microwave cavity perturbation method. The dielectric constant and dielectric loss factor of the CuCl residue have a linear correlation with the moisture content. With this relationship, we can realize on-line measurement of moisture content of the CuCl residue by dielectric properties. The microwave drying experiments with different thicknesses (1-5cm) of CuCl residue were carried out, and the heating curve of the materials under different microwave powers (250-450W) was made. From these parameters, an optimized operation parameter of microwave roasting technology, the thickness of CuCl residue filter cake to be microwave roasted should be 2cm , was deduced.

Introduction

Cl is a harmful element in zinc hydrometallurgical system. Thus, Cl- contained in the solution should be removed before electrodeposits[1,2]. In present, one of the method removing Cl from zinc sulfate solution is adding Cu^+ to form CuCl. So the CuCl residue is one of the most important intermediate material in zinc hydrometallurgy system. In order to realize the comprehensive utilization of resources, Kunming University of Science and Technology has developed a technique of removing chlorine of CuCl residue by microwave roasting. By this technique, the copper and chlorine can be recycled. Because of the initial moisture content of CuCl residue is about 36%, it must be dry. To meet the dechlorination requirement, the moisture content of CuCl residue should not exceed 5%.

In recent years, microwave irradiation has been widely investigated due to its unique heating advantage[3-10]. Microwave heat materials at molecular level, which leads to

* **Corresponding author:** JU Shao-hua, Associate Professor, PhD; Tel: +86-871-5147924; E-mail: shj_200801@126.com

homogeneous and quick thermal heating. Compared with traditional heating techniques, microwave heating provides additional advantages such as higher heating rates, selective heating, precise control of temperature, small equipment size, high heating efficiency, and so on[11-15]. The dielectric property （ε'、ε"、tanδ）of materials is an important physical indicator in the field of microwave chemistry[16-18]. It can largely describe the behavior characteristics of materials in microwave heating and drying process.

The study on the wave-absorption and temperature-rising behavior of CuCl residue at different moisture content is of great importance to the new process of drying CuCl residue by microwave roasting. By the method of microwave cavity perturbation, this paper measured the dielectric property（ε'、ε"、tanδ）of CuCl residue at different moisture content, providing the theoretical basics for the industrial application of drying CuCl residue by microwave roasting.

1 Dielectric measuring method

1.1 Raw material

The raw materials is the CuCl residue after dechlorination from a zinc hydrometallurgy plant in Yunnan Province. After chemical composition analysis, the Chemical composition of CuCl residue can be shown in Table1.

Table 1 The Chemical Elements of CuCl Residue from Zinc Hydrometallurgy Particles

Composition	Cu	Cl	S	Zn
Content(%)	54.68	17.27	4.74	5.63

The chlorine content is up to 17.27% and the main phase is CuCl and Cu_2O.

1.2 Measuring principles

The resonant cavity perturbation method is widely used as its high measurement accuracy and sphere of application[19-21].

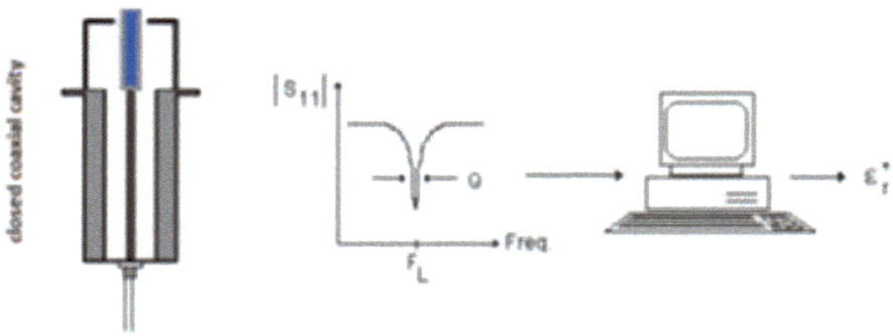

Fig. 1 Schematic diagram for the measuring principle of dielectric property measuring instrument

A microwave experiment is based on the cavity perturbation method, employing the single resonant mode. The change in the cavity characteristics in the presence of a sample is measured. If there is little sample in the resonant cavity, the principles can be expressed as follows:

$$\frac{\Delta\omega}{\omega} = -\omega_0(\varepsilon_r^{'} - 1)\int_{V_e} E_0^* \bullet E dv / (4W) \quad (1)$$

$$\frac{1}{Q}-\frac{1}{Q_0}=2\varepsilon_0\varepsilon_r^*\int_{V_e}E_0^*\bullet Edv/(4W) \quad (2)$$

$$W=\int_V[(E_0^*\cdot D_0+H_0^*\cdot B_0)+(E_0^*\cdot D_1+H_0^*\cdot B_1)]dv \quad (3)$$

where $\Delta\omega=\omega-\omega_0$, $\Delta\omega$ is the angular frequency deviation;ω_0 is the resonance frequency of cavity in the unperturbed condition;ω is the corresponding parameters of the cavity loaded with the sample; ε_r' and ε_r'' are the real and the imaginary part of the complex permittivity of the sample, respectively;V and V_e are the volumes of the cavity and the sample, respectively; E_0^*, D_0, H_0^* and B_0 are the fields in the interior of the sample; E is the field strength of cavity; d_v is the elemental volume; Q_0 is the resonance frequency of cavity in the unperturbed condition, respectively; Q is the corresponding parameters of the cavity loaded with the sample; W is the storage energy; D_1 and B_1 are the added value of the electric displacement and magnetic induction intensity.

1.3 Dielectric properties analysis result

Complex permittivity is the fundamental property that determines the microwave absorption of an ore [9]. Complex permittivity ε can be expressed as

$$\varepsilon=\varepsilon'-j\varepsilon'' \quad (1)$$

Where $j=\sqrt{-1}$; the real part (ε') is often referred as the dielectric constant, which reflects the ability of the material to store electromagnetic energy within its structure; the imaginary part (ε'') is the dielectric loss factor that characterizes the ability of the material to convert the stored electromagnetic energy into thermal energy. Another important parameter is loss tangent that describes how well the material dissipates stored energy into heat at a given frequency and temperature. Loss tangent (tanδ) can be expressed as

$$\tan\delta=\frac{\varepsilon''}{\varepsilon'} \quad (2)$$

The dielectric property（ε'、ε''、tanδ）of CuCl residue at different moisture content as a function of moisture content at room temperature is shown in Fig.2.

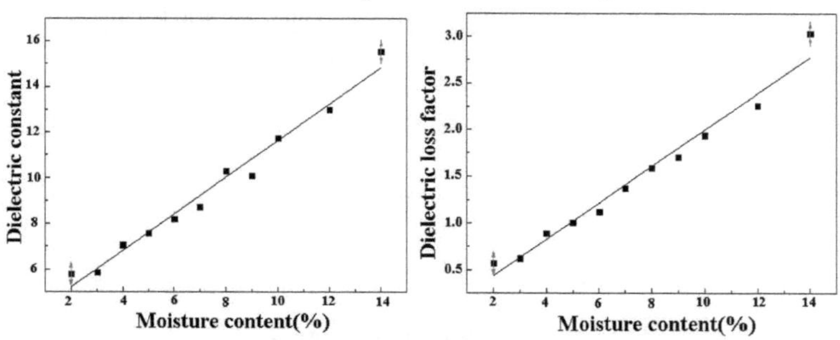

(a) Dielectric constant vs moisture content (b) Loss factor vs moisture content

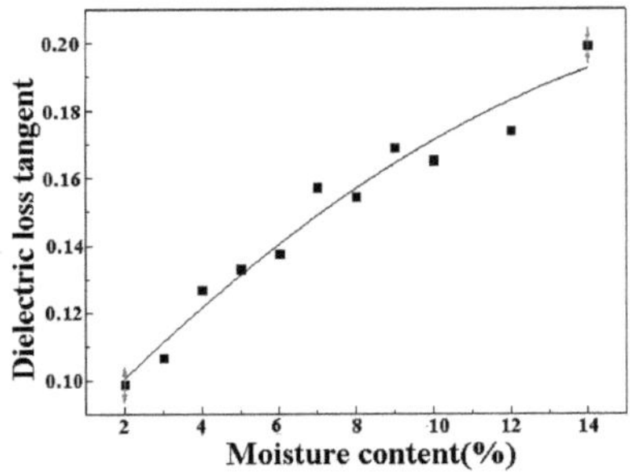

(c) Dielectric loss tangent vs moisture content

Fig.2 The relationship between dielectric properties and moisture content

Fig 2 shows that the impact of moisture content on the dielectric properties is significant. Fig. 2(a) and 2(b) show that the dielectric constant (ε') and dielectric loss factor(ε") of materials have a linear correlation with the increase of moisture content. The characteristics of dielectric constant increase with the moisture content of material are also present in most of the materials which has high moisture content[22-24]. By linear approximation of the curve, the linear equation of dielectric constant and dielectric loss factor about moisture content can be drawn and the linear correlation coefficient is 0.9873 and 0.9841, respectively. The result can be seen in Table2. By inverse algorithms of the equation, the linear equation that well describes moisture content about dielectric constant can be achieved. According to this linear equation, the on-line moisture content measuring instrument can be designed.

The dielectric loss angle tangent (tanδ) of the material as a function of moisture content at normal temperatures is shown in Fig.2(c). We can get the value of dielectric loss angle tangent by using the data of dielectric constant and dielectric loss factor. The variation in $\tan\delta$ with moisture content appeared to be linear but a quadratic equation relationship fitted the data. That is to say, the higher the moisture content is, the higher of the efficient for CuCl residue to convert microwave power into inner power.

Table2 is the Regression equations on moisture contents and dielectric parameters. The value of R^2 is not high enough. The causes of such a situation may be the influence made to the experimental data by the amount of the material and the compaction degree when filling materials into the containers.

Table2.Regression Equations on Moisture Contents and Dielectric Parameters

Name	Regression equations	R^2
Dielectric constant	ε'=3.6111+0.8016x	0.9873
Loss factor	ε''=0.0487+0.1948x	0.9841
Loss tangent	$\tan\delta$=0.0771+0.0123x-0.00029x^2	0.9855

2 Microwave drying

2.1 Microwave drying equipment

Microwave drying equipment was made by the Key Laboratory of Unconventional Metallurgy, has the ability to alter power intensity in the range of 0-3KW at the frequency of 2.45 GHz. It is equipped with a water-cooled condenser and a temperature controller to adjust the microwave power level for a preset temperature. The microwave system consists of two magnetrons, a waveguide and a multi-mode cavity. The dimensions of the multi-mode cavity body is 40mm, 35mm and 24mm, respectively. Moisture loss was measured by a thermogravimetric balance, which is connected to the computer. It is specifically demonstrated in Fig.3.

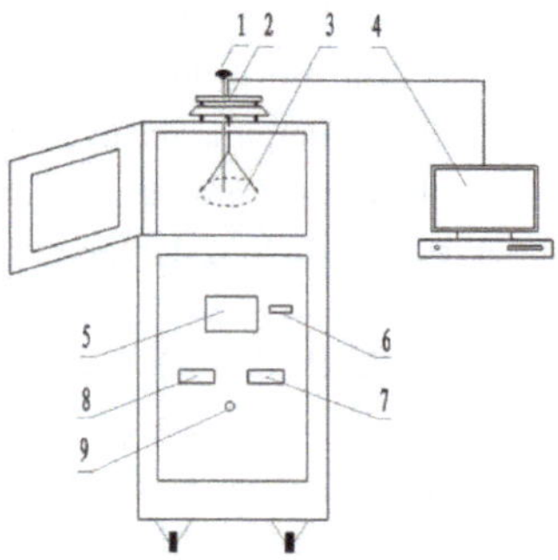

Fig. 3 The schematic of microwave drying system
1-thermocouple; 2-electronic balance; 3-container; 4-computer system; 5-temperature display instrument; 6-power controller; 7-ampere meter; 8-voltmeter; 9-start switch

2.2 Temperature-rising characteristics of CuCl Residue at different microwave power

A test was given to the heating curve of CuCl residue by microwave reactor, the power is setted at 250W, 350W and 450W, respectively. And the initial moisture content is setted at 18%, the gross weight of the material is 300g, and the material's thickness is 3cm. The favorable microwave absorbance characteristics of the CuCl compound is evidenced based on Fig. 4.

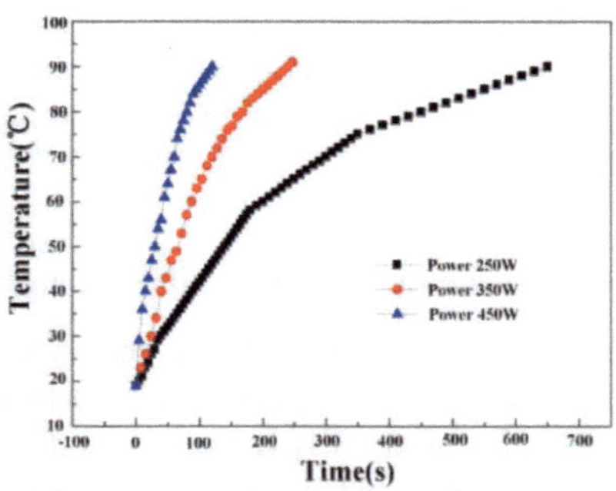

Fig. 4 Temperature-rising characteristics of CuCl residue at different microwave power

Temperature variation curve of CuCl residue by microwave roasting is made up by three stages, namely pre-heating stage, constant speed drying stage and falling-rate drying stage, but this paper only focuses on the relatively complex period of the pre-heating stage to do a research. In the initial phase of drying, due to the materials have relatively high moisture content and higher dielectric loss factor than dry material, large amount of microwave can be absorbed by CuCl residue. The microwave power absorbed by the material can convert quickly into heat warming the wet material, which makes the inner vapour pressure bigger than that of the air and water volatilizes gradually from the CuCl residue. With the decrease of the moisture rate, volatilization rate of the water gradually slow down. In Fig. 4, the curve for 250W appears to consist of four linear sections. Under the condition of low microwave power, material need long time to reach desired temperature. The water specific inductive capacity of 60-78 is much bigger than that of the common ores irradiated in a microwave field and the moisture content of the material is relatively high, so the water in the material has a more significant impact in the heating process. The influence of temperature on water dielectric constant will be observed under a long duration.

2.3 The results of microwave drying on different material thickness

The experiment procedures are as following: (1) 300g CuCl residue was putted into the container as a thin layer and placed the disk in microwave oven; (2) turned on the power of drying system and heated sample at a certain temperature with for a certain time; (3) data collection; (4) taken out the material. Drying experiment process conditions are as follows: the thickness of the materials is 1cm, 2cm, 3cm, 4cm, 5cm, respectively; the drying power is setted at 350W; the mass of the material is setted at 300g and the area of the sample changes over the thickness of the material. The total drying time is 5min.

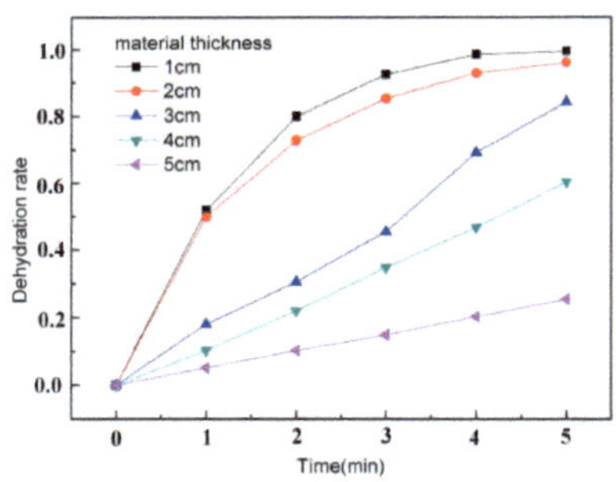

Fig.5 The results of microwave drying on different material thickness

The changes of the degree of dehydration with the time in microwave roasting can be shown in Fig.5. When the thickness is at 1-2cm, the degree of dehydration of the material can be above 98% in 5min. When the thickness is at 3-4cm, the degree of dehydration of the material can be around 70% in 5min; and when the thickness is at 5cm, the degree of dehydration is only 25%. In the follow-up process (dechlorination of CuCl residue by microwave radiation), having a certain amount of moisture content can enhance the effect of microwave heating and dechlorination. Therefore, the degree of dehydration by microwave within 4min when the material's thickness is 2cm meets the requirement.

Drying is a complicated heat and mass transfer process. The material thickness has a deeply influence on energy transmission and steam diffusion. The thicker the material thickness is, the greater the temperature gradient is. In this work, the mass transfer process will be weakened by the different direction of temperature gradient and steam diffusion, and this weakening effect will increase with the raise of material thickness.

3 Conclusion

1.The resonant cavity perturbation method is used by this paper to measure the dielectric constant and dielectric loss factor of CuCl residue with the moisture rate from 2-14 %.

2.Based on the fact that the dielectric constant and dielectric loss factor of the CuCl residue have a linear correlation with the moisture content, that the dielectric property can measure the moisture content of the material on line.

3.A series of microwave drying experiments were conducted. Based on the experimental results, a conclusion that the drying efficient is the highest when the thickness of materials is 2cm with other conditions are same as the premise was drawn.

Acknowledgements

The authors gratefully acknowledge the National Natural Science Foundation of China (NO. 51104073) for funding this work. The project was also funded as a Technology of People' s Republic of China.Yunnan Provincial Science and Technology Innovation Talents scheme-Technological Leading Talent (NO. 2013HA002)

References

[1]M.K Jha, V. Kumar and R.J. Singh, "Review of hydrometallurgical recovery of zinc from industrial wastes," *Resources, Conservation and Recycling,* 33 (2001), 1–22.

[2]S. Lu, Y. Xia and C. Huang, "Removing chlorine of CuCl residue from zinc hydrometallurgy by microwave roasting," *Journal of Central South University*, 21 (2014), 1290-1295.

[3]C.A. Pickles, "Microwaves in extractive metallurgy: Part 1 - Review of fundamentals," *Minerals Engineering,* 22 (2009), 1102-11111.

[4]C.A. Pickles, "Microwaves in extractive metallurgy: Part 2 - A review of applications," *Minerals Engineering,* 22 (2009), 1112-1118.

[5]J.W. Lekse, T.J. Stagger and J.A. Aitken, "Microwave Metallurgy: Synthesis of Intermetallic Compounds via Microwave Irradiation," *Chemistry of Materials,* 19 (2007), 3601-3603.

[6]S.W. Kingman, "Recent developments in microwave processing of minerals," *International Metals Reviews,* 51 (2006), 1-12.

[7]S.W. Kingman and N.A. Rowson, "Microwave Treatment of Minerals-A Review," *Minerals Engineering,* 11 (1998), 1081-1087.

[8]K.E. Haque, "Microwave Energy for Mineral Treatment Processes-A Brief Review," *International Journal of Mineral Processing,* 57 (1999), 1-24.

[9]D.A. Jones, S.W. Kingman and D.N. Whittles, "The influence of microwave energy delivery method on strength reduction in ore samples," *Chemical Engineering and Processing,* 46(2007), 291-299.

[10]P. Bergese, "Specific heat, polarization and heat conduction in microwave heating

systems: A no-equilibrium thermodynamic point of view," *Acta Materialia,* 54 (2006), 1843-1849.
[11]M.A. Janney, H.D. Kimrey and M.A. Schmidt, "Grain Growth in Microwave-Annealed Alumina," *Journal of the American Ceramic Society,* 74 (1991), 1675-1681.
[12]H.M. Tellez, J.P. Alquisira and A.C. Rius, "Comparative Kinetic Study and Microwaves Non-Thermal Effects on the Formation of Poly (amic acid) 4,4'- (Hexafluoroisopropylidene) diphthalic Anhydride (6FDA) and 4,4'- (Hexafluoroisopropylidene) bis (p-phenyleneoxy) dianiline (BAPHF) . Reaction Activated by Microwave, Ultrasound and Conventional Heating," *International Journal of Molecular Sciences,* 116 (2011), 2816-2824.
[13]S. Jyh, Z.W. Hong and L. Weihsing, "Study of dielectric constants of binary composites at microwave frequency by mixture laws derived from three basic particle shapes," *European Polymer Journal,* 45 (2009), 1316-1321.
[14]Z.M. Li, H.L. Du and X.K. Liu, "Synthesis and microwave dielectric properties of Si/C/B power," *Transactions of Nonferrous Metals Society of China,* 16 (2006), 470-473.
[15]C.S. Prasanth, H.P. Kumar and R. Pazhani, "Synthesis, characterization and microwave dielectric properties of nanocrystalline CaZrO3 ceramics," *Journal of Alloys and Compounds,* 464 (2008), 306-309.
[16]W. Li, J.H. Peng and L.B. Zhang, "Pilot-scale extraction of zinc from the spent catalyst of vinyl acetate synthesis by microwave irradiation," *Hydrometallurgy,* 92 (2008), 79–85.
[17]W. Li, J.H. Peng and L.B. Zhang, "Effects of microwave irradiation on the basic properties of wood-ceramics made from carbonized tobacco stems impregnated with phenolic resin," *Industrial Crops and Products,* 28 (2008), 143–154.
[18]A. Verma, A.K. Saxena and D.C. Dube, "Microwave permittivity and permeability of ferrite–polymer thick films," *Journal of Magnetism and Magnetic Materials*, 263 (2003), 228–234.
[19]L.B. Zhang, G. Chen and J.H. Peng, "Microwave absorbing properties of high titanium slag," *Journal of Central South University of Technology,* 16 (2009), 588-593.
[20]M. Huang, J.J. Yang and J.Q. Wang, "Microwave sensor for measuring the properties of a liquid drop," *Measurement Science and Technology,* 18 (2007), 1934-1938.
[21]R.G. Carter, "Accuracy of microwave cavity perturbation measurements," *Transaction on Microwave Theory and Techniques,* 49 (2001), 918-923.
[22]S. Timo, "Electrical properties of water in clay and silty soils," *Journal of Applied Geophysics,* 40 (1998), 73-88.
[23]G.P. Sharma and S. Prasad. "Dielectric properties of garlic (Allium sativum L.) at 2450 MHz as function of temperature and moisture content," *Journal of Food Engineering,* 52 (2002), 343-348.
[24]T.S. Ernest, M. Asce and M. Sundru, "Relationship of Soil Moisture to the Dielectric Property," *Journal of the Geotechnical Engineering Division,* 101 (1975), 755-770.
[25]C.H. Roth, M.A Malicki. and R. Plagge, "Empirical evaluation of the relationship between soil dielectric constant and volumetric water content as the basis for calibrating soil moisture measurements by TDR," *Journal of Soil Science,* 43 (1992), 1-13.

Drying, Roasting, and Calcining of Minerals
Edited by: Thomas P. Battle, Jerome P. Downey, Lawrence D. May, Boyd Davis, Neale R. Neelameggham, Sergio Sanchez-Segado, and P. Chris Pistorius
TMS (The Minerals, Metals & Materials Society), 2015

MOISTURE DEPENDENT DIELECTRIC PROPERTIES AND MICROWAVE DRYING BEHAVIOR OF ZIRCONIUM HYDROXIDE

Aiyuan Ma [1,2,3,4], Xuemei Zheng [1,2,3,4], Libo Zhang * [1,2,3,4],
Jinhui Peng [1,2,3,4], Shiwei Li [1,2,3,4], Yonggang Zuo [1,2,3,4]

[1] Yunnan Provincial Key Laboratory of Intensification Metallurgy, Kunming 650093, China;
[2] National Local Joint Laboratory of Engineering Application of Microwave Energy and Equipment Technology, Kunming, Yunnan 650093, China;
[3] Key Laboratory of Unconventional Metallurgy, Ministry of Education, Kunming 650093, China;
[4] Faculty of Metallurgical and Energy Engineering, Kunming University of Science and Technology, Kunming 650093, China.

Keywords: dielectric properties, Moisture, microwave drying, zirconium hydroxide

Abstract

The effect of moisture content on the dielectric properties and microwave drying behavior of zirconium hydroxide was studied by using an open-ended sensor dielectric measurement system at the frequency of 2.45 GHz. The experimental results showed that moisture content can significantly influence the variation of dielectric properties. Dielectric constant, loss factor and loss tangent all increase with the increasing moisture content. Compared with the conventional drying method, the microwave-assisted drying method resulted in a substantial decrease (81.8%) in the drying time, and the drying effect was better at the lower power level of 600 W. This study offers the useful information on dielectric properties of zirconium hydroxide, which can be used for developing microwave drying applications in mineral processing and designing better microwave sensors for measuring zirconium hydroxide moisture content.

Introduction

Zirconium hydroxide is the important intermediate products of zirconium. ZrO_2 powder obtained by calcining zirconium hydroxide is an important oxide due to its high strength and toughness, good corrosion and abrasion resistance, high melting point, refractory and low thermal conductivity, while preparation of zirconium

---

Author: Aiyuan Ma, male, PhD candidate, Tel: +86-15887855033; E-mail: may_kmust11@163.com

Foundation item: Project (2012HB008) supported by the Yunnan Province young academic technology leader reserve talents.

Corresponding author: Libo Zhang; Tel: +86-13888310177; Fax: +86-5138997; E-mail: lbzhang@kmust.edu.cn.

hydroxide is accomplished in the liquid [1-4]. So moisture content generally above 70%, the drying is important to the quality of the zirconium products, and the minimum moisture standard (below 10%) can only be achieved under specific controlled conditions. Conventional drying is still widely used for dewatering zirconium hydroxide, but this method is not economical. During the conventional drying, the heat transferred from outside to inside of samples, which is in the opposite direction to the steam diffusion, causing low drying efficiency. There are also other problems associated with conventional drying, such as environmental problems, long process and labor-intensive operation.

Thus searching for new techniques for drying of zirconium hydroxide is a subject long studied for improving the present drying techniques. Microwave as a kind of high efficient, clean green technology has been developing rapidly [5]. In general, microwave heating is a unique technology and offers a number of advantages over conventional heating, such as selective heating, rapid heating, and energy saving. In this study, microwave drying is chosen for the treatment of zirconium hydroxide.

The dielectric properties of a material are closely related to the interaction between the material and microwave radiation. Dielectric properties are defined in terms of complex permittivity (ε), composed of a real part (ε' dielectric constant) and an imaginary part (ε'' dielectric loss factor) by the equation: $\varepsilon=\varepsilon'- j\varepsilon''$ [6]. Loss tangent ($\tan \delta$), a parameter used to describe how well a material absorbs microwave energy, is the ratio of dielectric loss factor and the dielectric constant ($\tan\delta= \varepsilon''/ \varepsilon'$) [7]. A material with a higher loss tangent will heat faster under microwave field as compared to a material with a lower loss tangent. Dielectric constant and loss tangent are functions of measurement frequency, material homogeneity and anisotropy, moisture, and temperature in the material.

However, the researches on the metallurgical materials, especially on the zirconium hydroxide mixed with water have not been reported. The objectives of the present work are: 1) to assess the effects of moisture content on dielectric constant, loss factor and loss tangent. 2) to assess the dehydration rate on conventional drying and microwave drying.

Materials and equipment

Experiment Materials

The zirconium hydroxide used in this study was obtained from an enterprise company in Jiangxi Province of P.R.China. The moisture in zirconium hydroxide is almost all free water, with the initial moisture content the range of 76.6%.

Dielectric Measurement System

Dielectric parameters measuring device of zinc oxide dust was shown in Figure 1. Dielectric properties testing equipment from Germany Puschner company dielectric parameter tester (Dielectric kit for Vials). Microwave power source generates microwave excitation signal and the directional coupler divides the signal into two types of reflected and transmitted signals. The microwave signal receiver of AD-8320 integrated circuit can detect the signal amplitude and phase. Resonator was used to hold in the analyse cavity. Test control unit via a USB data cable was connects with the computer test software, which can obtain the measured dielectric parameters and output test results. Test sample was placed on the size of the cavity rather small bottle and the test sample vials will be took into the cavity resonator through the opening holes. Microwave source excitation signal interacts with the sample. By comparing the resonance frequency and quality factor the before and after, according to the test cavity perturbation theory can calculate the dielectric parameters of the tested sample.

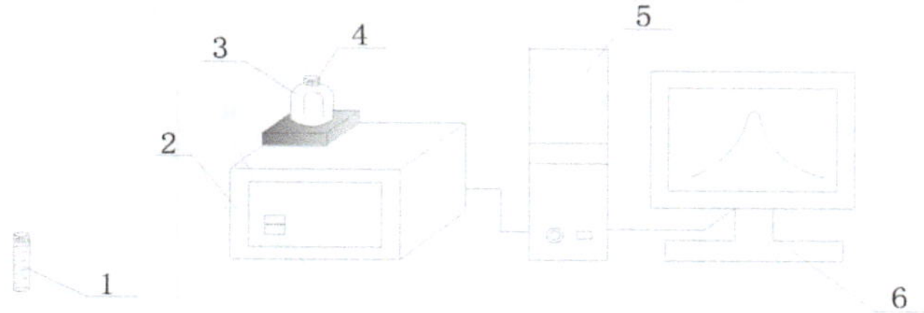

Figure 1. Schematic diagram of dielectric constant measurement device
1-sample; 2-Sensor; 3-Cavity resonator; 4-Standard vial; 5,6-PC/Date collection system

Microwave drying equipment

The main experiment equipment was a 3kW box-type of microwave reactor, which was designed by Key Laboratory of Unconventional Metallurgy, Ministry of Education, Kunming University of Science and Technology, P.R.China. The experimental device connection diagram was shown in Figure 2.

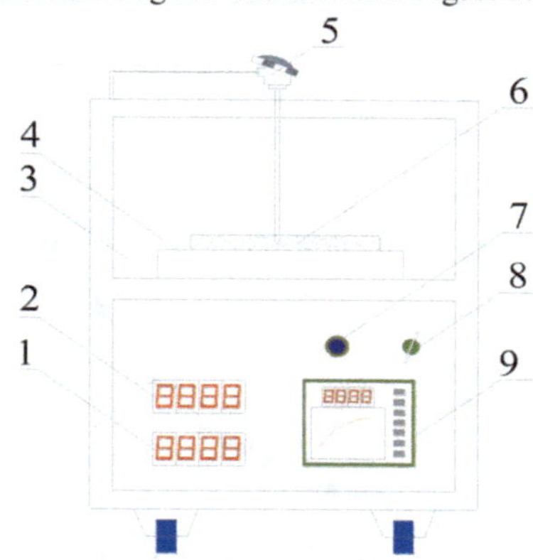

Figure 2. Schematic diagram of microwave heating system
1-Voltage measuring devices; 2-Current measuring devices; 3-Microwave cavity; 4-Glass plate; 5-Thermocouple; 6-Zirconium hydroxide sample; 7-Power button; 8-Adjusting knob; 9-Temperature controller

The microwave heating device contains an automatic temperature control, and its microwave heating frequency is 2450MHz. Its power is from 0 to 3kW, and it is continuously adjustable. A thermocouple was used with shielded sleeve to measure the temperature in range from 0 to 1300℃, and the material supporting body was a glass plate with good wave-transparent property and heat shock property.

Results and discussion

Effects of Moisture on Dielectric Properties

The effect of moisture on the dielectric constant, dielectric loss and loss tangent, respectively were shown in Figure 3, Figure 4 and Figure 5.

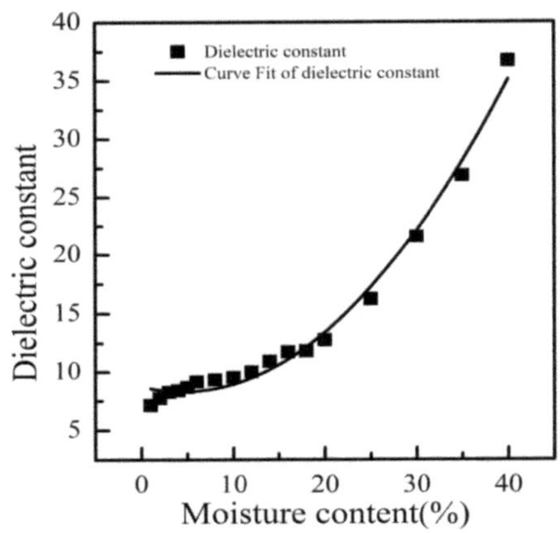

Figure 3. Effect of moisture content on dielectric constant.

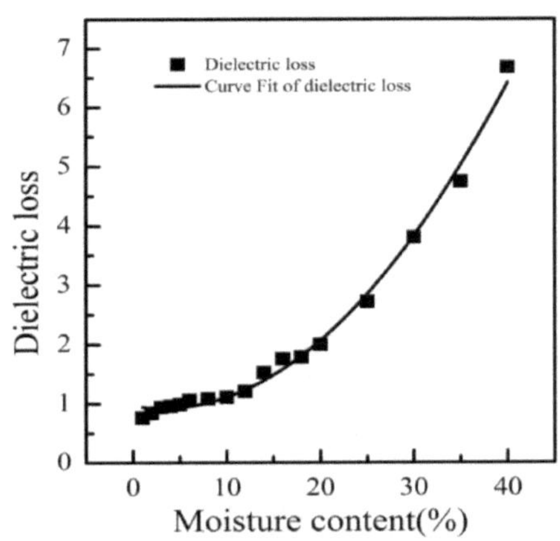

Figure 4. Effect of moisture content on dielectric loss.

Figures 3 and 4 show that the relationship between the value of ε′ and ε″ and the different zirconium hydroxide moisture content at 2.45 GHz. 2.45 GHz is one of the ISM (Industrial, Scientific and Medical) bands, which is the most industrially-used frequency. The relationship found is in good agreement with the values reported by Hotta et al. [8]. It can be seen that the relation in this study are in good agreement with the reported relation. The relation between the values (ε′) and the moisture content (ρ) can be expressed by the following equation (1),

$$\varepsilon'=0.0213\rho 2-0.1928\rho+8.7411\ (R2=0.9871). \quad (1)$$

The relation between the values (ε″) and the moisture content (ρ) can be expressed by the following equation (2),

$$\varepsilon''=0.0040\rho 2-0.0264\rho+0.9655\ (R2=0.9926). \quad (2)$$

Figures 3 and 4 show that the dielectric constant (ε′) and loss factor (ε″) stay at a relatively low level and the change trend is not obvious when the moisture content is small (< 20%). With the increasing of the moisture content, the dielectric characteristics value of the zirconium hydroxide（ε′ and ε″）significantly increases.

Figure 5 shows that the value of $\tan\delta$ does not change significantly with increasing moisture content when the moisture content is small (<10%), and then the value of $\tan\delta$ varies significantly. For zirconium hydroxide that contains abundant free water, the loss tangent may either stay constant or decrease when the moisture content is small enough [9-11]. It is found that the values of dielectric constant (ε′) and dielectric loss (ε″) keep constant when the moisture content is small (<10%), as described in Fig. 3 and Fig. 4, resulting in the ratio of dielectric loss (ε″) and dielectric constant (ε′), the loss tangent staying constant.

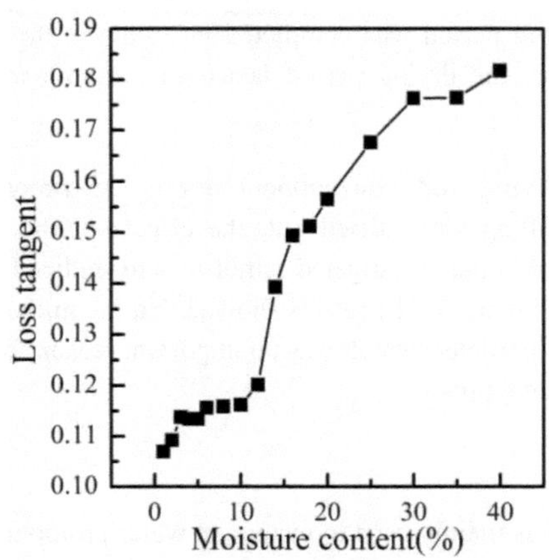

Figure 5. Effect of moisture content on loss tangent.

Thermal behavior (TG/DSC)

The thermal decomposition of zirconium hydroxide sample was analyzed through thermogravimetric and differential scanning calorimetric analyses. The TG - DSC curves obtained from zirconium hydroxide sample are shown in Figure 6. As seen, the total mass loss within a wide temperature range between 0 and 300℃, which can be attributed to the release of moisture. The DSC curve displays a remarkable peak around the endothermic temperature, 122.1℃, showing the presence of an appreciable weight loss. The exothermic peak situated at around 450.9℃ could be attributed to a precursor of monoclinic zirconia replaced by the amorphous zirconia [12].

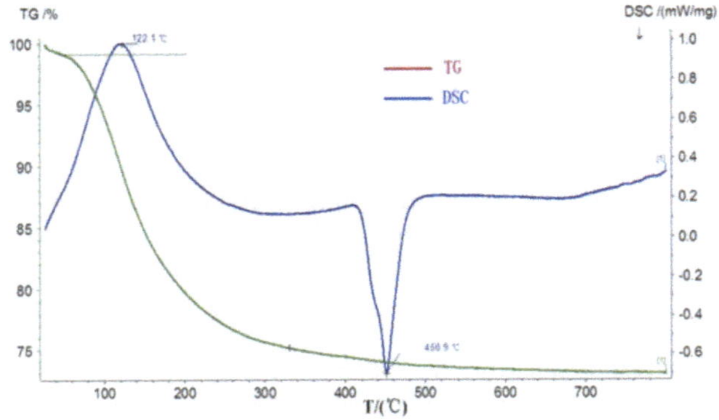

Figure 6. TG—DSC curves of as-received zirconium hydroxide sample.

<u>The study on conventional drying and microwave drying</u>

The effects of two different drying methods on the drying of zirconium hydroxide were evaluated based on the drying parameters such as the drying time, and the drying rate, which is shown in Figure 7. The zirconium hydroxide of 150g were dried at 95℃. The microwave drying period was completed in 20min at the microwave power of 600 W. Nevertheless, the drying period lasted for 110min in conventional-drying method.

After microwave drying and conventional drying of zirconium hydroxide, the treatment by ball-milling was carried out, the effects of the modifications on the morphological change in the investigated samples were studied by SEM. The obtained images are shown at Figure 8. The results showed that the microwave drying can ease the agglomeration of particles, which was an important reason for the high drying rate in the microwave drying process.

The absorbed energy is mainly used to evaporate water promoting drying. The ability of absorbing microwave energy of zirconium hydroxide decreases with the decrease of the moisture content at 2.45GHz, as a result, the drying rate of zirconium hydroxide is faster at the beginning of drying than that at the ending of microwave dying. The advantage of microwave drying system is attributed to the uniform moisture content of the entire sample [13,14].

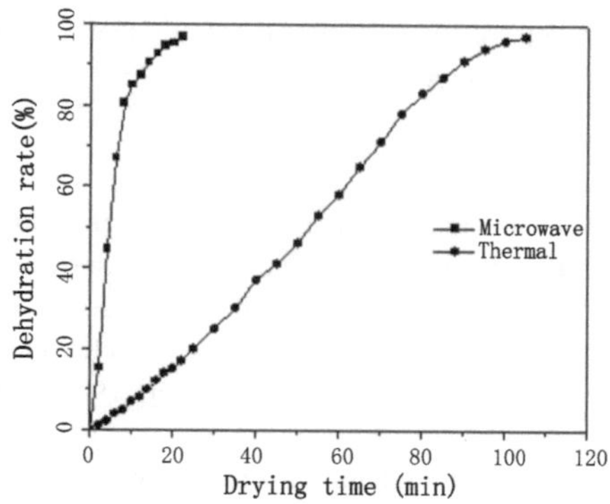

Figure 7. Comparison of conventional drying and microwave drying

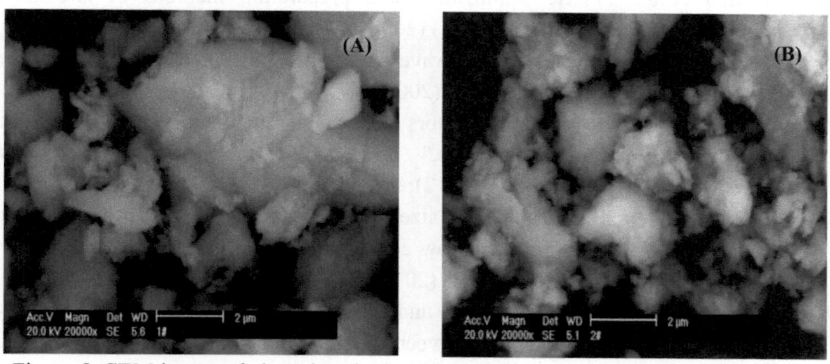

Figure 8. SEM imges of zirconium hydroxide dried by conventional and microwave: (A)- conventional drying; (B)- microwave drying.

Conclusions

1) In this paper, covering different moisture content levels at a frequency of 2.45GHz was measured with an open-ended coaxial probe dielectric measurement system. The experimental results showed that moisture content to be the significant influence on the variation of dielectric properties. Dielectric constant, loss factor and loss tangent all increase with the increasing moisture content.

2) Compared with the conventional drying method, microwave-assisted drying resulted in a substantial decrease (81.8%) in the drying time, and the drying effect was better at the lower power level of 600 W. Microwave drying can ease the agglomeration of particles.

3) The dielectric properties model could be applied to predict microwave drying characteristics as a continuous function of changing moisture content.

Acknowledgements

The authors are grateful for financial supports by the Yunnan Province young academic technology leader reserve talents (2012HB008).

References

1. T.M. Suzuki, and J.O. Bomani, "Preparation of porous resin loaded with crystalline hydrous zirconium oxide and its application to the removal of arentic," *Reactive and Functional Polymers,* 43(1-2,13) (2000), 165-172.
2. L.Z. Wen et al., "Preparation of ultrafine zirconia particles and it's drying methods," *Journal of Guangdong University of Technology,* 19(1) (2002): 63-69,105. (in Chinese)
3. Y.P. Li et al., "Preparation of zirconium hydroxide with huge specific surface area," *Chemical Industry and Engineering*, 24(2) (2007): 150-153,177. (in Chinese)
4. J.L. Mao et al., "Phase stabilization and preparation methods of zirconia," *Inorganic Chemicals Industry*, 40(1) (2008): 5-7. (in Chinese)
5. D.K. Xia and C.A. Picklesi "Microwave caustic leaching of electric arc furnace dust," *Minerals Engineering,* 13(1) (2000): 79-94.
6. S.O Nelson. and S. Trabelsi, "Factors influencing the dielectric properties of agricultural and food products," *Journal of Microwave Power and Electromagnetic Energy*, 46(2) (2012): 93-107.
7. N.G. Patil et al., "Effect of the load size on the efficiency of microwave heating under stop flow and continuous flow conditions," *Journal of Microwave Power and Electromagnetic Energy*, 46(2) (2012): 83-92.
8. M. Hotta et al., "Complex permittivity and permeability of SiO_2 and Fe_3O_4 powders in microwave frequency range between 0.2 and 13.5 GHZ," *ISIJ International*, 49(9) (2009): 1443-1448.
9. M.E. Martín-Esparza et al., "Dielectric behavior of apple (var. Granny Smith) at different moisture contents," *Journal of Food Engineering*, 77(1) (2006): 51-56.
10. C.M. McLoughlin, W.A.M. McMinn, and T.R.A. Magee, "Physical and dielectric properties of pharmaceutical powders," *Powder Technology,* 134(1–2) (2003): 40-51.
11. A. Navarrete, R.B. Mato, and M.J. Cocero, "A predictive approach in modeling and simulation of heat and mass transfer during microwave heating. Application to SFME of essential oil of Lavandin Super," *Chemical Engineering Science,* 68(1) (2012): 192-201.
12. G.Y. Guo, Y.L. Chen, "A nearly pure monoclinic nanocrystalline zirconia," *Materials Science and Processing*, 178(5) (2005): 1675-1682.
13. A. Hussain et al., "Microwave Drying of Ginger by Online Aroma Monitoring," *Drying Technology,* 28(1) (2010): 42-48.
14. N. Therdthai, W.B. Zhou, and K. Pattanapa. "Microwave vacuum drying of osmotically dehydrated mandarin cv. (Sai-Namphaung)," *International Journal of Food Science & Technology,* 46(11) (2011): 2401-2407.

Drying, Roasting, and Calcining of Minerals
Edited by: Thomas P. Battle, Jerome P. Downey, Lawrence D. May, Boyd Davis, Neale R. Neelameggham, Sergio Sanchez-Segado, and P. Chris Pistorius
TMS (The Minerals, Metals & Materials Society), 2015

Optimization of Microwave Drying of Salt with Response Surface Methodology

Bao-bao Wang[1,2,3], Li-bo Zhang[1,2,3],Bing-guo Liu[1,2,3], Jin-hui Peng[1,2,3], Hong-ying Xia[1,2,3]*, Zhi-qiang Li[1,2,3]

1National Local Joint Laboratory of Engineering Application of Microwave Energy and Equipment Technology, Kunming 650093, China
2 Key Lab. of Unconventional Metallurgy, Ministry of Education, Kunming 650093, China
3 Faculty of Metallurgy and Energy Engineering, Kunming University of Science and Technology, Kunming 650093, China

Keywords: microwave heating , industrial salt , response surface methodology (RSM)

Abstract

Based on optimization using response surface methodology designed by Box- Behnken, to study the interaction of wet industrial salt with moisture drying temperature (50-90°C); duration (10-50s); and the weight of the materials (20-60g). The results show that: the experimental parameter conforms to the quadratic equation model to influence the moisture content and the dehydration speed., in analyzing the significance and the interaction of various factors, the optimum by the Design Expert software conditions for drying with microwave radiation include temperature of 90°C, duration of 10s, and sample mass of 26g. Verified by experiment, the moisture content after drying is less than 0.25% and the dehydration rate reached 1.37g/min, between the predicted value and the actual value confirmed good fitting degree.

Introduction

Titanium metallurgy industry, high titanium slag blended with petroleum coke and the sodium chloride output of titanium tetrachloride [1]. Due sodium chloride is a strong absorbent material will absorb the moisture in the storage phase, water is generally up to 1 percent, to ensure that when entering sodium chloride process does not react with chlorine to produce hydrogen chloride gas corrosion of equipment, chlorine sodium in the water must be removed to less than 0.5%.

Conventionally known drying methods, drum drying, spray drying, fluidized bed drying [2-4] and so on. The heat transfer from the surface to the inside of this drying method, the material in the presence of a temperature gradient is inevitable, so the drying material is uneven, internal material containing more water, the external material overheating, take advantage of low energy [5]. Therefore, the development of the new salt drying process is important. Selective use of microwave heating, the material does not conduct heat, and only the moisture in the material is heated, the water in order to achieve diffusion of material from the inside outwards, thus greatly reducing the drying time [6]. To prevent the material during drying due to high temperature caused by the loss of energy, In this paper response surface optimization method, microwave temperature drying of salt discussed under the temperature conditions of the drying temperature, drying time and sample mass relative dehydration ratio.

Experimental

Materials

The industrial salt was supplied by a smelter in Yunnan Province. Before the experiment, three simples were placed in the oven to bake at 80°C for 24h. Then their exact moisture was determined. The results showed that the industrial salt was with moisture of 1.13%.

Experimental Equipments and Methods

A microwave drying system (0.75kW; 2450 MHz) developed independently by the unconventional metallurgy ministry of education key laboratory , electronic balance (FA2004) manufacture by Shanghai Shangping Instruments Corporation , electric blast oven (101A-3) manufacture by Shanghai Chongming Experimental Instrument Factory.

According to the requirement of experiment , a certain amount of industrial salt m_g (g) put in the container is dried for a period time by microwave. Then, the mass m_t (g) of industrial salt dried is weighed. The dehydration rate W_t is calculated according to the next formula.

$$W_t = \frac{m_g - m_t}{m_g \times 0.0113} \times 100\%$$

Experimental Design

Experiment design based on three levels of the Box-Behnken, inspection three factors: drying temperature (χ_1), drying time (χ_2), sample mass (χ_3), the test factor level and coding shown in Table 1.

Table 1 Independent variables and their levels used for Box-Behnken design

Factor	Coding	Coding level		
		-1	0	1
Drying temperature (°C)	χ_1	50	70	90
Drying time (s)	χ_2	10	30	50
Sample mass (g)	χ_3	20	40	60

Selected the dried sample moisture content (Y_1) and the loss rate (Y_2) as a response, the experimental group, experiments were done to 15, including the three groups of center point experiment was repeated. Each response values Y_1, Y_2 and experimental factors X_1, X_2, X_3 of the mutual relations (1) of the quadratic polynomial obtained [7,8].

$$Y = \beta_0 + \sum_{i=1}^{n} \beta_i \chi_i + \sum_{i=1}^{n} \beta_{ii} \chi_i^2 + \sum_{i<j} \beta_{ij} \chi_i \chi_j \quad (1)$$

Where Y is the response predicted, β_0 is the constant, β_i, β_{ii}, β_{ij} are the linear, quadratic and cross product coefficients, respectively.

Result and Discussion

Microwave drying results shown in Table 2, using Design-Expert software equation (1) for multiple regression fitting [9], moisture content and loss rate of linear and quadratic models were found, which were fitted equation shown in Eq2, 3.

$Y_1 = 0.72 - 0.15\chi_1 - 0.088\chi_2 + 0.16\chi_3 + 0.02\chi_1\chi_2 - 0.068\chi_1\chi_3 - 0.065\chi_2\chi_3 - 0.15\chi_1^2 -$

$$0.087\chi_2^2 - 0.14\chi_3^2 \quad (2)$$

$$Y_2 = 0.33 + 0.20\chi_1 - 0.44\chi_2 + 0.14\chi_3 - 0.19\chi_1\chi_2 + 0.11\chi_1\chi_3 + 0.038\chi_2\chi_3 + 0.23\chi_1^2 + 0.36\chi_2^2 - 0.037\chi_3^2 \quad (3)$$

Table 2 Experimental design matrix and results

Run	Drying temperature χ_1 (°C)	Drying time χ_2(s)	Sample mass χ_3(g)	Moisture Y_1(%)	Dehydration rate Y_2 (g / min)
1	50	10	40	0.71	1.04
2	90	10	40	0.33	1.95
3	50	50	40	0.58	0.27
4	90	50	40	0.28	0.41
5	50	30	20	0.34	0.32
6	90	30	20	0.22	0.37
7	50	30	60	0.76	0.45
8	90	30	60	0.37	0.93
9	70	10	20	0.38	0.90
10	70	50	20	0.25	0.21
11	70	10	60	0.86	1.02
12	70	50	60	0.47	0.48
13	70	30	40	0.69	0.35
14	70	30	40	0.71	0.33
15	70	30	40	0.75	0.30

Development of Regression Model

Table 3 Analysis of variance (ANOVA) for response surface quadratic model for moisture

Source	Squares	df	Mean square	F value	Prob> F
Model	0.65	9	0.07	16.17	0.0034
χ_1	0.18	1	0.18	40.15	0.0014
χ_2	0.06	1	0.06	13.89	0.0136
χ_3	0.2	1	0.2	45.73	0.0011
$\chi_1\chi_2$	1.60×10^{-3}	1	1.60×10^{-3}	0.36	0.5732
$\chi_1\chi_3$	0.02	1	0.02	4.13	0.0977
$\chi_2\chi_3$	0.02	1	0.02	3.83	0.1076
χ_1^2	0.09	1	0.09	20.01	0.0066
χ_2^2	0.03	1	0.03	6.35	0.0532
χ_3^2	0.07	1	0.07	16.32	0.0099
Pure Error	0.02	5	4.41×10^{-3}	-	-

R^2= 0.97; R^2_{adj}= 0.91; Adequate precision = 11.41 (> 4).

It can be seen that the model used in this experiment Prob> F is 0.0034 <0.0500, show a significant experimental model used, the model correlation coefficient R^2 of 0.97, 24.77 precision greater than 4, indicating that the water content of the model equation can be used (2) descriptively, R^2_{adj}= 0.91 showed that only about 9% of water content variation cannot be explained by the model. Equation (2) of the regression coefficient is significantly testing indicated: factor χ_1,χ_2,χ_3 On moisture content rate of all signification, while χ_1^2 and χ_3^2 interaction of the moisture content rate of also significantly. This is because the evaporation temperature is the important factor in whether or not water, up to a certain temperature, moisture was able to evaporation, drying time to decide whether the evaporation of water completely, and the feed weight related to the water in the drying time is completely evaporated, the sample is dried is uniform moisture content of the main factors. Figure 1 is the moisture content of the experimental and predicted value graph, we can see better predictive value close to the experimental values, indicating that the quadratic model is suitable for describing experimental factors and moisture content relevance.

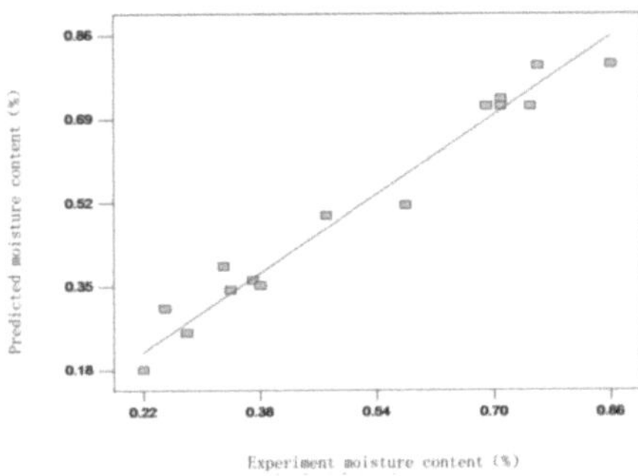

Fig.1 Predicted vs experimental moisture content

Analysis of RSM

Figure 3 showed the microwave drying temperature , drying time and the relationship between sample moisture content of three-dimensional graphics, the sample moisture decreases with increasing temperature or the drying time extended, wherein the drying temperature effect on the moisture content of the most significant, when the drying time and marerial weight are10s and 40g, the drying temperature is 50°C and 90°C , the moisture content of the sample was 0.71% and 0.33%, which is due to the drying temperature was necessary for evaporation of water, the higher temperature, evaporation of water was faster, the moisture content of the sample is smaller, the same drying temperature, extended the drying time, increase evaporation of water, the moisture content of the sample is smaller. Therefore, to meet the requirement to obtain the moisture content of the salt product, experiments should be selected by higher drying temperatures and a longer drying time.

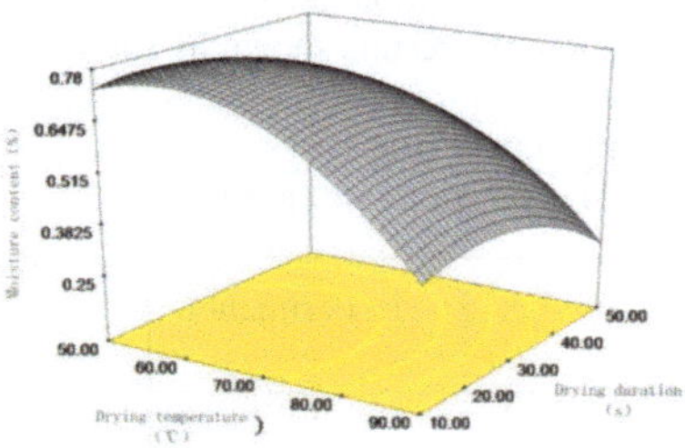

Fig.3 Three-dimensional response surface plot of moisture (effect of drying temperature and time, weight = 40g)

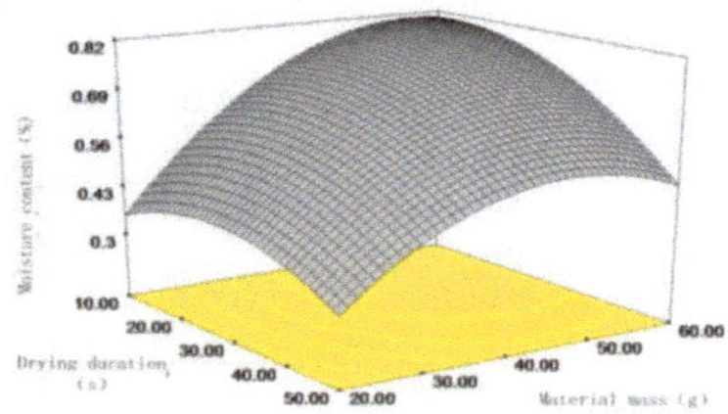

Fig.4 Three-dimensional response surface plot of moisture (effect of drying time and mass, temperature=70°C)

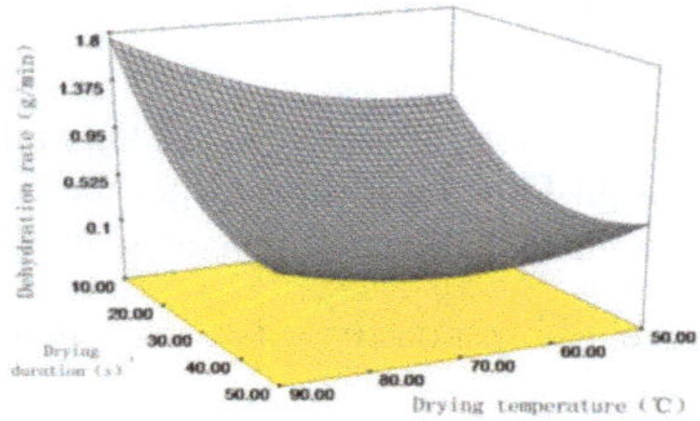

Fig.5 Three-dimensional response surface plot of drying rate (effect of drying temperature and time, weight =40g)

Figure 4 shows the sample mass influenced the drying time was significant, because the salt was not the absorbing material, during the drying process, the microwave can only heating the water. The sample mass determined the sample moisture content, if the sample mass is less, it will used for a long time for heating the material from room temperature to 70°C, when sample mass is decreased, it will shorten the heating time from room temperature to 70°C.
Figure 5 shows the three-dimensional response surfaces of the combined effect of sample mass and drying time on the dehydration rate, the microwave temperature on the salt dehydration rate was not significant, as the dehydration rate to extend the drying time is reduced. When the microwave temperature and sample mass were 90°C and 40g, the drying time of the 50s, the dehydration rate is 0.41g/min, however, when drying time was 10s, the dehydration rate is high to 1.95g/min, is 4.76 times too when 50s. This is because the same weight of water contained in the same material, in the beginning of drying, heating rate causes the water molecules absorbing large initial burst of water evaporation in the drying in the water evaporation unit time increases, the larger loss rate; with the drying time extended, the water moisture of the sample decreases as the material in the water is reduced to a certain extent, the water evaporation unit time decreases, resulting in dehydration rate becomes smaller.

Process Optimization

Taking into account of the economic benefits of drying experiments and energy saving, expect to get a lower moisture content of the sample, and make the drying process has a higher rate of dehydration, using Design-Expert Parameter Optimization Module on the drying temperature, time and material weight were re-optimization, the results shown in Table 5.

Table 5. The optimum conditions for the process

Drying temperature (℃)	Drying time (s)	Sample mass(g)	Dehydration rate (g/min)	
			Predicted	Experimental
90	10	26	1.42	1.37

Conclusions

(1) RSM is used to optimize the process conditions for microwave drying industry salt.The effects of operating parameters of drying duration, material thickness and drying temperature dehydration rate are investigated using analysis of variance (ANOVA), to identify the significant factor. The experimental data on dehydration rate is found to agree satisfactorily with the model predictions.

(2) In order to obtain a moisture content of less than 0.3%, while a high rate of water loss experimental parameters, the Design-Expert software to get the best optimization parameters: drying temperature is 90°C, drying time is 10s and sample mass is 26g.

(3) The technology of microwave drying industry salt is efficient and non-polluting compared to traditional drying methods.

Acknowledgments

The authors are grateful for the financial support from the National Natural Science Founda tion (No.51104073), the National Technology Research and Development Program of China (N o. 2013AA064003), the Yunnan Province Young Academic Technology Leader Reserve Talen ts (2012HB008), and the Yunnan Provincial Science and Technology Innovation Talents scheme - Technological Leading Talent (NO. 2013HA002)

Reference

1. W. MO, *Titanium metallurgy* (Metallurgical Industry Press, 1998).
2. G.-Q. Shen, *Drum drying* (Encyclopedia of Agricultural, Food, and Biological Engineering, 2003), 211.
3. K. Masters, *Spray drying handbook* (Spray drying handbook, 1979).
4. S. Hovmand, *Fluidized bed drying* (Handbook of industrial drying, 1995) 195-248.
5. X. MA et al., *A method of microwave drying sodium chloride*, in.
6. C.H. Liu et al., "STUDY ON MICROWAVE ABSORPTION PROPERTY AND MICROWAVE DRYING PROCESS OF SODIUM CHLORIDE", *Mining and Metallurgy*, 2 (2012) 011.
7. D. Gangadharan et al., "A. Pandey, Response surface methodology for the optimization of alpha amylase production by Bacillus amyloliquefaciens", *Bioresource technology*, 99 (2008) 4597-4602.
8. I. Tan, A. Ahmad, and B. Hameed, "Preparation of activated carbon from coconut husk: optimization study on removal of 2, 4, 6-trichlorophenol using response surface methodology",

Journal of hazardous materials, 153 (2008) 709-717.
9.P. Lim, "Optimization of the rough cutting factors of impeller with five-axis machine using response surface methodology", *The International Journal of Advanced Manufacturing Technology*, 45 (2009) 821-829.

Drying, Roasting, and Calcining of Minerals
Edited by: Thomas P. Battle, Jerome P. Downey, Lawrence D. May, Boyd Davis, Neale R. Neelameggham, Sergio Sanchez-Segado, and P. Chris Pistorius
TMS (The Minerals, Metals & Materials Society), 2015

THE IMPACT OF CALCINATION CONDITIONS ON PRODUCTION OF MAGNESIUM BY THE MAGNATHERM PROCESS

James C Sever[1]

[1]Alpha Omega Engineering, Inc.
5614 S. Leeway Ct., Spokane, WA 99223

Keywords: magnesium, calciner, dolime, specific surface area, magnetherm, vacuum furnace

Abstract

The mode of operation of dolomite kilns at a magnesium reduction plant using the Magnatherm process was designed to minimize operation cost and maximize kiln refractory life. Analysis of downstream operation issues shows the quality metric for dolime changed from loss on ignition to specific surface area. This produced a complete change in kiln operation; the net results being major plant productivity increases decrease in operating cost, and increases in both kiln operating costs and reline frequency.

Introduction

A kiln appears to be a simple piece of equipment to operate; however, it is deceivingly complex, as it is dependent on many variables and their interrelationship. Some of these variables are readily adjustable, but others, such as lining refractory type, are difficult and costly to change. In this instance, the parameters that did not reflect the demands of the downstream process were used to specify the original equipment. By using these parameters to control the kiln operation, the major process, using the kiln product, behaved in an unpredictable and uncontrolled fashion. Identifying the cause and changing the key parameter resulted in kiln operation outside its design limits. Consequently, the kiln required higher and more costly maintenance but the plant operation significantly improved overall.

Owing to the plant being closed for over 12 years, no quantitative data is available, thus the following is an anecdotal presentation.

Process Description

Northwest Alloys produced primary magnesium ingot using a silicothermic process, the Magnatherm process. [1] The chemistry for this process is the same as that of the Pidgeon process only the furnace design and operating conditions differ.

$$2CaOMgO \quad + \quad Fe\,Si_{75\%} \quad \underset{1350^{\circ}\,C\ @\ 0.1\ atm.}{\longrightarrow} \quad Ca2SiO4 \quad + 2Mg\uparrow \quad + \quad FeSi_x$$

The vacuum reduction furnace was heated electrically using the slag as the resistor. The raw materials were fed gradually to the furnace. The furnace was designed such that a pressure gradient was to progress from the raw material containers, through the furnace and into the condenser. This was to result in the magnesium vapor, the product of the reduction reaction, to flow into the condenser and be held until transferred to the casting department.

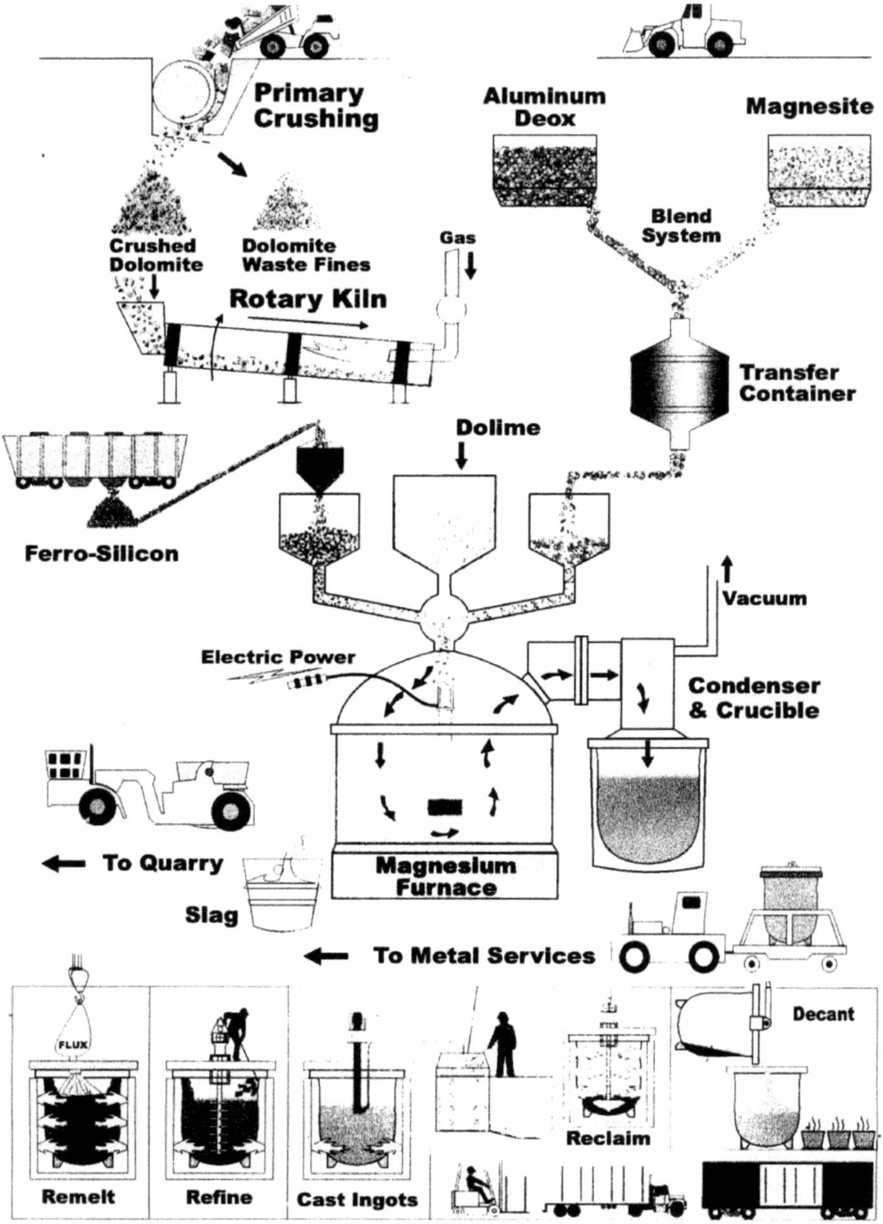

Figure 1. Northwest Alloys Process Schematic: Note the three separate feed material tanks connected to the reduction furnace.

Problem Described

Unfortunately, there were periods during the operation of the magnesium reduction furnace where the internal pressure unpredictably surged to such a degree that the condenser could not remove the magnesium vapor quickly enough in order to maintain the desired pressure gradient and magnesium vapor flow. It was determined that the reduction reaction did not occur uniformly and consistently but rather fluctuated unpredictably. The net result of these fluctuations was that the magnesium vapor was pushed into the feed system where it condensed. As it condensed, it cemented particles of feed flowing through the piping, which blocked the feed system and interrupted the process.

In order to clear the blockage, argon was used to purge the furnace and bring it to atmospheric pressure. The power input was simultaneously reduced to a level just sufficient to balance the heat loss from the furnace. The furnace feed system was then opened and the blockage cleared using an air hammer and long chisel bit. Once the blockage was removed, the furnace was re assembled and brought back to operation.

Unfortunately, the process of opening the furnace to atmosphere allowed air to enter into the furnace and condenser. The air reacted with the magnesium producing both oxides and nitrides on the active surface of the condenser. The pressure surge not only reduced the amount of time the furnace had available to produce magnesium but also compromised the condenser by forming an internal insulating layer, thus reducing the ability to condense as much magnesium as it had prior to the blockage. This in turn reduced the quantity of raw materials that could be fed to the furnace. These conditions persisted until the condenser was changed, part of a scheduled service.

The presence of magnesium oxides and nitrides in the condenser resulted in the generation of excess sludge when the magnesium was processed into ingot. When the condensed metal was re-melted and cleaned in the casting department, molten salt was stirred into the melt to remove nonmetallic impurities. This mixture of molten salt, oxides and nitrides comprised the waste material, sludge.

Search for a Cause

To determine the origin of these back pressure incidents, the operating conditions of the furnace were studied. These conditions appeared to be within the design specifications and no material or operation was shown to cause a variation sufficient to produce the high reaction rate and cause a blockage. A statistical factorial experiment was then developed to assess the impact of conditions surrounding the furnace. The net result of the experiment was to identify the dolime as the major variable contributing to the pressure fluctuations.

Up until this point, the dolime was characterized and the kiln controlled based on the loss on ignition (LOI) of the calcined material at 1000°C. An alternate attribute of the dolime was selected for assessment: specific surface area of the dolime. In performance of the assessment a number of samples were taken at the kiln discharge over a 24 hour period. These samples showed significant variation in the specific surface area of the dolime.

Hypothesis

The variation in the specific surface area impacted the rate of reaction within the furnace. A high surface area dolime particle, when delivered to the furnace, resulted in a large area of dolime exposed to the liquid reductant. This resulted in a large amount of magnesium oxide being exposed and reacting at one time which produced the high pressure condition. Normally, the dolime which had a lower surface area and the rate of reaction was controlled by the smaller amount of magnesium oxide available followed by dissolution of the dolime particle's surface before a new magnesium rich surface was exposed.

It was deemed necessary to minimize the variation in the specific surface area of the dolime. To accomplish this, the kiln operation was evaluated to determine the variables that could be adjusted to reduce the product dolime surface area variation.

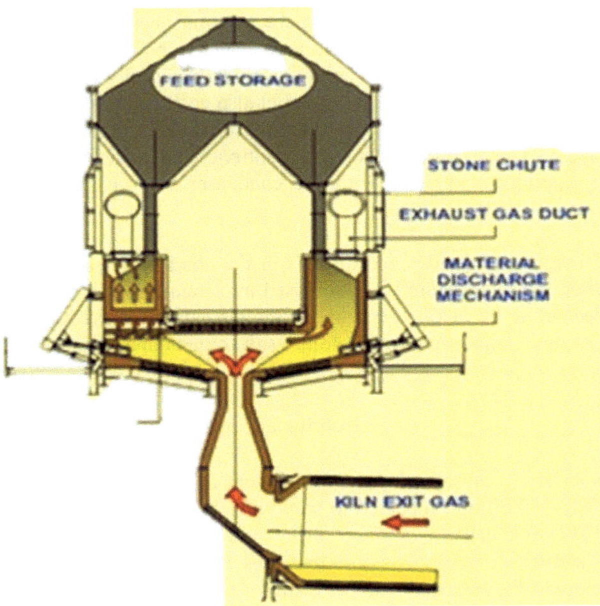

Figure 2. Cross Section of Kiln Preheater [2]

Kiln

Two kilns were needed to produce sufficient dolime for plant operation; both were designed and constructed by Kennedy Van Saun. They were preheater rotary kilns and were selected as being the most energy efficient calciners available at the time of plant design. They consisted of a preheater and a rotary barrel into which a burner was inserted. The preheater maintained an annulus filled with dolomite through which the kiln off gas flowed prior to exiting through a bag house. (See Figure 2) The dolomite would be pushed off a ledge by hydraulic rams, which were arranged around the total circumference of the kiln. The preheated dolomite then cascaded into the rotary barrel, where it was slowly conveyed through the length of the kiln as the barrel rotated.

The design criterion was: to deliver aggregate dolime at the discharge temperature to the reduction feed tank at the lowest cost. The target LOI was 0.5% +/ - 0.1%. To meet this target the temperature of the burn zone was set at 1250°C and the kiln length defined to provide a typical residence time of <3 hours.

The variables available for adjustment in order to influence the physical attributes of the discharged product were defined as follows:

1. Feed Rate – As the volume of the rotating kiln is relatively fixed, the amount of material entering the kiln over time was a main variable controlling the residence time of material within the kiln. The ability to vary this parameter was limited due to the necessity to match the amount of dolime required by the reduction furnaces.
2. Rotation Rate of the Kiln Barrel – This controlled the bed depth and combined with the feed rate, influenced the residence time. The range of variation of this variable was limited as too slow a rotation would create an unstable bed leading to segregation and irregular calcination of the product. Too fast a rotation would lower the bed depth and reduce the retention times to below that which was needed to produce a fully calcined product.
3. Fuel Feed – The amount of fuel fed to the burner is the primary adjustment of kiln operating temperature. The hottest portion of the kiln is termed the "burn zone". Too little fuel will result in product that is insufficiently calcined. Too much fuel and the refractory lining of the kiln will fail.
4. Feed Particle Size – The maximum particle size defines the amount of time at temperature the particle must be retained to sufficiently calcine the product. This is dependent on the heat capacity of the material and the heat transfer rate of the material. The particle maximum size was limited to 1 3/8 inch diameter to prevent bridging in the furnace feed system. It was not desirable to reduce this top size as the preheater gas flow resistance was at maximum and a smaller top size feed would result in kiln flame-out due to back pressure within the preheater.
5. Feed Particle Size Distribution – The smallest size particle was limited to 3/8 inch diameter. This was found to be the smallest size possible to prevent excessive back pressure in the preheater. In addition, too many fines could result in the formation of a core of fine material that would not receive sufficient calcination. This occurs because the rotating bed acts as an inertial classifier with the coarse particle carried to the outside and the finest particles concentrated in the center of a "kidney shaped" core. [3] The bottom or smaller size could not be raised to make the feed more uniform in size due to the fact that over one third of the good dolomite mined was being wasted as it was deemed to be too small for kiln feed.

Corrective Action Taken

After lengthy study and testing the following occurred:

1. The control parameter used to set kiln operating conditions was changed from LOI to specific surface area of the dolime.
2. The specific surface area target was set at 2 m^2/gm +/- 0.5 m^2/gm.
   (Note: The target was set to accommodate the natural variation in specific surface area that occurred over time due to operational changes such as increased production to meet furnace consumption demand or the slight change in the screen curve in the preheater due to variations within the feed pile. Normally the specific surface area remained within 5% of the previous day's average readings.)

To meet these desired parameters the following actions were taken:

1. The temperature of the burn zone was gradually raised until the desired specific surface area was reached. This finally settled at a point 200°C higher than the design burn zone temperature.
2. Kiln rotation was reduced to increase residence time of the dolomite within the kiln.
3. The feed particle distribution variation was reduced by accessing two feed piles of dolomite rather than one. During the construction of a feed pile the discharge of rock onto the apex of the pile results in a classification of the material with the coarse rock falling to the bottom outside and the fines concentrated at the core. By accessing two piles - one of which is new and the other half consumed - coarse material can be taken from the newer pile and fine material from the older pile. As the piles are consumed, the end of the old pile will consist of coarse material and the new pile will have the center of fines exposed.

Outcome

With the changes made in kiln operation, the following occurred:

1. Unexpected pressure surges within the reduction furnaces decreased.
2. Episodes of magnesium furnace feed blockage were reduced dramatically. This resulted in a net increase in the overall plant productivity due to uninterrupted furnace runs and the ability to feed material at a consistently high rate due to good condenser heat transfer.
3. The recovery of good metal from crude metal increased while the amount of sludge decreased. By keeping the furnace sealed throughout the cycle, less magnesium was lost through burning and correspondingly, the amount of oxides and nitrides were reduced.
4. Increasing the burn zone temperature had detrimental consequences as well:
   a. The higher temperature produced a scale formation within the kiln. When pieces of this material broke free they occasionally resulted in blockage of the kiln discharge.
   b. The burn zone brick were consumed due to formation of low melting point compounds from sintering of the contaminants contained in the dolomite. As time permitted, the type of brick was changed from a high alumina brick to a high magnesite brick. This greatly reduced, but did not eliminate, the loss of brick thickness in the "burn zone".

c. The time span between kiln rebuilds decreased from an average of 24 months to an average of 10 months.
d. The cost for operating the kiln increased due to the increased consumption of fuel.

Overall the increased productivity and lower amount of waste more than offset the negative impact the changes had on kiln operating costs.

20–20 Hindsight

If specific surface area was used as the key product parameter for the kiln discharge, the design of the kiln would have been substantially different. The kiln barrel would have been increased in length to provide the necessary sintering time at the lower burn zone temperature. With the correct design, the objective of low cost kiln operation could have been achieved together with stable reduction furnace operation.

References

1. J. Sever and M. Ballain, "Evolution of the Magnetherm Magnesium Reduction Process," *Magnesium Technology,* ed. N. Hort, S. N. Mathaudhu, N. R. Neelameggham and M. Alderman (San Antonio, Texas: TMS, 2013) 69-74.

2. Sarna,Satyendra Kumar, web page http://ispatguru.com/calcination-of-limestone/

3. Akwasi A. Boateng, *Rotary Kilns, Transport Phenomena and Transport Processes*, (New York, NY: Butterworth-Heinemann and imprint of Elsevier, 2008), 105-127.

Drying, Roasting, and Calcining of Minerals
Edited by: Thomas P. Battle, Jerome P. Downey, Lawrence D. May, Boyd Davis, Neale R. Neelameggham, Sergio Sanchez-Segado, and P. Chris Pistorius
TMS (The Minerals, Metals & Materials Society), 2015

CALIX CALCINER: A GREEN APPLICATION IN THE PRODUCTION OF MAGNESIUM

James C Sever[1]

[1]Nevada Clean Magnesium
Suite #602 15216 North Bluff Road, White Rock, British Columbia V4B0A7

Keywords: magnesium, Calix, calciner, dolime, Nevada Clean Magnesium, ScanMag

Abstract

Two green field magnesium reduction plants are under consideration using dolomite as the raw material. The Calix calciner was independently selected for both plants. The unique design of this calciner is detailed and the advantages defined for the specific application. The low carbon foot print of this unit combined with the ability to capture food grade carbon dioxide makes this an environmentally proactive approach to carboniferous ore calcination.

History

In 2009, while searching for gold, Molycor Gold Corporation discovered a very large deposit of high grade, high purity dolomite near Ely, Nevada. [1] It was decided to pursue development of the property with the goal of establishing a commercial primary magnesium smelter at - or near - the site. In the course of developing the project the company's name was changed to Nevada Clean Magnesium, Inc. to better reflect the new nature of the business.

At about the same time, a major downturn in the production of solar panels produced in the West, resulted in closure of a number of companies at Glomfjord, Norway. A search was initiated to find a suitable business that would provide long-term employment and utilize established assets. Eventually it was decided to produce magnesium metal using a local source of dolomite. This resulted in the formation of ScanMag. [2]

Both companies independently defined the process and equipment needed to establish productive, profitable operations. In the course of this work, both companies selected a silicothermic reduction method and also determined that the Calix calciner would not only provide dolime in a suitable form, but would also allow capture of the pure carbon dioxide liberated from the ore on heating.

Eventually both companies learned of the existence of the other and it was deemed beneficial to enter into a joint venture to assist one another in the establishment of the magnesium reduction plants.

Decision Logic

The attributes leading to selection of the Calix technology [3] were as follows:

1. Energy Efficiency – There are no moving parts in the calciner, so it is possible to make full use of insulation to minimize heat loss.
2. Fuel Flexibility – Both projects are in the preliminary stage of development and the ability of the Calix calciner to use electric power or a variety of fuels as the energy source for the calciner is a positive attribute.
3. Capture of Food Grade Carbon Dioxide – By design, the combustion gas is confined to a combustion circuit which is physically separate from the calcination chamber. (See Figure 1) This results in the evolved carbon dioxide to be free from contaminants and maintained at a high concentration (typically > 98% CO_2). These attributes allow the carbon dioxide to be readily recovered and converted into a solid or liquid for ease of transport and use off site.
4. Specific Surface Area Control – By varying the assembly of equipment used in the calciner, namely the calcination tubes and holding tanks, the calcined particle's specific surface area can be custom tailored to support the optimum desired condition needed for the downstream operations. The specific surface area can vary from $< 10\ m^2/gm$ to $> 10^4\ m^2/gm$.
5. Minimal In-process Inventory – Conventional calciners have a large mass of material contained in the shaft or rotating tube. These conditions result in holding times of hours and a large mass of material treated at a given set of conditions. This is not the case with the Calix calciner. The process occurs in the time it takes for the particle to fall through the calcination tube, typically 3 to 7 seconds. Consequently, the "in-process" inventory is minimal.
6. Fine Particle Feed – The dolomite fed to the kiln is ground to less than 200 microns. This means that all unit operations to prepare for use in a silicothermic reduction operations are completed prior to calcining and the resulting dolime is delivered in a uniform particle size suitable for blending and compaction into the feed briquette. The small particle precludes an unreacted core, as in the case of shaft calciner product.
7. Carrier Fluid is Superheated Steam – To transport the dolomite particle into the calciner chamber, it is introduced into a stream of superheated steam. Use of steam allows exclusion of any other gaseous component and improves the heat transfer between the hot calciner tube and the dolomite particle and catalyzes the de-carbonization operation allowing it to occur at a lower temperature. [4]

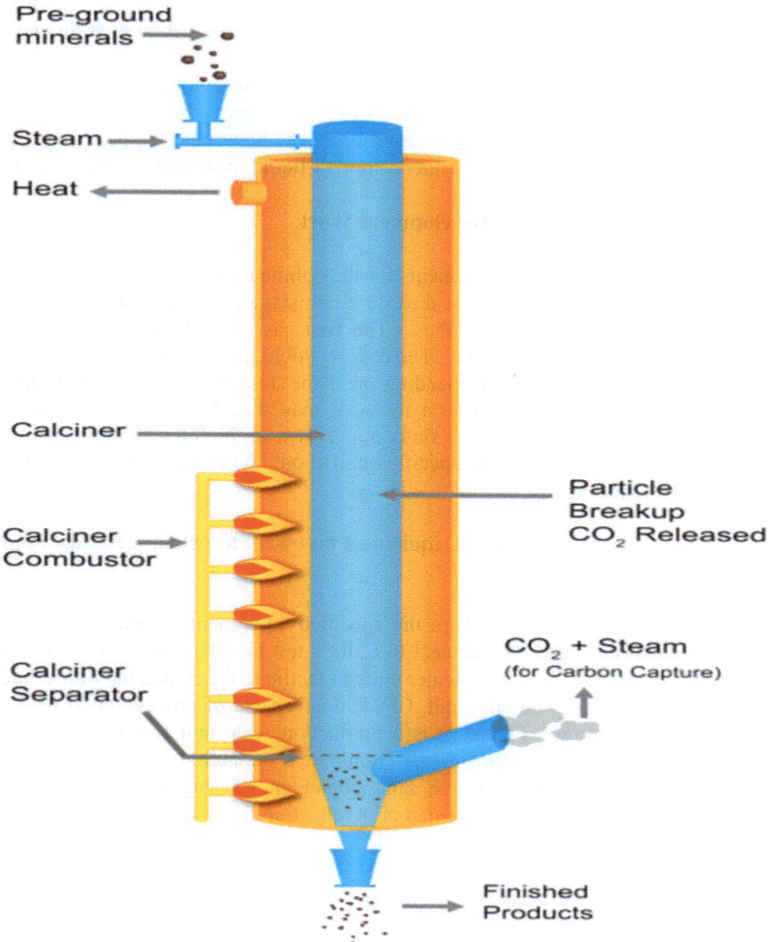

Figure 1. Simplified view of Calix Calciner showing steam transport system and separate combustion and calcination chambers.

Equipment Description and Operation

Crushed dolomite is delivered to a feed tank. A feeder at the tank's bottom allows a constant stream of dolomite to enter a pipe containing superheated steam. The steam flow elutriates the fine solids and carries them through a preheater, then up to the top of the calciner tube where they are tangentially injected. As the particles travel radially through the tube, the wall (which is heated to the desired calcination temperature) radiates and conducts via the steam. Thermal energy is transferred to the particle allowing it to disassociate into carbon dioxide and dolime. The particle settles into the bottom of the calciner tube and it is then removed for further processing. Less than ten micron particles are carried in the exiting gas stream; these are removed via cyclone. Heat exchangers are used to transfer energy from the gas stream to the incoming dolomite. The steam is then condensed back into water and the nearly 100% carbon dioxide stream is sent for processing into either a solid or liquid for transport and handling.

Development Work

ScanMag provided Calix with the requirements for the dolime to be used in their project. Based on the need for full removal of all carbon dioxide a two stage calcination unit is proposed. [5] This would consist of five calcination tubes. The first pass would consist of three tubes to liberate the majority of the carbon dioxide. The dolime would then travel through a second stage consisting of two calcination tubes. Depending on if the desired specific surface area can be attained by passing through the calcination tubes, it may then be necessary to have a high temperature holding tank to facilitate final sintering. Test work on the feed material will enable better definition of the equipment. The specified unit should be capable of processing 350,000 tonnes per annum of dolomite.

Additional work is required to define the equipment necessary to obtain the desired preheat of 300 to 500 deg. C.

When the project reaches the stage where the specified equipment is ready for performance verification, 20 tons of dolomite is to be sent to Calix's test facility near Melbourne, Australia. [6] Part of this facility consists of two calciner units to facilitate the evaluation of the calcination process on specific materials. The pilot unit, CFC850 is capable of processing 20 to 100 kg/hr. The full scale CFC 15000 (See Figure 2) is a production scale unit that is currently used to produce a variety of products. With the provided dolomite test material, Calix's engineers will be able to demonstrate the capability of the kiln, obtain engineering parameters and provide sufficient calcined dolime for testing in a small pilot reduction furnace.

Figure 2. Calix CFC15000 Calciner located at Bacchus Marsh near Melbourne, Australia

Carbon Dioxide Usage

It is the intention of Nevada Clean Magnesium to capture and sequester the liberated carbon dioxide via deep well injection. This is in accordance with the mission of the company to have a major impact on the elimination of greenhouse gas. By producing magnesium without emitting carbon dioxide, the reduced emissions attained by use of the magnesium product in aircraft or automotive applications will be fully realized.

ScanMag has decided to provide the food grade carbon dioxide as a product. To this end they have already identified a number of outlets for the product.

Conclusion

While the cost of operation and capital costs are equivalent to other calciner types, [7] the quality of calcined product is ideally suited to the proposed magnesium reduction process and the ability to capture the liberated carbon dioxide as a pure product enables a multiplicity of uses with minimal downstream processing.

While Scan Mag and Nevada Clean Magnesium are early adopters of this technology, the expected benefits make the Calix calciner a major positive attribute for future smelters.

References

1. Nevada Clean Magnesium, 43-101 Report, "Preliminary Economic Assessment and Technical Report of the Tami-Mosi Magnesium Project, Nevada", rev. 2014.

2. D4K Consulting, *ScanMag Project Metal – investor presentation,* October 2014.

3. Calix web site, www.calix.com/au

4. W. H. MacIntyre and T. B. Stansel, "Steam Catlysis in Calcinations of Dolomite and Limestone Fines," *Industrial and Engineering Chemistry*, 45 (1953), 1548-1555.

5. Mark Skeats communication to Scan Mag via Vulcan Inc. 2014.

6. Andrew Okely, Calix, Calciner Testwork Proposal submitted to ScanMag AS, January 2013

7. J. Dinsdale, P Tirzins, M. Sceats, "Flash Calcining Technology for Cementitious Binders and Additives" (Paper presented at Future Cement Conference, London, UK, 8 February 2011), 17.

Drying, Roasting, and Calcining of Minerals
Edited by: Thomas P. Battle, Jerome P. Downey, Lawrence D. May, Boyd Davis, Neale R. Neelameggham, Sergio Sanchez-Segado, and P. Chris Pistorius
TMS (The Minerals, Metals & Materials Society), 2015

MODELING AND DESIGN OF EXPERIMENT IN CALCINATION OF MAGNESITES

B.K.Chakrabarti
Department of Materials & Metallurgical Engg.
National Institute of Foundry & Forge Technology
Ranchi-834003. INDIA.
e-mail: bijoy_2000in@yahoo.com.
Phone:91-9431767648

Keywords: Modeling & DOE, Dead Burning of Magnesite, Rotary kiln, Energy Conservation.

ABSTRACT

Modeling and a screening design of experiment (DOE) was performed with the process variables to identify the greatest effect of variables on the response quality in the dead burning of magnesite refractory in a rotary kiln. Among various interdependent variable factors, the design variables chosen were raw magnesite (RM) feed rate, kiln rotation time i,e, seconds per rotation(SPM) and burning zone temperature (BZT). The response variables chosen were exit gas temperature and density of dead burnt magnesite(DBM). ANOVA and normal plots were used to evaluate the effects.This work had enabled to correlate the BZT, RPM, RM feed with the exit gas temp and density of magnesite produced. The result opens up an avenue to look into the optimum region of operations within the ranges of variables considered in order to conserve energy and to minimize the exhaust gas temperature and to maximize the density of the DBM produced.

1. Introduction

The ceramic and refractory industries are the big energy consumers [1,2,3] , where the thermal energy constitutes an important part of the manufacturing costs. Proper utilization and conversion of energy plays an important role in the energy management of any system. DBM production in rotary kiln involves consumption of an alarming amount of energy, where thermal energy constitutes an important part of the manufacturing costs. The success of the dead burning process of magnesite rests on proper controlling of the large number of variable factors responsible for the manufacture of DBM. However in order to visualize optimum out put it is necessary to segregate the important factors which may be directly and greatly responsible for the success of the process. If conservation of energy of the process is of concern then obvious step would be to aim at higher quality of DBM at a lesser energy consumption. An energy campaign of the kiln system shows [4,5,6] that the potential heat loss areas are through kiln shell, chamber exit gas, clinker exit etc. Due to high exit gas temperature and high volume the heat loss, the exit gas temperature was considered as one of the response variables. The quality of DBM may be visualized in terms of its sintering, grain growth, crystalline size, density etc. In the present work the density had been chosen as the other response variable.

Multivariate Data Analysis refers [7,8,9] to statistical technique used to analyze data that arises from more than one variables. This essentially models in reality where each situation, product, or decision involves more than a single variable. Despite the quantum of data available, the ability to obtain a clear picture of what is going on and make intelligent decisions is a challenge. The available information is stored in database tables containing rows and columns, multivariate data analysis can be used to process the information in a meaningful fashion.

Factorial design is an approach to perform any experiment for analysis. DOE(design of experiment) & ANNOVA(analysis of variance), are various techniques used to analyze multivariate data and depict response patterns.

UNSCRAMBLER software package of CAMO Software Ltd. was utilized to create factorial design and analyze effect of variables on responses.

Experimental design is aimed precisely at solving the problems in a situation where experiments are costly and determines which variables are important & provides more information, and thus will have to be performed with higher priority. The advantage of designing experiments is that one gains control over data distribution, and ensures further analysis.

2. Data Acquisition & Kiln Description

Description of the kiln system and data acquisition.

Process	:	Dry process.
Type	:	Rotary.
Kiln Length	:	84 meters
Inclination	:	3 %
Fuel Type	:	LSHS (Low sulfur heavy stock) .
Calorific Value of the Fuel	:	10800 K.cals / Kg.
Fuel / DBM	:	0.185.
Average Exit gas temperature	:	235 C.
Atmospheric temperature	:	30 C.

Data acquisition and utilization of the data needs proper attention, since there are certain variable factors which influences the results very much.

3. Experimental

There are many variable factors involved in dead burning of magnesite in a rotary kiln.. However, based on data acquisition and the work as reported [4,5,6] the variable factors chosen in this work were Raw magnesite (RM) feed rate, kiln rotation time (SPR) and burning zone temperature (BZT) and estimated their effect on two responses viz, Density of DBM and Temperature of the stack exit gas. Density of magnesia influences the quality of product and exit gas temperature influences the energy efficiency of the process. UNSCRAMBLER statistical software package was used to design the experiment. Two level (upper & lower limits) experimental design was made with two centre sample.

Table I: Lower & Upper limit of the input variables.

Variables	Lower limit	Upper limit
Raw material feed (RM Feed) kg/hr	10080	10320
Seconds per rotation of the kiln (SPR)	60	80
Burning zone temperature ^{0} C (BZT)	1550	1650

Table II: The Data table of Design matrix.

RM Feed(Kg/hr)	SPR(sec)	BZT ^{0}C	Exit Temp ^{0}C	Density (Kg/m^3)
10080	60	1550	230	3.30
10200	70	1600	225	3.39
10320	60	1650	225	3.45
10200	70	1600	235	3.38
10080	80	1550	240	3.35
10200	70	1600	230	3.38
10320	80	1550	225	3.25
10320	80	1650	240	3.46
10080	60	1650	234	3.46
10320	60	1550	200	3.32
10080	80	1650	244	3.40

Table III: Effect summary for the Response Exit Temp

	Effect Summary 1		
Effects summar	Significance	Effect value	p-value
RM_FEED (A)	-	-14.5000	0.0113
SPR (B)	++	15.0000	0.0096
BZ_Temp (C)	+	12.0000	0.0257

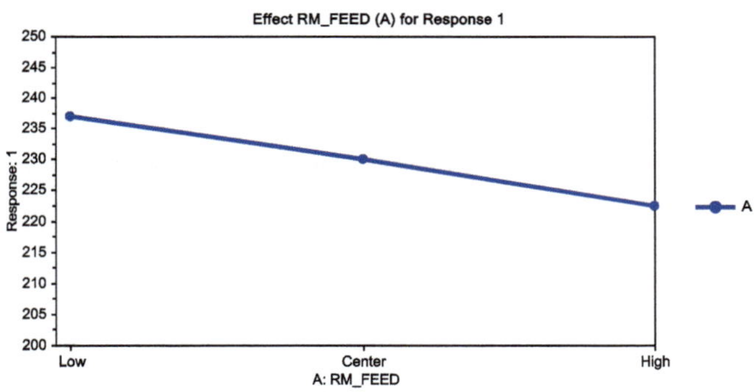

Figure1: Effect of RM Feed on the exit temperature.

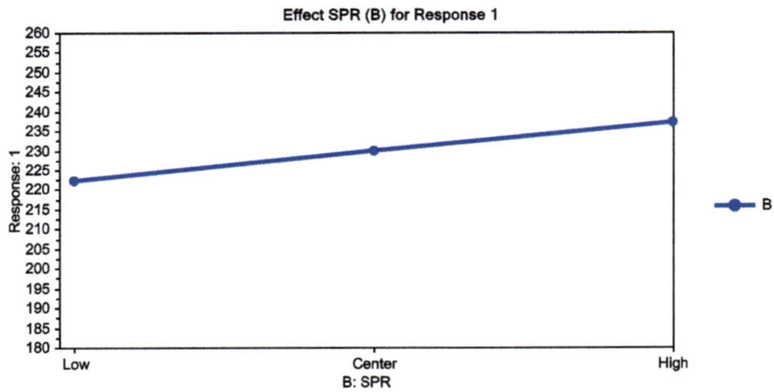

Fig 2: Effect of SPR on Exit Temp

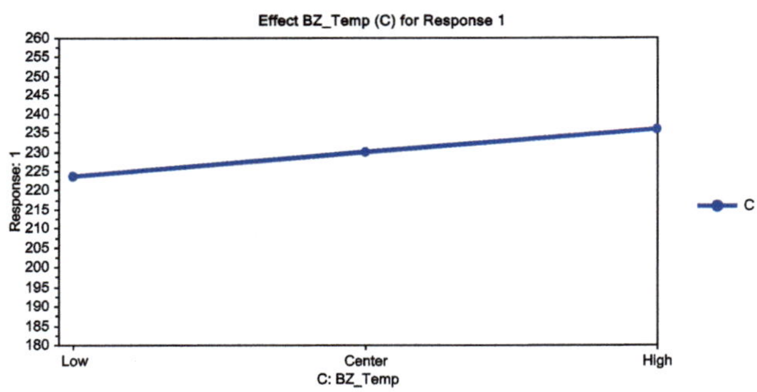

Fig 3: Effect of Burning Zone Temp(BZT) on Exit Temp.

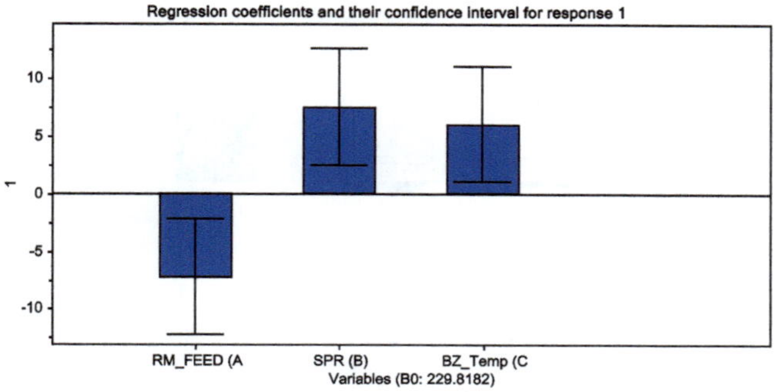

Fig. 4 : Regression Coefficients and their confidence interval for the response exit temp.

Table IV: Effect summary for the Response variable Density

	Effect Summary 2		
Effects summar	Significance	Effect value	p-value
RM_FEED (A)	NS	-0.0075	0.7525
SPR (B)	NS	-0.0175	0.4690
BZ_Temp (C)	+++	0.1375	0.0005

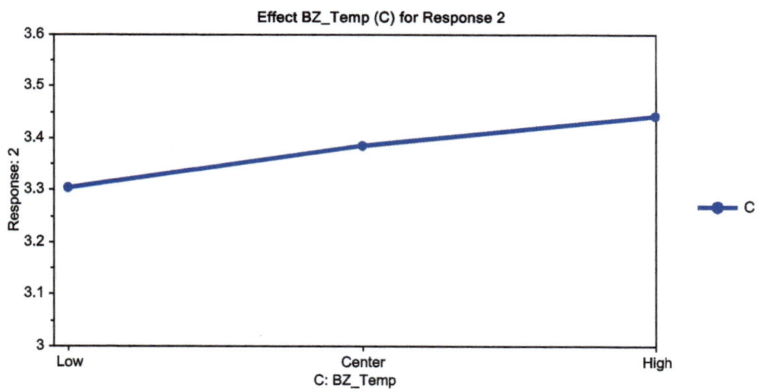

Fig 5: Effect of Burning Zone Temp on the response Density of Magnesite.

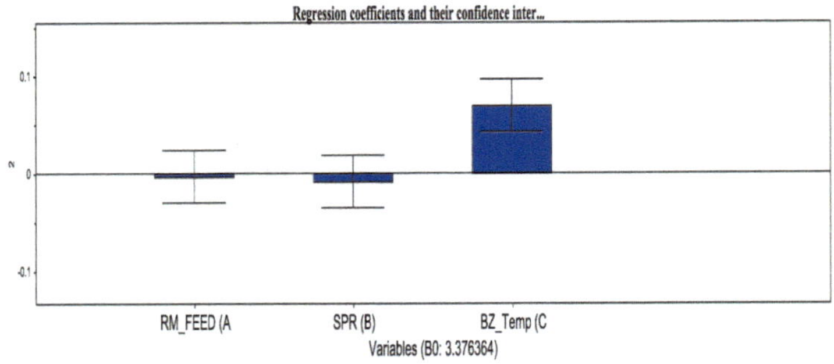

Fig 6: Regression coefficients and their confidence interval for response Density

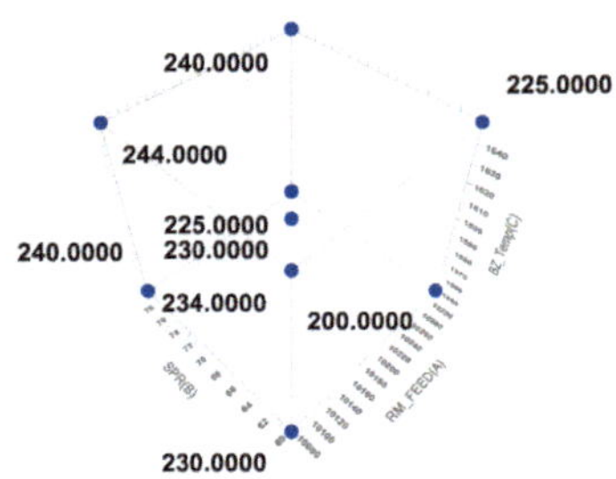

Fig. 7: Cube plot of RM Feed, SPR and BZT vs response exit gas temp.

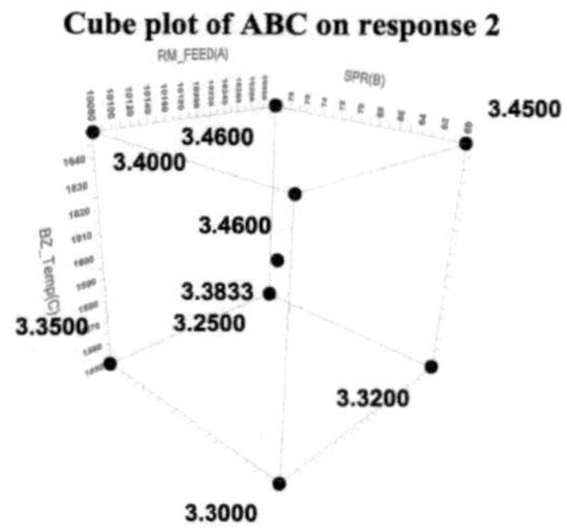

Fig. 8: Cube plot of RM Feed, SPR and BZT vs response Density of DBM.

4. Results & Discussions

Table-I describes about the lower & upper limit of the data acquisitioned for the design variables RM Feed, SPR and BZT. Based on which the design matrix table was constructed. Table-II depicts about experimental design matrix, where in each row of the data table correlates various input design variables with the responses.
Main effect plot depicts :

- Exit gas temperature decreases as the raw material feed increases as shown in fig.1. Exit temp has also been found to increase as the time of rotation of the kiln increases as shown in fig.2.
- Fig.3 illustrates the effect of input variable BZT on response exit gas temperature, and shows that exit temp. increases with increase in burning zone temperature.
- Fig. 5 illustrates, the effect of input variable BZT on response density and illustrates that the density of DBM increases as the burning zone temperature increases.
- Fig.4 & fig.6. shows the volumetric representation of the regression coefficients and describes about the regression coefficients & their intervals. It also correlates visually the effect of input variables on the response variables.
- Fig.7 & Fig.8 are the 3-D cube plots, which illustrates the effect of all the three input variables simultaneously on each of the response variable. Fig.7 illustrates the effect of RM feed, SPR and BZT on exit temperature. Similarly, Fig.8 illustrates the effect of three input variables on the response variable, i,e density of DBM.
- Table III and Table IV describes the effect summary for the response Exit Temp & Density respectively.

5. Conclusion

Factorial design was an approach to perform an experiment for analysis. With the use of process data we could predict various effects of variables on responses which consequently helped in defining and monitoring optimized solutions to the process. There were many process variables for dead burning of magnesite in rotary kiln. However, due to complexity involved in data acquisition & experiments, only three input variables and two response variables were taken into account. Numerous other possibilities and combinations of variables and responses may be aimed at.

It may be concluded that results obtained in the work was an outcome of systematic approach towards data acquisition and utilization of the same. The Experimental Design, ANNOVA & Response surface , main effect plots etc. given an insight into the interactions among variable factors and into the operation of the kiln in the dead burning of magnesite. This work had enabled to correlate the design variables burning zone temperature (BZT), Kiln rotation time (SPR), RM feed with the response variables exit gas temp and density of magnesite produced. The details of interactions & their effects of the data correlating various input factors and responses were described & discussed in results & discussions section through design matrix table (Table.II), main effect plots (Fig.1, Fig.2 Fig.3 & Fig.5), volumetric representation of regression coefficient (Fig. 4 & 6), 3-D cube plots (Fig.7 & 8). A proper monitoring of the kiln is extremely essential so that the above factors could be regulated properly. The work provides data and appropriate information's to the management of the company about how the units have been functioning. It also guides and enables the industry to set up realistic targets of energy utilization & economics of the process. And to identify the critical areas where & how conservation measures could be implemented.

6. **References:**

1. John D. Morgan, Mc Kelvey, Ceramics: Energy Update, a report from 79th Annual Meeting of American Ceram. Society, J. Am. Ceram. Soc.Bull,56 (1977) p-586.

2. Clifford Storey, Energy and the refractories industry, Refractories Journal, 6 (1977) p-26-28.

3. I.K.Smithson, Fuel and energy economy in industry, Tans. British Ceram. Soc. 74 (6) 1975, p-203.

4. B.K.Chakrabarti " Energy conservation & management in refractory burning units.''-, Presented in the technical session and published in the proceedings of the 65th annual session of the Indian Ceramic Society, 2002.

5 . B.K.Chakrabarti " Investigations on heat loss through the kiln shell in magnesite dead burning process- a case study''Journal of Applied Thermal Engineering, 22(2002)1339-45.

6. B.K.Chakrabarti "Studies on the heat transfer behaviour in a magnesite kiln and its assessment'' Journal of High Temperature Materials & Processes, London, Volm.21, No.4, 2002. P187-194.

7. Montgomery, D. C. "*Design and Analysis of Experiments*", Wiley, NY (2004)

8. .Dean A. and Voss D. (1999) "*Design* and analysis of experiment" Springer, NEW YORK.

9. T. W. Anderson, "*An Introduction to Multivariate Statistical Analysis*", 2nd Edition (New York: John Wiley & Sons, 1984).

Drying, Roasting, and Calcining of Minerals
Edited by: Thomas P. Battle, Jerome P. Downey, Lawrence D. May, Boyd Davis, Neale R. Neelameggham, Sergio Sanchez-Segado, and P. Chris Pistorius
TMS (The Minerals, Metals & Materials Society), 2015

STUDY ON EFFECT OF UNTREATED AND CALCINED OLIVINE ON LOW SILICON PELLET PRODUCTION PROCESS AND QUALITY

Gele Qing[1,2], Keng Wu[1], Shaoguo Chen[1], Qihang Liu[1], Li Ma[2]

[1]University of Science and Technology Beijing, 30 Xueyuan Road, Beijing 100083, P. R. China;
[2]Shougang Research Institute of Technology, 69 Yangzhuang Road, Beijing 100043, P.R.China
E-mail：wukeng1030@sina.com.cn; qinggele_68@163.com

Keywords: untreated olivine, calcined olivine, low silicon pellet

Abstract

Olivine contains higher MgO element, which can be added into blast furnace burden to improve slag stability and fluidity. Effect of calcined olivine and untreated olivine on low silicon pellet production process and quality was studied in this paper. The results shows, calcined olivine has no effect on bursting temperature of green pellet, however the bursting temperature of green pellet will decrease with increase of untreated olivine ratio. Untreated olivine has slight effect on green pellet strength than calcined olivine while the ratio is higher than 2%. The compressive strength of low silicon pellet will increase first and decrease afterwards with increases of calcined olivine ratio under firing temperature of 1250°C, whereas, the changing sequence is reverse fired under 1280°C. The result of untreated olivine pellet compressive strength is same as calcined olivine pellet. However, the compressive strength and metallurgical property of calcined olivine pellet is slightly higher than that of untreated olivine pellet under same condition.

Introduction

Flux as dolomite was charged into blast furnace directly to supply MgO for slag in the past; however it has brought negative effect on blast furnace operation like endotherm, etc. Presently, most of the MgO in slag is from sinter which contains around 2% of MgO[1,2]. The metallurgical properties as reduction, swelling and smelting of pellet can be improved by increased of MgO content[3-5]. Usually reduction swelling of acid pellet is higher, so ratio of acid pellet in blast furnace burden is lower than fluxed pellet. On the other hand, softening and smelting property of acid pellet and high basicity sinter is different, which will affect the blast furnace operation. So the ratio of acid pellet in blast furnace burden is around 20% in domestic. Therefore, when some of MgO in sinter is diverted into pellet, not only the quality of sinter can be increased, also the metallurgical property of pellet will be improved, which is beneficial blast furnace operation.

Generally, control of firing temperature of MgO-bearing pellet has more difficulty than acid pellet, and various MgO-bearing fluxes have different influence on pelletizing, firing and quality of pellet. Many investigations on different fluxes as dolomite, light-burned dolomite, magnesium oxide and olivine used in pelletizing process have been reported[6-9], it is found that olivine has the extensive use in pellet[10-12]. Olivine is from natural, and it can be pretreated to become calcined olivine. The properties of untreated olivine and calcined olivine are different, which have different effect on pelletizing, firing and quality. So effect of untreated and calcined olivine on low silicon pellet production process and quality were studied in the present paper.

Experimental method and material performance

Experimental method

First, raw materials were mixed on the basis of blending schemes, and then green pellet was agglomerated in pelletizing disc with diameter of Φ800mm, rotation speed of which was 18r/min, pelletizing time was for 10min. After pelletizing green pellets with 10~12.5mm was screened out for dropping strength, compressive strength, moisture and bursting temperature test. Bursting temperature test of Green pellet as following procedure: put 20 green pellets into the hanging basket through 1.8m/s hot wind, observed the integrity of pellets after 5min, if the green pellets were not burst, raise the hot wind temperature and repeat the test until 10% of green pellets were burst, and record the temperature as bursting temperature. High temperature tube furnace was used in the green pellet preheating and firing experiment, Compressive strength of pellet was tested by microcomputer control electronic presser, Scanning electron microscopy(SEM) was used for observation of pellet microstructure.

18 pellets with diameter of 10~12.5mm were selected for reduction swelling test. The pellets were reduced at 900℃ for 1 hour by gas of CO and N 2 (volume 30/70), with flow rate 15±1L/min. 500g pellets with diameter of 10~12.5mm were screened out for reduction test and reduced by same gas and flow rate for 3h. that was as same experimental temperature, reducing gas compositions and flow rate as reduction swelling test, except for reduction time of 3 hours. Low temperature reduction disintegration index, 500g pellets with diameter of 10~12.5mm were reduced at 500℃ for 1 hour by gas of $CO/CO_2/N_2$ (20/20/60), with flow rate 15±1L/min, after cooling samples were put into tumbler machine for 300r, and then pellets were screened by 6.3mm, 3.15mm and 0.5mm screens.

Material performance

Two kinds of olivine were used in the experiment, one was untreated olivine, and another is calcined olivine. Bentonite was used as binder for pelletizing. The chemical composition of raw materials is given in Table 1, iron ore was low silica magnetite with iron grade 69.32% and SiO2 content 1.72%, Table 2 shows grain size of iron ore. Grain size of iron ore concentrate less than 74μm content is 88.2%, Before experiment, two olivines have been ground fine with size below 74μm content up to 100%. There is not much difference except burning loss in chemical composition of two kind of olivine, the burning loss of calcined olivine is low, which is 1.06% and burning loss of untreated olivine is high, which is 7.83%.

Physical properties of bentonite are shown in Table 3, 2h water absorption is 456%, swelling index is 18ml/2g.

Table 1 Chemical composition of raw material (%)

Ingredients	TFe	SiO_2	Al_2O_3	CaO	MgO	P	S	K_2O	Na_2O	loss
Iron Ore	69.32	1.72	0.32	0.43	0.45	0.004	0.16	0.023	0.090	-2.76
Calcined olivine	5.17	36.76	4.51	2.97	44.70	0.15	0.22	1.2	0.14	1.06
Untreated olivine	5.65	36.84	0.60	1.21	45.01	0.039	0.025	1.17	0.11	7.83
Bentonite	--	61.22	14.52	2.28	2.59	0.001	0.037	0.015	0.008	13.75

(Note: Components in mass fraction)

Table 2 Grain size of iron ore (%)

Size range (mesh)	+80	80~100	100~120	120~160	160~180	180~200	-200
Iron ore	0.2	1.44	1.25	1.26	3.09	4.56	88.2

(Note: mass fraction)

Table 3 Physical properties of bentonite

Name	Glial price ml/15g	Blue absorption capacity g/100g	Swelling Index ml/2g	2hWater absorption %	<0.074 μ m Content %
Bentonite	38	32.2	18	456	100

Results and Discussions

Effects of calcined and untreated olivine on pelletizing and green pellet quality

To clarify the influence of calcined and untreated olivine on pelletizing and green pellet quality, pelletizing tests were carried out by adding different proportion of calcined and untreated olivine to iron ore, respectively, 0%、0.5%、1.0%、1.5%、2%、2.5%, 3.0%. Bentonite proportion was 1.5%. Figure 1 shows the result of green pellet dropping strength. Figure 2 shows the compressive strength of green pellet changes with olivine proportion. When calcined olivine is below 1.5%, dropping and compressive strength of green pellet is similar with benchmark experiment of 0% proportion olivine, while exceed 2.0% the dropping strength start to decrease, when is 3.0% it slows down to 4.7times/0.5m, but compressive strength get to be improved up to 11.2N/P. when untreated olivine was changed from 0.5% to 3.0%, it is observed that slight change is caused between dropping and compressive strength. The results of bursting temperature of green pellet are shown in Table 4. The bursting temperature of green pellet with calcined olivine is higher, which exceed 650℃, however the bursting temperature of green pellet is decreased with increase of untreated olivine ratio.

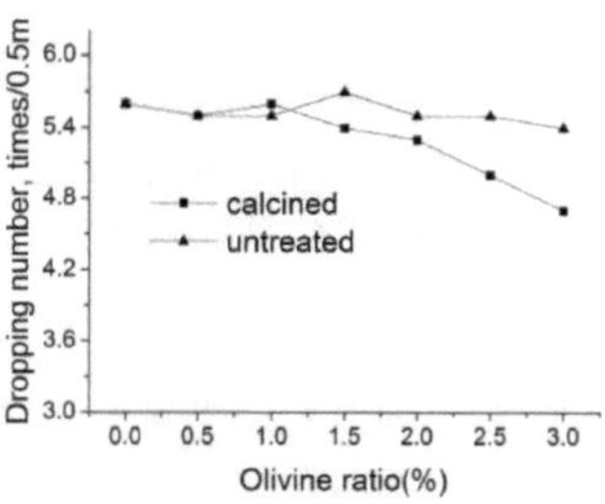

Figure 1. Dropping strength of green pellets changes with olivine ratio

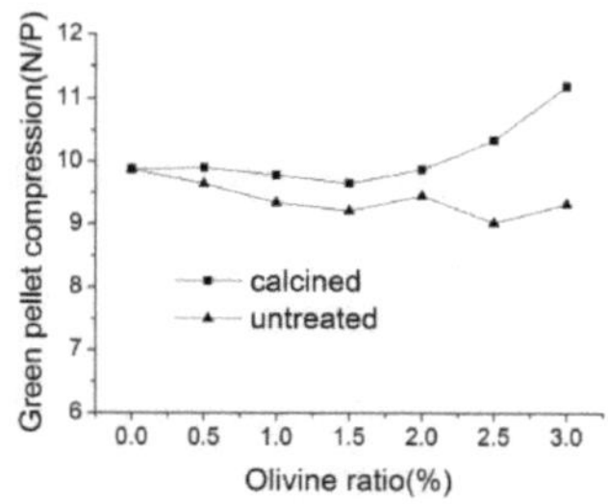

Figure 2. Compressive strength of green pellets changes with olivine ratio

Table 4 Bursting temperature of green pellet with different olivine

Name	Calcined olivine							untreated olivine					
Ratio,%	0	0.5	1	1.5	2	2.5	3	0.5	1	1.5	2	2.5	3
Green pellet moisture,%	8.8	8.7	8.6	8.6	8.7	8.6	8.5	8.8	8.8	8.9	8.7	8.6	8.8
Bursting temperature,℃	650	650	650	650	650	650	650	610	610	610	610	600	600

Influence of calcined and untreated olivine on pellet compressive strength

High temperature tube furnace was used in the firing experiment and temperature was separately designed as 1250°C and 1280°C, firing time is 20min. Before firing, all green pellets were dried at

120°C for 2 hours and preheated at 950°C for 20min. Compressive strength result of pellet using calcined olivine is shown as Fig 3, the result of pellet using untreated olivine is show as Fig 4. It can be found that both the calcined and the untreated proportion are below 1.5% at 1250°C, pellet compressive strength improves with the increase of olivine content, the highest is up to 2655N/P, while proportion exceeds 2% the downward trend is observed to compressive strength. Under the same olivine proportion, the compressive strength of pellet using calcined is higher than one using untreated. When firing temperature was improved to 1280°C, compressive strength first decreased and then increased, up to 3134N/P at adding 3% calcined olivine to pellet. It is seemed to indicate that at high temperature condition, the calcined olivine is beneficial to improve compressive strength of pellet. Although, changing tendency of pellet strength with untreated olivine is the same as calcined olivine. However the compressive strength of calcined pellet is higher than untreated olivine pellet under same condition. Figure 5 shows SEM images of fired pellet with 1%, 3% of calcined and untreated olivine. Little liquid was formed in pellet with 1% of olivine, even the temperature is raised to 1280℃. The main phase is magnetite oxidation and recrystallization of Fe2O3. While the proportion of olivine was raised to 3%, there appeared monticellite and forsterit, so internal structure of pellet become tighter and the compressive strength increase significantly.

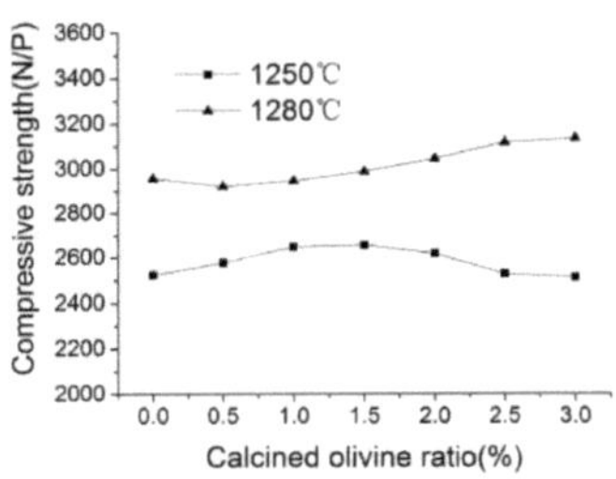

Figure 3. Compressive strength of pellet with calcined olivine

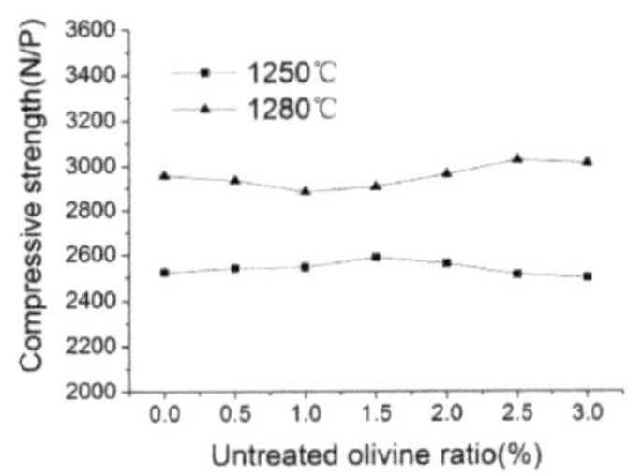

Figure 4. Comressive strength of pellet with untreated olivine

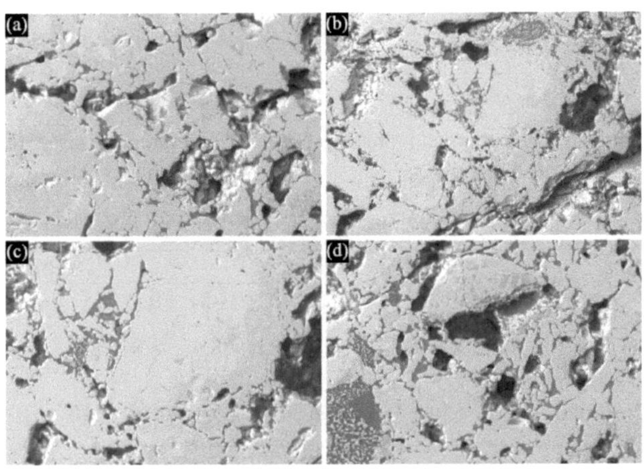

(a: 1% untreated olivine; b: 3% untreated olivine; c:1% calcined olivine; d:3% calcined olivine)

Figure 5. SEM images of fired pellet

Effect of calcined and untreated olivine on pellet metallurgical properties

Table 5 shows the pellet chemical composition of schemes designed as separately adding 1%、2% and 3% of calcined and untreated olivine. Iron grade of Pellet without olivine is 66.33%. the pellet added 1% calcined olivine is 65.61%,the decreasing amount is 0.75%, however the MgO content was increased from 0.47% to 0.93%. While proportion of calcined olivine was increased to 3%, the MgO content of pellet will increased to 1.81%. The chemical composition of pellet with untreated olivine is almost same as pellet with calcined pellet under same ratio.

Table 5 Pellet chemical composition(%)

No.		TFe	FeO	SiO_2	CaO	MgO	Al_2O_3
1	--	66.33	0.34	2.50	0.45	0.47	0.47
2	1%calcined olivine	65.61	0.43	2.86	0.47	0.93	0.52
3	2%calcined olivine	65.02	0.38	3.21	0.50	1.37	0.55
4	3%calcined olivine	64.39	0.41	3.56	0.52	1.81	0.61
5	1%untreated olivine	65.69	0.48	2.87	0.45	0.93	0.48
6	2%untreated olivine	65.17	0.36	3.23	0.46	1.38	0.48
6	3%untreated olivine	64.46	0.47	3.59	0.47	1.83	0.46

The reduction swelling index of pellet with different proportion of olivine fired at 1280℃ was shown in Figure 6. Form figure it was seen, the reduction swelling index of pellet will decrease with increase of both calcined and untreated olivine proportion. The reduction swelling index of pellet without olivine is too high to satisfy the blast furnace operation due to low SiO_2 content, which is only 2.55%. While SiO_2 content is low it is difficult to suppress the development of the iron whisker during pellet reduction. With increase of both calcined and untreated olivine proportion, the amount of liquid phase increase and the magnesium ferrite formed inside pellet, which restrain the development of iron whisker effectively during pellet reduction to reduce swelling. At same proportion, the reduction swelling index of pellet with calcined olivine is lower than that of pellet with untreated olivine. While proportion was 2.0%, the calcined olivine pellet reduction swelling index is 19.56%, untreated olivine pellet reduction swelling index is 20.57% , while the proportion of untreated olivine increased to 2.5%, the reduction swelling index of pellet will decreased to 19.48%.

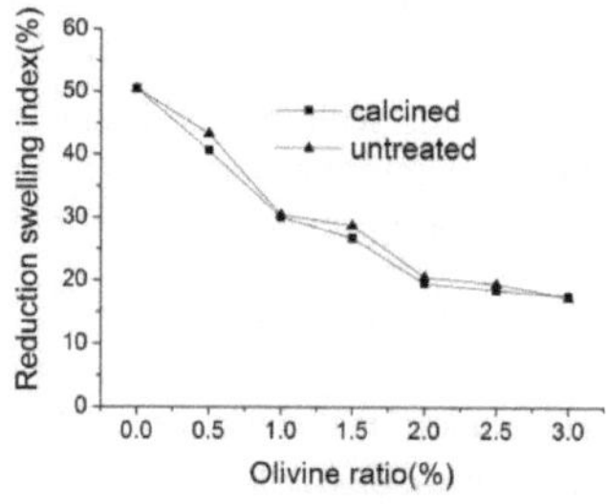

Figure 6. Pellet reduction swelling index changes with olivine ratio

The result of pellet reducibility changes with olivine proportion is shown as Figure 7. The pellet reducibility was continuously improved with the increase of olivine ratio. High smelting phase such as forsterite ($2MgO.SiO_2$) and metasilicate magnesium ($MgO.SiO_2$) can be formed when appropriate amount of MgO was added in pellet which improves the reducibility. The smelting point of forsterite and metasilicate magnesium is 1890℃and 1557℃ respectively. The reduction index of pellet without olivine is 67.6%, pellet with 2% calcined olivine is increased to 72.56%, while proportion of calcined pellet increased to 3%, the reduction index is increased to 74.56%. Pellet with 3% untreated olivine is also reached 73.41%. which illustrate that calcined olivine and untreated olivine both will improve the pellet reducibility.

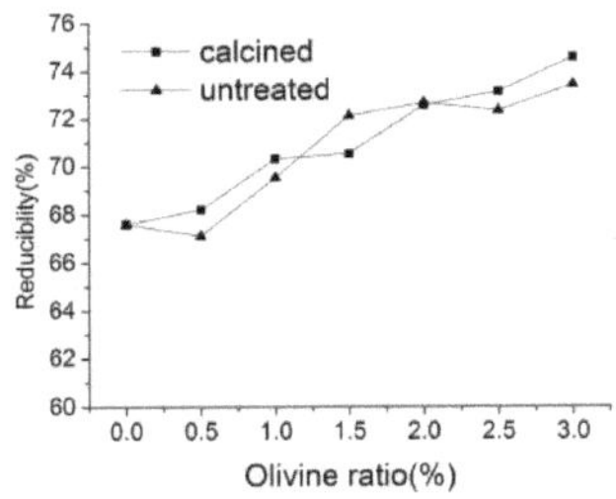

Figure 7. Pellet reducibility changes with olivine ratio

Conclusions

Calcined olivine has slight effect on green pellet strength while proportion is less than 2.0%, however, the dropping strength of green pellet decreases after calcined olivine proportion exceed 2%. The compressive strength of green pellet is enhanced with increase of calcined olivine ratio and bursting temperature is higher than 650℃.

Untreated olivine has less effect on green pellet dropping strength and compressive strength, however, bursting temperature of green pellet is decreased with increase of untreated olivine ratio.

The compressive strength of pellet is raised with increase of calcined olivine ratio, while firing temperature is 1250℃. While firing temperature is 1280℃, the compressive strength of pellet with high ratio calcined olivine is higher than pellet with low ratio calcined olivine. The variation compressive strength of untreated olivine pellet is same with calcined olivine pellet; however, the compressive strength of calcined pellet is higher than untreated olivine pellet under same condition.

Both untreated and calcined olivine will decrease the reduction swelling index of low silicon pellet. The reduction swelling index can be decreased to lower than 20% when calcined olivine ratio increased to 2.0%, and untreated olivine ratio increased to 2.5%.

The reducibility of pellet is improved both with the addition of untreated or calcined olivine. Calcined olivine will improve the low temperature reduction disintegration index of pellet, however, untreated olivine will decrease the low temperature reduction disintegration slightly.

Acknowledgments

The authors would like to express their thanks for the support by National Natural Science Foundatio China (51274026), Independent research project (41603003) of State Key Laboratory of Advance Metallurgy.

References

1. Y.Z. Zhang, Z.W. Lu, "Effects of basicity and contents of SiO2 and MgO on sinter production and quality", *Sintering and Pelletizing*, 5(2001),47-50.
2. H.Y. He, "the smelting of low silicon pig iron and the research of influence of MgO content in slag", *Journal of Wuhan University of Science and Technology*, 5(2002),1-5.
3. H.Q. Zhang, "Production and development of fluxed pellet", *China Mining Magazine*,4(2009),89-92.
4. Paulo F, Richard, Blast furnace burden softening and melting phenomena-melting onset in acid and basic fluxed pellets, *Iron making Conference Proceedings*,(2002),585-595.
5. Kenneth G.Oja, Test work to determine the effect of varying percent SiO2,CaO/SiO2 and MgO/SiO2 ratio on iron ore pellet quality using design of experimental software,(Paper presented at AISTech meeting, Pittsburgh,5 May 2008),131-142.
6. E.Eliasson, P.L Hooey, H.Annesten. Formation of potassium slag in olivine fluxed blast furnace pellets, *Iron making and Steelmaking*, 34(5)(2007), 422-430.
7. Ulrika Leimalm, Lena Sundqvist, Okvist Bo Bjorkman. Effect of simulated PCI rate on olivine pellet reduction,(Paper presented at The 4th international congress on the science and technology of ironmaking meeting, Osaka, 26 November 2006),201-204.
8. Ulrika Leimalm, Lena Sundqvist Okvist, Anna Brannmark, "Correlation between pellet reduction and some blast furnace operation parameters",(Paper presented at AISTech meeting,Cleveland, 1 May 2006),39-50.
9. B. Xu, et al, "Study on fluorine-bearing iron concentrate pellets added with dolomite", *Iron and Steel*, 43(2008),9-15.
10. H.X. Fan, et al, "Influence of flux additives on iron ore oxidized pellets", *Journal of Central South University*, 17(2010), 732-737.
11. G.S. Wu, M.Z. Bian, F.M. Shen, "Application of basic pellets contained magnesium and research for the reasonable structure of burden", *iron and stee*,41(12)(2006),19-22.
12. T.J. Chen, et al, "Experimental study on application of forsterite in iron ore pelletizing", *Sintering and pelletizing*, 4(2)(2005),5-8.

DRYING, ROASTING, AND CALCINING OF MINERALS

FLUIDIZATION, REDUCTION ROASTING, AND MICROWAVE TREATMENT

Session Chairs:
Jerome P. Downey
Lawrence D. May

Drying, Roasting, and Calcining of Minerals
Edited by: Thomas P. Battle, Jerome P. Downey, Lawrence D. May, Boyd Davis, Neale R. Neelameggham, Sergio Sanchez-Segado, and P. Chris Pistorius
TMS (The Minerals, Metals & Materials Society), 2015

Fluidization and Magnetization Roasting Technology Research on Siderite

Wen Chen[1],Zhenhong Liao[1],Xiaoyin Liu[1],Ligang Zhang[1],Jialin Li[1],Xinghua Liu[1]

[1]Changsha Research Institute of Mining and Metallurgy Co.,Ltd.; 966 Lushan South Road; Changsha, Hunan, 410012, China

Keywords: Siderite; Fluidization; Magnetization Roasting;Kinetics

Abstract

A research-scale fluidization and magnetization roasting device was developed to evaluate a processing option that might offset the complexity and high cost of the conventional magnetization roasting process on siderite. In this study, the operating parameters of the fluidization and magnetization roasting device for siderite were evaluated through cold-flow tests and magnetization roasting tests. The reaction mechanism functions of the fluidized roasting process are revealed with kinetic studies.

Test Sample and Equipment

Test Sample

The siderite ore sample used for this study was from Dechang,Sichuan,China. The crude ore was purified by low intensity magnetic separation (LIMS) and high intensity magnetic separation (HIMS) after crushing and grinding. The graph of XRD and the chemical phase of iron are listed in Figure 1 and Table 1, respectively.

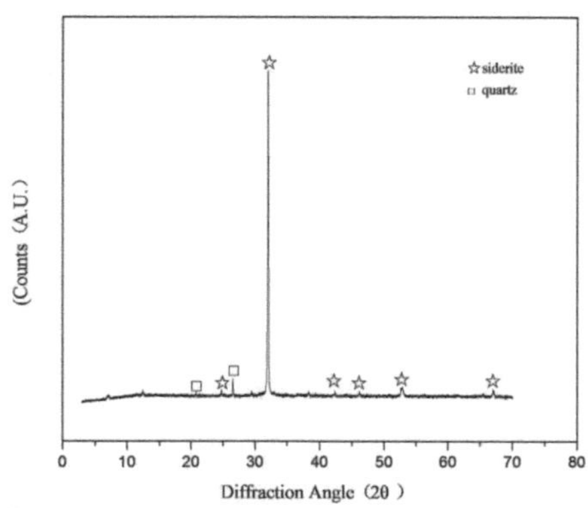

Figure.1. X-ray Diffraction Spectra of Siderite

Table 1. Chemical Phase Analysis of Iron, Wt.%

Iron Phase	Fe in magnetite	Fe in limonite (or hematite)	Fe in carbonate	Fe in sulfide	Fe in silicate	Total
Content, Wt.%	0.9	0.2	28.44	3.71	3.92	37.17
Distribution, Wt. %	2.42	0.54	76.51	9.98	10.55	100.00

Test Equipment

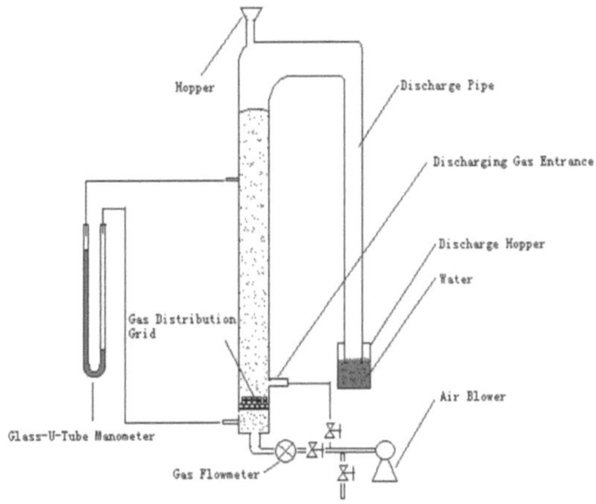

Fig.2. Cold-Flow simulations equipment

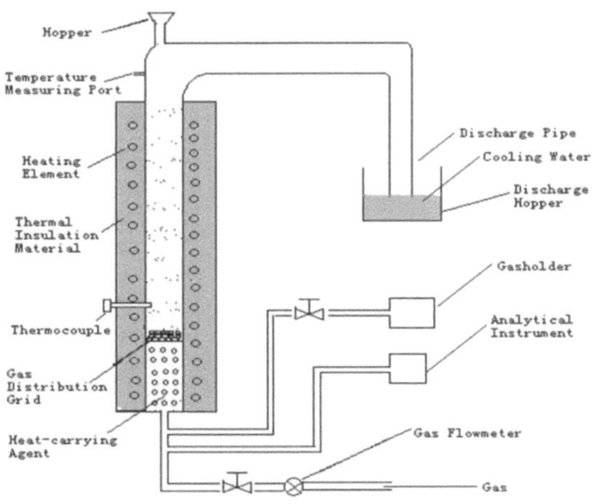

Fig.3. Fluidization and magnetization roasting equipment

Cold-Flow Simulation Tests

In order to determine the siderite size fraction for fluidized roasting and the best fluidization velocity, different size fractions of siderite were used to conduct the cold-flow simulation tests. The results are shown as Figure 4.

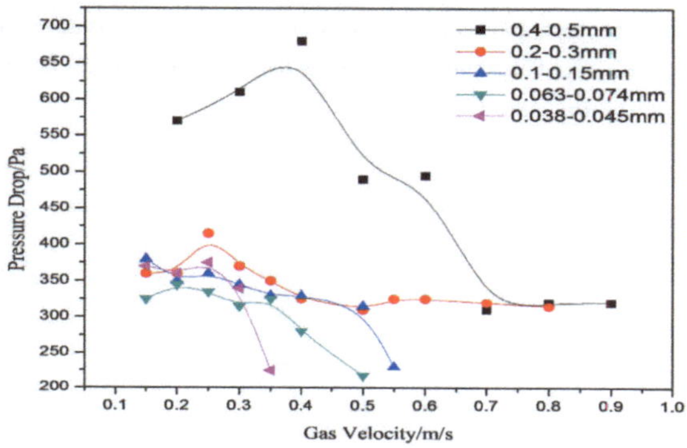

Fig.4. Cold-Flow simulation tests of different size fraction

Under proper fluidization, the bed pressure drop for the various sizes of siderite remains about the same. With increased gas velocity, some particles will elutriate and escape the bed, and the bed pressure drop will decrease. The results in Figure 2 show that most small particles of siderite will escape the bed when the gas velocity reaches 0.3m/s. In order to avoid the siderite of -0.038mm (or smaller size) elutriating from the bed, the biggest size of siderite chosen for fluidization was 0.3mm. In order to verify the results of cold-flow simulations, siderite sized to -0.3mm was used to conduct another cold-flow simulation test. The results are shown in Figure 5.

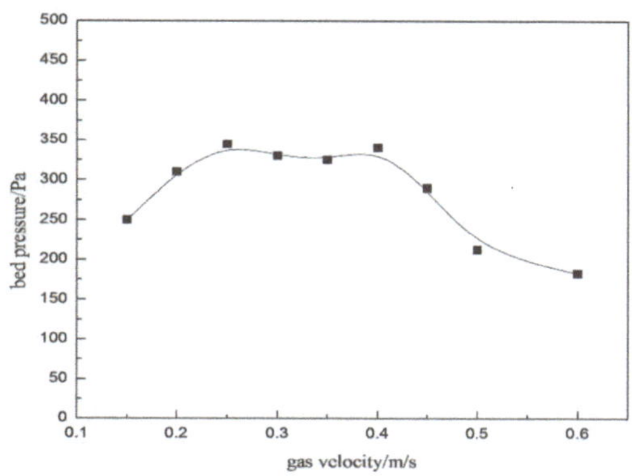

Fig.5.Cold simulation tests of -0.3mm Siderite

It can be seen that gas velocity between 0.2 m/s to 0.4 m/s gas is more suitable for the fluidization of -0.3mm whole size fraction of siderite. But with higher gas velocity, the possibility of smaller siderite particles leaving the bed increases. Therefore, a gas velocity of 0.25 m/s was chosen as the best gas operating parameter for fluidization and magnetization roasting.

Magnetization Roasting Tests

The selected temperatures for magnetization roasting were 550°C, 600°C, 650°C, 700°C, and 750°C. For each temperature, different roasting times were conducted. After reduction roasting, low intensity magnetic separation (LIMS) tests were carried out on roasted ore under the magnetic field intensity of 2000 Oe. The results are shown in the plots provided in Figure 6.

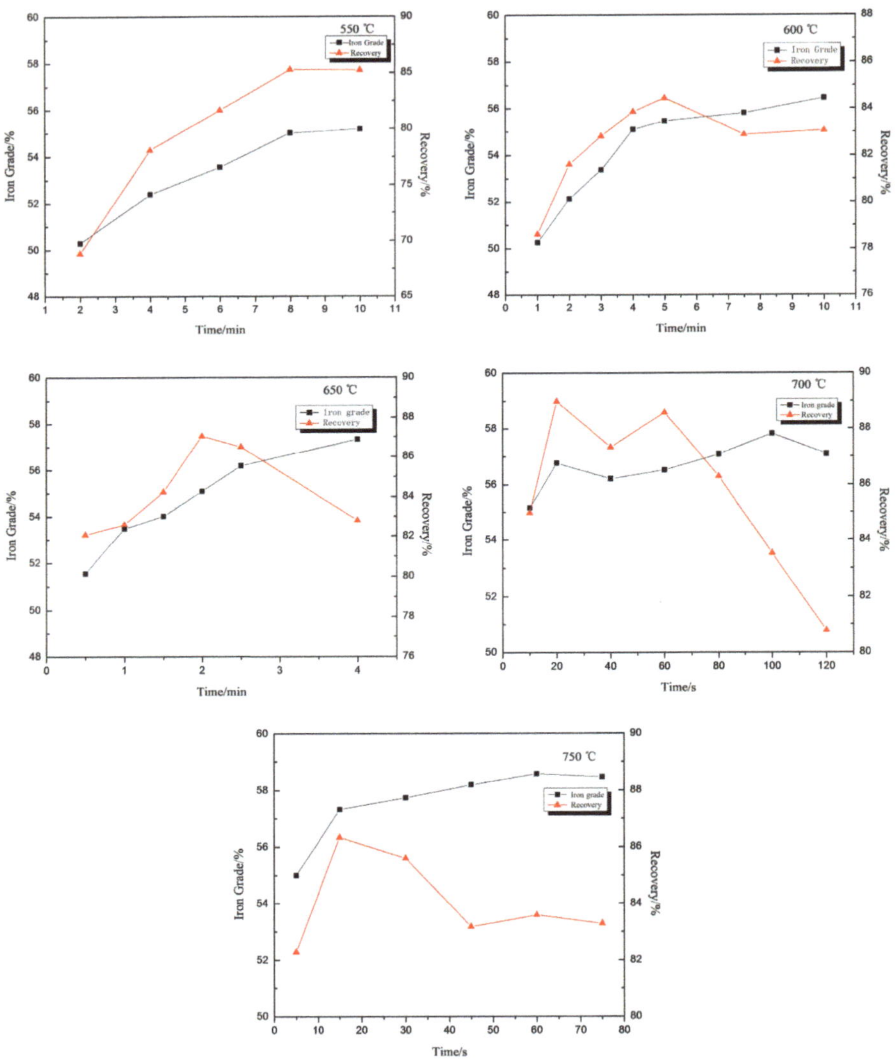

Fig.6.Results of Magnetization Roasting Tests for Different Times at Different Temperatures

The changes of chemical phase of iron

In order to find out the change of chemical phase of iron at each temperature, the roasted ore samples were studied by XRD. The results are shown in Figure 7.

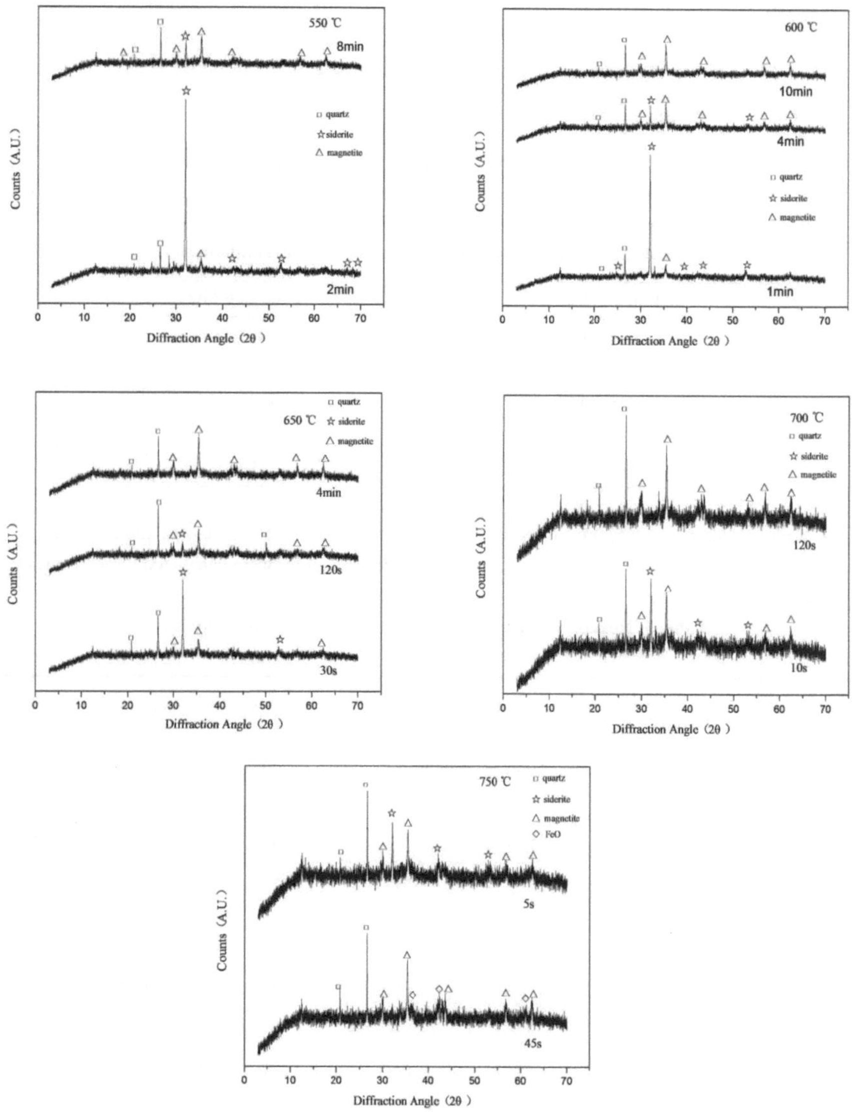

Fig.7. Changes of chemical phase of iron

The figures show that with the increase of time more siderite is converted into magnetite at the same temperature. And with the increase of temperature, less time is required to complete the conversion. Of special note is that FeO appeared when siderite was roasted for 45s at 750°C, which indicates that over-reduction occurred in the siderite. FeO is feebly magnetic and its existence is harmful to fluidization and magnetization roasting.

Kinetics

The reaction mechanism function of isothermal kinetics is shown as follow:

$$G(a) = \int_0^t A\exp(-\frac{E}{RT})dt = kt \qquad (1)$$

a: percent conversion of iron in siderite; k: reaction rate constant

There are 5 kinds of reaction mechanism functions of isothermal kinetic suitable for fluidization and magnetization roasting on siderite. They are shown in table 2.

Table 2. reaction mechanism functions of isothermal kinetic

reaction mechanism	Functions G(a)
convergent geometry R3	$1-(1-\alpha)^{1/3}$
convergent geometry R2	$1-(1-\alpha)^{1/2}$
Random nucleation A2	$1-(1-\alpha)^{1/2}$
Reaction order F1	$-\ln(1-\alpha)$
Diffusion D3	$[1-(1-\alpha)^{1/3}]^2$

Putting a at temperature 550°C and 650°C into G(a) shown in Table 2 get G(a)-t curves shown in Figure 8. Reaction rate constants and coefficient of determination were calculated through linear fitting. Results are shown in Table 3.

Table 3. Reaction rate constants and Coefficient of determination

Temperature	reaction mechanism $g(a)$	R3	R2	A2	F1̃	D3
550°C	Reaction rate constant k (s^{-1})	5.68576×10^{-4}	7.93591×10^{-4}	0.0021	0.00198	1.50485×10^{-4}
	Coefficient of determination	0.98027	0.975	0.92495	0.98924	0.98977
650°C	Reaction rate constant k (s^{-1})	0.00212	0.00289	0.00704	0.00778	7.17059×10^{-4}
	Coefficient of determination	0.87052	0.85788	0.81064	0.81064	0.96178

From these plots, it can be seen that D3 is more suitable for the reaction mechanism function of fluidization and magnetization roasting on siderite.

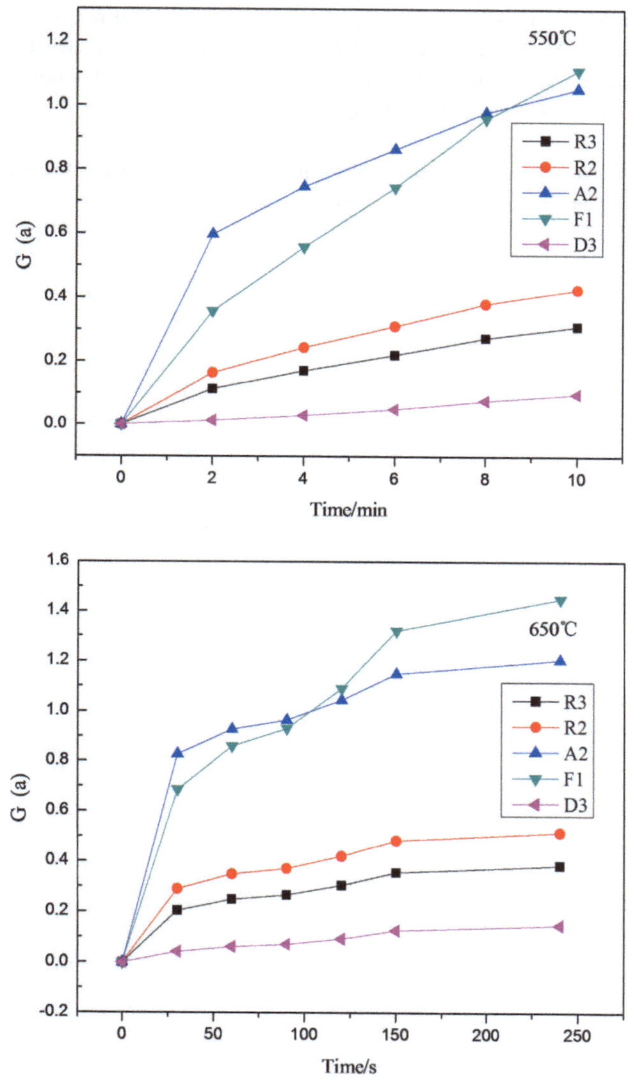

Fig.8. Curves of G(a)-t

Conclusions

(1) In this study, a -0.3mm size fraction of siderite was most suited for fluidization and magnetization roasting. For this size fraction, a gas velocity of 0.25 m/s was chosen as the best gas operating rate for fluidization and magnetization roasting.

(2) With the increase of time and temperature, the majority of siderite is converted into magnetite. At a temperature of 750°C, the conversion can be achieved in about 1 minute.However, when the temperature is 750°C, FeO can be formed, which is harmful to fluidization and magnetization roasting on siderite.

(3) D3 is more suitable for the reaction mechanism function of fluidization and magnetization roasting on siderite. The chemical reactions are controlled by diffusion.

Drying, Roasting, and Calcining of Minerals
Edited by: Thomas P. Battle, Jerome P. Downey, Lawrence D. May, Boyd Davis, Neale R. Neelameggham, Sergio Sanchez-Segado, and P. Chris Pistorius
TMS (The Minerals, Metals & Materials Society), 2015

THE TEMPERATURE BEHAVIOR AND MICROWAVE THERMO GRAVIMETRIC ANALYSIS CHARACTERISTIC OF AMMONIUM PARATUNGSTATE IN A MICROWAVE FIELD

Guo Sheng-hui[1,2,3], Cheng Fang[1,2,3], Peng Jin-hui[1,2,3], Zhou Lie-xing[1] [1,2,3]

[1]National local joint engineering laboratory of engineering Applications of Microwave Energy and Equipment Technology, Kunming 650093, Yunnan, China;
[2]Key Laboratory of Unconventional Metallurgy, Ministry of Education, Kunming University of Science and Technology, Kunming 650093, Yunnan, China;
[3]Faculty of Metallurgical and Energy Engineering, Kunming University of Science and Technology, Kunming 650093, Yunnan, China;

Keywords: ammonium paratungstate; microwave; calcined; thermo gravimetric analysis; behavior of temperature

Abstract

This article discusses the behavior of temperature and microwave thermo gravimetric analysis characteristic of ammonium paratungstate in a microwave field. Experimental results show that when 12g of material is place in a microwave with 900W of power, paratungstate's temperature rises at 70°C/min. This results in paratungstate's temperature being 980°C after 14 minutes. Compared with thermo gravimetric analysis characteristic, former two-step process of paratungstate microwave calcined completed ahead of schedule, after two-step process reaction is more intense,Paratungstate warming is more rapidly, the reaction time is much shorter; According to XRD and SEM, the product is high purity, the crystal particle size is uniform, and the grain size is fine.

Introduction

Ammonium paratungstate is an important raw material for the manufacturing of tungsten products and alloys. It is also one of the important export goods of China, as well as being a main trade good on the international tungsten market [1]. Tungsten trioxide is usually obtained by calcining ammonium paratungstate [2], and is mainly used in the production of tungsten powder and tungsten carbide powder. Traditionally,

Author: Guo Sheng-hui, male, Professor, Tel: +86-13888778655; E-mail: shguo78@hotmail.com.
Foundation item: Project (2013HB) support by the Yunnan Provincial Science and Technology Innovation Talents Scheme-Technological Leading Talent.
Corresponding author: Cheng Fang, Tel:+86-15808897400; E-mail: 1079299682@qq.com

to produce tungsten trioxide, ammonium paratungstate powder is placed in either a fixed or rotary furnace for calcining[3].This production method has the disadvantages of long production time, high production cost, high energy consumption, low thermal efficiency of the furnace, and the product purity and granularity is difficult to guarantee.

Microwave heating is a green and efficient heating method. Not only does microwave heating have the advantages of selectively heating material, a fast heating rate, and high heating efficiency, but it can also reduce the reaction temperature, shorten the reaction time, and reduce energy consumption [4].Microwave thermo gravimetric analysis is a new thermo gravimetric analysis method that uses microwave heating. This method is mainly used to study the basic theory of microwave and thermo gravimetric characteristics of materials in a microwave field.

The purpose of this analysis was to determine the temperature behavior of ammonium paratungstate when exposed to a microwave field. The results were compared to the conventional thermo gravimetric characteristics for ammonium paratungstate. The experiment was carried out in a homemade single-mode cavity thermal analysis device. The phase composition and crystal morphology of the product were analyzed by XRD and SEM to provide further knowledge of the materials reaction mechanism in a microwave field.

Experiment

Experimental materials

The amount of WO_3 in the ammonium paratungstate used in this experiment was 90.06%. The material was calcined in microwave at 720°C, resulting in a weight loss of 9.94%. The maximum amount of impurity elements in ammonium paratungstate is shown in table 1.

Table.1. Maximum content of ammonium paratungstate impurity elements($\times 10^{-6}$)

Mo	Fe	Si	Ca	Ti	V	As	Sb	K	Na	Al	Mn
30	8	8	8	8	8	8	8	6	6	6	6
Mg	Ni	Cr	P	Co	S	Cu	Pb	Bi	Sn	Cd	
6	6	5	5	5	4.2	3	1	1	1	0.6	

Experimental device

The experiment used the homemade single-mode cavity microwave thermal analysis device. Operating parameters of device is shown in Table 2. The single-mode cavity microwave device is shown in Figure 1.

Table.2. Operating parameters of device

Frequency	Power	Temperature range	Weighing range	Atmosphere	Crucible
2.45GHz	0～3KW	20～1000°C	0.1～300g	Air, Nitrogen	Corundum - mullite

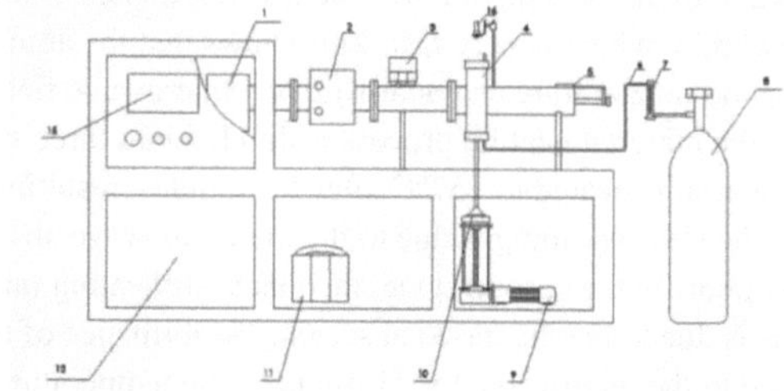

Figure.1. Single-mode cavity microwave thermal analysis device

Put in figure caption. Listed as (1) magnetron (2) circulator (3) three pin regulator (4) single-mode cavity quartz tube reaction chamber (5) motor control valve (6) sulfur coil (7) rotor flow meter (8) cylinders (9) lifting mechanism (10) electronic balance, (11) transformer (12) electrical cabinet (13) sample crucible (14) water load (15) control system (16) infrared thermometer.

Experimental procedure

The thermo gravimetric experiment was carried out in a microwave with a power of 900W. 12 g of material was placed in the microwave for 25 minutes and heated to 1000°C at a ramp rate of 10°C /min in an argon atmosphere. The products were analyzed for content and morphology with XRD and SEM.

Results and Discussion

Temperature rising behavior of ammonium paratungstate in a microwave field

The temperature curve of ammonium paratungstate in a microwave field is shown in Figure 2.

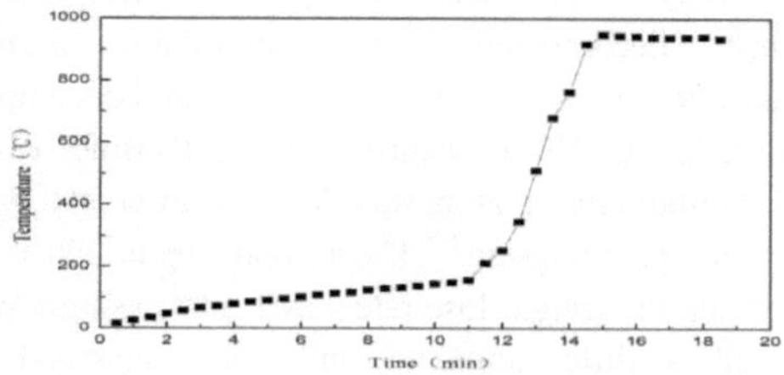

Figure.2. Temperature curve of ammonium paratungstate in a microwave field

As shown in Figure 2, when 12 g of ammonium paratungstate is heated for 14 minutes in a 900W microwave field, the temperature reaches 980°C. This is an average heating rate of 70°C/min, which is very fast. This shows that the ammonium paratungstate has good temperature characteristic in a microwave field. As shown in heating curve, that the material heating process is divided into three stages. In the initial stage, the material is heated to 157°C after 11 minutes, resulting in a heating rate of 12°C/min. The slow warming is due to the materials wave absorbing performance being poor. In the second stage, the materials heating rate is much higher at 247°C/min. This is due to the increased absorbing performance of the material. After being exposed to the microwave for 11 minutes, the temperature increased rapidly. Material up to the reaction temperature and release a large amount of heat: $5(NH_4)_2O \cdot 12WO_3 \cdot 5H_2O \rightarrow 12WO_3 + 10NH_3\uparrow + 10H_2O\uparrow$ [5] ; The third stage, the constant segment, occurs when the ammonium paratungstate is completely decomposed into tungsten trioxide and the temperature becomes stable at 980°C.

The conventional thermo gravimetric characteristics of ammonium paratungstate

The differential thermal – thermo gravimetric curve for ammonium paratungstate heated to 1000°C at a rate of 10°C/min in an argon atmosphere is shown in Figure 3.

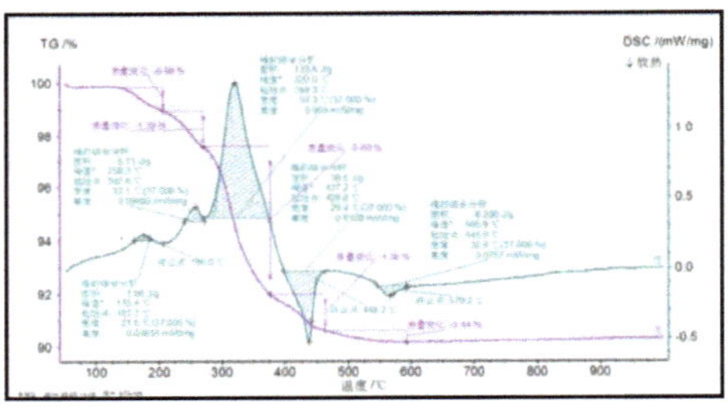

Figure.3. Differential thermal–thermo gravimetric curve

Figure 3 shows that under routine conditions of ammonium paratungstate are consistent. The five peaks of the differential thermal curve are corresponding to the weight loss steps of TGA curve. From the literature, ammonium paratungstate under routine conditions mainly experienced 4 stages of weight change: removal of water and ammonium, decomposition into amorphous ammonium tungsten bronze, crystallization into ammonium tungsten bronze and decomposition into WO_3 [6, 7, 8]. The first step in 137°C~200°C,occurred an endothermic reaction, the weight lose rate was 0.88%, ammonium paratungstate lost part of crystal water changed into a single water ammonium paratungstate[9].The second step in 200°C~262°C, occurred an endothermic reaction, the weight lose rate was 1.39%, a single water ammonium paratungstate removed a little ammonia and water changed into amorphous ammonium tungsten bronze (AATB).The third step in 262°C~375°C, occurred an intense endothermic reaction, weight lose rate was 5.60%, this is due to amorphous ammonium tungsten bronze (AATB) losing part of ammonia and water changed into ammonium tungsten bronze (ATB).The final step in 375°C~600°C, occurred an exothermic reaction, the weight lose rate was 1.82%,at this step material lost all ammonia and water completely decomposition into WO_3.

Ammonium paratungstate microwave thermo gravimetric characteristics

Using a microwave with 1000W of power, 12 g of material was heated for 25 minutes. The thermo gravimetric (TG) and derivative thermo gravimetric (DTG) curves for the ammonium paratungstate is shown in Figure 4.

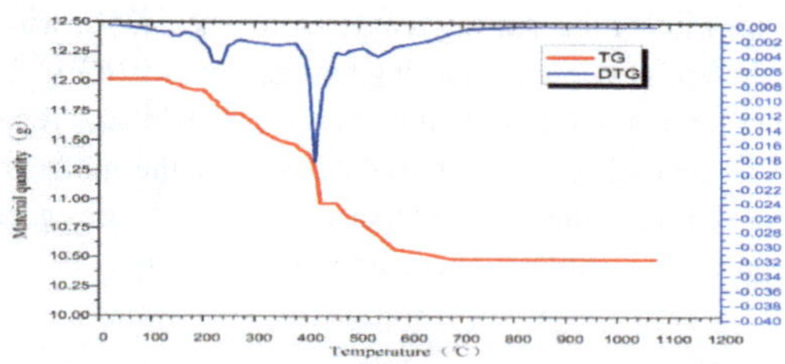

Figure.4. TG and DTG curve of ammonium paratungstate

Figure 4 shows that the thermo gravimetric (TGA) and derivative thermo gravimetric (DTG) curves of ammonium paratungstate are consistent in a microwave field. The five peaks of the derivative thermo gravimetric (DTG) are corresponding to the weight loss steps of thermo gravimetric (TGA) curve. A calculation result from the two curves of ammonium paratungstate, the process of thermal decomposition of ammonium paratungstate is divided into four steps. The first step in 123°C~183℃, the weight lose rate was 0.75%.The reason for this step is ammonium paratungstate loses part of crystal water was changed into a single water ammonium paratungstate. The second step in 200℃~248℃, the weight lose rate was 1.67%.This is due to a single water ammonium paratungstate losing a little ammonia and water, because they changed into amorphous ammonium tungsten bronze (AATB).The third step in 273℃~426℃, the weight lose rate was 6.48%, This is due to amorphous ammonium tungsten bronze (AATB) losing part of ammonia and water into ammonium tungsten bronze (ATB).The final step in 457~679℃, the weight lose rate was 4.38%.This is due to ammonium tungsten bronze (ATB) losing all ammonia and water completely decomposing into WO_3.Paralysis process of ammonium paratungstate in a microwave field is shown in table 3.

Table.3. Microwave paralysis process of ammonium paratungstate

temperature	123℃	183℃	248℃	426℃	679℃
material	ammonium paratungstate→	a single water ammonium paratungstate→	AATB →	ATB→	WO_3

The comparison of microwave thermo gravimetric and conventional thermo gravimetric of ammonium paratungstate

Microwave thermo gravimetric characteristics of ammonium paratungstate were compared to conventional thermo gravimetric analysis. It was determined by the first step of microwave thermo gravimetric that the starting temperature was 123°C, which is lower than the conventional thermo gravimetric starting temperature of 137°C. This may be attributed to water having a good absorption performance in a microwave [10]. The dehydration reaction occurs very quickly, resulting in the early removal of the water of crystallization from a microwave field. From third step of microwave thermo gravimetric, it is shown that the starting temperature was 273°C, which is higher than the conventional thermo gravimetric starting temperature of 262°C. In these steps, the material temperature increased rapidly in a microwave field, and removed a large number of NH_3. The second and final step of microwave thermo gravimetric is fairly consistent with conventional thermo gravimetric, at temperatures greater than 600°C, ammonium paratungstate completely decomposed into WO_3.

Product XRD analysis

Using XRD, the microwave calcined product was analyzed. The results are shown in Figure 5.

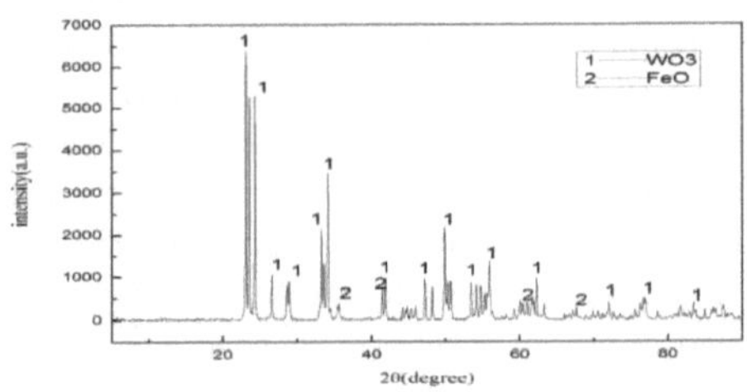

Figure.5. Calcined product XRD

Figure 5 shows that the structure of ammonium paratungstate was changed, the water and ammonia was lost, and the final product, WO_3, was left as the single solid component. The peaks of the XRD pattern are removing aligned with the standard card, showing that ammonium paratungstate decomposition completely in the microwave field. The results indicate a high purity, crystalline product with a small grain size. By analysis, it was determined that the product's crystal systems were monoclinic and orthogonal, respectively.

Product SEM analysis

In order to explore the microscopic properties of the microwave calcined ammonium paratungstate product, SEM was used at 500x and 10000x magnification. The results are shown in Figures 6 and 7.

Figure.6. SEM of 500x

Figure.7. SEM of 10000x

Figure 6 shows that there are very fine particles, with a rough surface and uniform grain size. This is helpful to get fine granularity and uniform grain tungsten trioxide. This figure also indicates that ammonium paratungstate decomposed completely, and most of ammonium paratungstate changed into WO3.

Figure 7 shows that particle size of WO3 is small, with an average particle size of 1.7μm, and uniform with a surface grain size of 0.5-3μm. From Figure 7, it can be determined that the crystal shape of WO_3 is block–like. This indicates that calcination in a microwave field yields better morphology of the WO_3 crystal.

Conclusions

The purpose of this analysis was to determine the temperature behavior of ammonium paratungstate when exposed to a microwave field. The results were compared to the conventional thermo gravimetric characteristics for ammonium paratungstate. The experiment was carried out in a homemade single-mode cavity thermal analysis device. The phase composition and crystal morphology of the product were analyzed by XRD and SEM to provide further knowledge of the materials reaction mechanism in the microwave field. The results are as follows:

1) When 12g of ammonium paratungstate is placed in a 900W microwave field and heated to 980°C after 14 minutes, the average heating rate is 70°C/min. This shows that the ammonium paratungstate has good temperature characteristics in a microwave field.

2) Ammonium paratungstate under routine and microwave conditions remove experienced four stages of weight change: removal of water and ammonium, decomposition into amorphous ammonium tungsten bronze (AATB), crystallization into ammonium tungsten bronze (ATB) and decomposition into WO_3.But ,the Weight lose steps are different.

3) When compared to TGA characteristics, the microwave thermo gravimetric reaction of ammonium paratungstate is shorter and warming occurs faster.
4) XRD indicates that ammonium paratungstate, after microwave calcining, changes structure, loses all water and ammonia, and results in the final product of WO_3 as a single solid component. The peaks of the XRD pattern are well aligned with the standard card. The crystal systems of the product are monoclinic and orthogonal, respectively.
5) The SEM image at 500x magnification displays fine particles with a rough surface and uniform size. The SEM image at 10000X magnification displays a small, uniform particle size of 1.7μm with a surface grain size of 0.5-3 μm. The crystal shape of WO_3 is completely developed and takes on a block-like structure.

References

[1]Fu Xiao-ming, Zhong Yun-bo, Ren Zhong-ming. Rapid thermal decomposition of monoclinic ammonium paratungstate. Powder Metallurgy Technology[J], 2006，24（6）

[2]Guo Shen-ghui, Peng Jin-hui, Fan Xing-xiang. Microwave calcining ammonium paratungstate preparation of tungsten trioxide. Chinese Journal of Rare Metals [J], 2002, 7： 26-4.

[3]Ye Wei-hong, Wang Chong-jing. Tungsten--resources, metallurgical properties and application.

[4]Peng Jin-hui, Yang Xian-wan. New applications of microwave energy[M].Kun Ming: Yunnan science and technology press, 1997， 75-78.

[5]Liu Shi-jun, Chen Qi-yuan, Zhang Ping-min. Thermo chemical determination of thermal decomposition of monoclinic ammonium paratungstate. Acta Physico-Chemica Sinica[J]. 2000,11:11（16）

[6] Van Put J W.Crystallisation and processing of ammonium paratungstate (APT). 1995, 61-76.

[7]Zeng Wen-ming, Chen Qi-yuan, Li Yuan-gao. The thermal decomposition kinetics research of ammonium paratungstate. Chinese Journal of Rare Metals [J], 1997,1:1（18）

[8]Wang Xue-qing, Yang Jian-shen, Liu Shan-yu. The thermal decomposition processes research of ammonium paratungstate. China Tungsten Industry[J], 2014,4:2（29）

[9] Fu Xiao-ming, Zhong Yun-bo, Ren Zhong-ming. Slow thermal decomposition processes research of monoclinic ammonium paratungstate. Cemented Carbide[J], 2005，9:3（22）

[10]Jin Qin-hang, Dai Shu-shan. Microwavechemical[M],BeiJing:Science press, 1999：30-50.

Drying, Roasting, and Calcining of Minerals
Edited by: Thomas P. Battle, Jerome P. Downey, Lawrence D. May, Boyd Davis, Neale R. Neelameggham, Sergio Sanchez-Segado, and P. Chris Pistorius
TMS (The Minerals, Metals & Materials Society), 2015

GREEN EVALUATION OF MICROWAVE SHAFT FURNACES

Jin Chen [1,2], Guo Chen [1,3*], Jinhui Peng [1,3]

[1]Key Laboratory of Unconventional Metallurgy, Ministry of Education, Kunming University of Science and Technology, Kunming 650093, PR China
[2]City College, Kunming University of Science and Technology, Kunming 650093, PR China
[3] Yunnan Mingzu University, Kunming 650500, PR China

*Corresponding author: guochen@kmust.edu.cn

Keywords: Microwave shaft furnaces; Green evaluation; analytic hierarchy process; Resource consumption; Environment impact

Abstract

In this study, the influences of the green evaluation variables on the characterization of microwave shaft furnaces were systematically investigated. These variables include: resource consumption, environment impact, production cost, processing time and quality of microwave shaft furnaces. A greenness evaluation index, evaluation system and evaluation model of microwave shaft furnaces were established. The analytic hierarchy process and fuzzy comprehensive evaluation were applied to the evaluation process. Resource consumption and environment impact had a statistically significant effect on the green evaluation of microwave shaft furnaces. The results indicated that green evaluation techniques can be applied effectively and efficiently to the process manufacturing of microwave shaft furnaces.

Introduction

The microwave part of the electromagnetic spectrum corresponds to frequencies between 300 MHz and 300 GHz [1]. However, most research and industrial activities involve microwaves at either 2.45 GHz or 915 MHz frequencies. Microwave heating technology is one effective route in realizing energy-savings and green production. Compared with conventional heating techniques, microwave heating has the additional advantage that it has both internal and volumetric heating [2]. Because no thermal conduction is involved in the energy conversion, the heating is very rapid, uniform, and highly energy efficient; these advantages result in energy savings and shorten the processing time [3].

As a new method of green metallurgy and the preparation of materials, microwave heating has developed into an outstanding field [4]. Many domestic and foreign scholars have conducted research in the fields of characteristics of the minerals in microwave heating, pretreatment of the refractory mineral, microwave milling, microwave drying depth, carbon reduction, calcinations, microwave leaching, etc. [5]. The results showed that microwave heating can effectively solve the question of the existence of "cold center" by conventional heating, improving the response rate, lowering the response temperature, and reducing the energy consumption of the reaction. With broad prospects for development in metallurgical research, microwave technology is of great significance to the innovation process of the mining and metallurgy industries [6]. Compared with conventional heating, microwave heating has several incomparable advantages, but some issues still exist, such as poor comparability and experimental reproducability based on their respective devices employed in most of the research studies, at home and abroad. Because of the absence of a specialty microwave reactor, the basic theory and practical application of research in microwave metallurgy is weaker, which makes it difficult to promote industrialization [7].

Green evaluation is a collection of experimental strategies, mathematical methods, and statistical inferences that enable an experimenter to make efficient empirical explorations of the system of interest [8-11]. Therefore, green evaluation is searching for the best route of the objective function to optimize a manufacture process. For green evaluation, the analytic hierarchy process is the most commonly used approach for evaluation of experiments [12-15]. The fuzzy decision-making method is efficiently utilized for selecting the most significant factors that are to be further analyzed in an analytic hierarchy process [16-17]. Recently, the analytic hierarchy process and fuzzy decision-making methods have been successfully applied to different green evaluations for optimizing manufacture conditions using experimental designs [18-19].

The objectives of this study were attained using the analytic hierarchy process and fuzzy decision-making method to optimize the green evaluation of microwave shaft furnaces.

Features of microwave shaft furnaces

The microwave shaft furnaces typically consist of a magnetron, a wave guide, a resonant cavity, a power controller and a temperature measurement system. The picture in Figure 1 shows the microwave shaft furnaces. The equipment used in present study contained the following:

1. A magnetron is used to generate microwave power, which is most widely used for the industrial microwave heating applications due to its availability and low cost. A magnetron is an oscillator used to convert electric power, usually in the form of high-voltage, and direct current into high-frequency radiant energy. The polarity of the emitted radiation oscillates from negative to positive at high frequencies causing the material within the radiation field to heat up. The microwave power of the microwave shaft furnaces is supplied by two magnetrons at 2.45 GHz frequency and 1.5 kW power, which are cooled by water circulation.

2. Wave guides transfer the generated microwave energy from the magnetron to the resonant cavity. Wave guides differ in their geometry, which can improve microwave energy distribution, depending on the wave guide design.

3. The resonant cavity is the most important part of the microwave shaft furnaces, the size of which is accurately calculated by microwave theories.

4. A ceramic cavity, 420 mm (length)×420 mm (breadth)×300 mm (height), is positioned at the center of the microwave oven by drilling holes on the side faces, with ends projecting on both sides. Raw materials are loaded in a ceramic cavity by an automatic screw feeder.

5. The temperature measurement system consists of an infrared pyrometer and an automatic control system. Infrared pyrometers provide temperature information to the automatic control system, which controls the temperature of the sample during the microwave heating process.

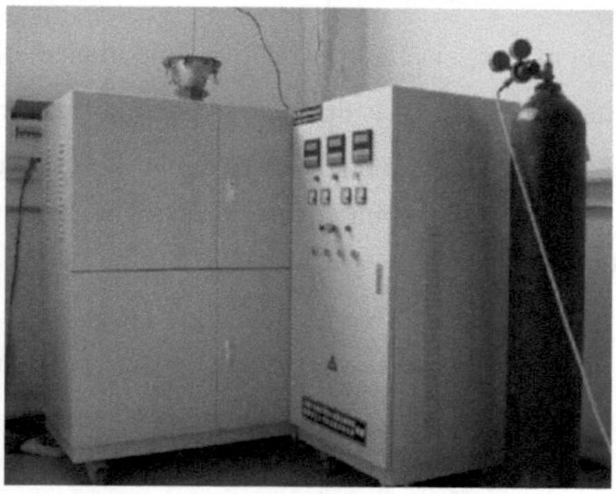

Figure 1 The microwave shaft furnaces employed in the present study

Methodology

Structure of assessment

The effects of variables such as: material resources, energy utilization and surplus energy on the resource consumption were systematically investigated. The effects of air pollution, liquid pollution, solid pollution and other pollution on the environment were also investigated. Hierarchical structure of green evaluation of microwave shaft furnaces is shown in Figure 2.

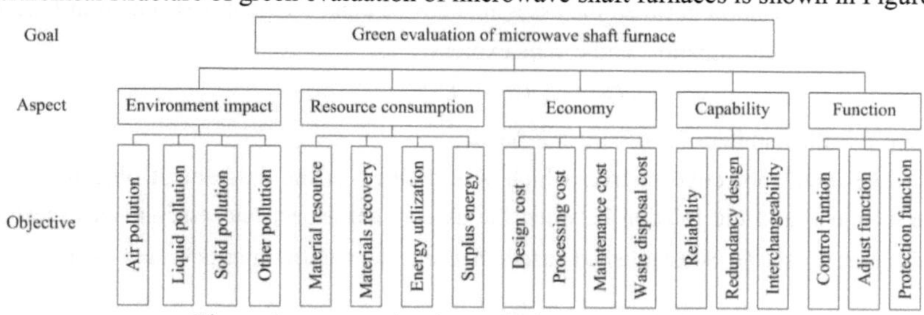

Figure 2 green evaluations of microwave shaft furnaces

Assessment model of green evaluation

For each criterion and weight of the hierarchy structure all the associated elements are compared in matrices as follows:

$$P = \begin{bmatrix} 1 & \frac{w_1}{w_2} & \cdots & \frac{w_1}{w_n} \\ \frac{w_2}{w_1} & 1 & \cdots & \frac{w_2}{w_n} \\ \vdots & \vdots & \ddots & \vdots \\ \frac{w_n}{w_1} & \frac{w_n}{w_2} & \cdots & 1 \end{bmatrix} \quad (1)$$

Where P is the comparison matrix; w_1 is the relative importance of criterion 1; w_2 is the relative importance of criterion 2; and w_n is the relative importance of criterion n.

The Eigenvalue method and Gauss-Seidel iterative matrix are used to calculate the relative weights of elements in each pairwise comparison matrix. The largest Eigenvalue is obtained from the following equation:

$$\det[P - \lambda_{max} I] = 0 \quad (2)$$

Where λ_{max} is the largest Eigenvalue of matrix A; I is the unit matrix. The relative weights (A) of matrix (P) are obtained from the following equation:

$$(P - \lambda_{max} I) \times A = 0 \quad (3)$$

The largest Eigenvalue (λ_{max}) is obtained by solving the Eigenvalue, Eq. (2), and showed in Eq. (3). The relative weights (A) is obtained by solving and normalizing the Eigenvalue Eq. (4).

$$\lambda_{max} = \frac{1}{n} \sum_{i=1}^{n} \left\{ \frac{\sum_{j=1}^{n} a_{ij} \times w_j}{w_i} \right\} \quad (4)$$

It is crucial to select assessment parameters which are representative, rational and accurate in order to form an assessment aspect matrix set U, which was defined as the assessment aspect of the goal, and the relation between the assessment aspect and the goal was shown in Figure 2. U is based on the actual situation, and can be expressed as:

$$U = \{U_1, U_2, \cdots, U_i\} \quad (5)$$

Where i is the number of assessment aspects.

The assessment objective matrix, which was defined as the assessment objective of the assessment aspect, and the relation between the assessment aspect and assessment aspect was shown in Figure 2, set U_i is established based on assessment aspect matrix U and expressed as:

$$U_i = \{U_{i1}, U_{i2}, \cdots, U_{ij}\} \quad (6)$$

Where j is the number of assessment aspect U_i.

The weight reflects the significance degree of each indicator, and the value is obtained as follows:

$$A = \{a_1, a_2, \cdots, a_i\},\ A_i = \{a_{i1}, a_{i2}, \cdots, a_{ij}\} \quad (7)$$

where A is the weight matrix of the assessment aspect, A_i is the weight matrix of the assessment objective, a_i is the ith weight of the assessment aspect, and a_{ij} is the jth weight of assessment objective of the assessment aspect a_i.

Generally, a weight number of assessment model, a_i and a_{ij} should satisfy the normalization and non-negative conditions, respectively.

$$\begin{cases} 0 < a_i \le 1 \\ \sum_{i=1}^{i} a_i = 1 \end{cases}, \begin{cases} 0 < a_{ij} \le 1 \\ \sum_{j=1}^{j} a_{ij} = 1 \end{cases} \tag{8}$$

Fuzzy assessment matrix in the first stage, can be expressed as:

$$R_{ij} = \begin{bmatrix} r_{i11} & r_{i12} & \cdots & r_{i1n} \\ r_{i21} & r_{i22} & \cdots & r_{i2n} \\ \vdots & \vdots & \vdots & \vdots \\ r_{im1} & r_{im2} & \cdots & r_{imn} \end{bmatrix} \tag{9}$$

Where r_{imn} is the fuzzy subset membership function between the mth assessment objective and the nth assessment objective.

Fuzzy assessment matrices in the second stage come from the synthesis of all objectives in the first stage. A fuzzy assessment matrix in the second stage is calculated using the fuzzy comprehensive algorithm, and can be expressed as:

$$R_i = A_i \circ R_{ij} \tag{10}$$

Where $\circ$ is the fuzzy relation comprehensive algorithm.

Using the methods mentioned above, fuzzy assessment matrices in the third stage can be obtained as follows:

$$R = A \circ R_i \tag{11}$$

In order to obtain better results of the life cycle assessment model, the following expression is used to represent the green grade matrix of each assessment aspect, which can be described as excellent, good, ordinary, poor and bad, respectively.

$$V = \{1.0, 0.8, 0.7, 0.6, 0.5\} \tag{12}$$

Green degree describes the degree of environmental effect or the environmental friendly level of an enterprise's various activities. Based on the mathematics theory, the green degree results can be computed by matrix multiplication, which can be derived from Eq. (11) and Eq. (12). Then, the green degree results can be expressed as:

$$Y = VR^T \times 100 \tag{13}$$

Results and discussion

The usability of microwave shaft furnaces is evaluated under the established green evaluation with the analytic hierarchy process and fuzzy comprehensive evaluation. The weight values and assessment matrices of microwave shaft furnaces are shown in Table 1. The comparison of these results can be considered as a decision making model since it guides decision makers to choose the best alternative for manufacturing projects and processes.

Table 1: Green evaluation of microwave shaft furnaces

Aspects		Objective		Fuzzy assessment matrices				
U_i	a_i	U_{ij}	a_{ij}	r_{ij1}	r_{ij2}	r_{ij3}	r_{ij4}	r_{ij5}
Environment impact	0.25	Air pollution	0.3	0.3	0.3	0.2	0.1	0.1
		Liquid pollution	0.3	0.3	0.2	0.3	0.1	0.1
		Solid pollution	0.3	0.4	0.2	0.1	0.2	0.1
		Other pollution	0.1	0.5	0.2	0.1	0.1	0.1
Resource consumption	0.25	Materials resource	0.2	0.4	0.1	0.3	0.1	0.1
		Materials recovery	0.1	0.2	0.2	0.1	0.1	0.2
		Energy utilization	0.1	0.3	0.1	0.3	0.1	0.2
		Surplus energy	0.2	0.5	0.1	0.2	0.1	0.1
Economy	0.2	Design cost	0.2	0.3	0.2	0.3	0.1	0.1
		Processing cost	0.1	0.2	0.3	0.3	0.1	0.1
		Maintenance fee	0.2	0.5	0.2	0.1	0.1	0.1
		Waste disposal fee	0.3	0.6	0.1	0.1	0.1	0.1
Capability	0.15	Reliability	0.2	0.4	0.3	0.1	0.1	0.1
		Redundancy design	0.1	0.5	0.2	0.2	0.1	0.1
		Interchange ability	0.2	0.3	0.3	0.2	0.1	0.1
Function	0.15	Control function	0.3	0.4	0.2	0.2	0.1	0.1
		Adjustment function	0.2	0.3	0.3	0.1	0.1	0.2
		Protection function	0.2	0.4	0.3	0.1	0.1	0.1

From the above theoretical work, the calculation processes can be carried out according to the analytic hierarchy process and fuzzy comprehensive assessment method. The life cycle assessment results are shown in Table 2. The green degree of microwave shaft furnaces is 88.56, a relatively high value indicating that microwave shaft furnaces have a good green degree, especially in: energy consumption, processing time and environmental protection. Thus, microwave shaft furnaces can fulfill the aim of green manufacturing.

Table 2: Green evaluation results for the microwave shaft furnaces

Species	Green degree
microwave hot air systems	88.56

Conclusions

In this study, the analytic hierarchy process and fuzzy comprehensive evaluation method were applied to green evaluation of microwave shaft furnaces. The hierarchical structure consists of assessment aspect and assessment objective. The fuzzy matrix, a pairwise comparison scale to determine the weights of aspects and objectives was used as the mathematical model. The fuzzy assessment matrices are calculated using the Eigenvalue method and Gauss-Seidel iterative matrix. The objectivity of the method increases as more objective decisions are made. The assessment results show that microwave shaft furnaces have a good green degree. The case study shows that the green evaluation is applicable as an evaluation technique for green manufacturing.

Acknowledgements

Financial supports from the Key Projects in the financial supports from the National Scientific Foundation of China (No: 51404114), the National Science & Technology Pillar Program during the Twelfth Five-year Plan Period (No.: 2015BAB17B00), Specialized Research Fund for the Doctoral Program of Higher Education (No.: 20125314120014), the International Science & Technology Cooperation Program of China (No.: 2012DFA70570), the Yunnan Provincial International Cooperative Program (No.: 2011IA004), and the Applied Foundation Fund of Yunnan Province of China (No.: 2012FD015)

References

[1] Amankwah R. K. and Pickles C. A. (2009) 'Microwave roasting of a carbonaceous sulphidic gold concentrate', *Minerals Engineering*, Vol.22, pp.1095–1101.

[2] Popescu S., Misawa T., Ohtsu Y., Fujita H. and Sanematsu S. (2008) 'New microwave reactor for paper-based waste neutralization', *Resources, Conservation and Recycling*, Vol.52, pp.671–677.

[3] Grossin D., Marinel S. and Noudem J. G.(2006) 'Materials processed by indirect microwave heating in a single-mode cavity', *Ceramics International*, Vol.32, pp.911–915.

[4] Zhao Y. F. and CHEN J. (2008) 'Applications of microwaves in nuclear chemistry and engineering', *Progress in Nuclear Energy*, Vol.50, pp.1–6.

[5] Pan R. J., Wu Y. C., Wang Q. P. and Hong Y. (2009) 'Preparation and catalytic properties of platinum dioxide nanoparticles: A comparison between conventional heating and microwave-assisted method', *Chemical Engineering Journal*, Vol.153, pp.206–210.

[6] Clemens J. and Saltiel C. (1996) 'Numerical modeling of materials processing in microwave furnaces', *International Journal of Heat and Mass Transfer*, Vol.39, pp.1665-1675.

[7] Li W., Peng J. H., Zhang L.B., Zhang Z.B., Li L., Zhang S.M. and Guo S.H.(2008) 'Pilot-scale extraction of zinc from the spent catalyst of vinyl acetate synthesis by microwave irradiation', *Hydrometallurgy*, Vol.92, pp.79–85.

[8] Zhu J. Y. and Deshmukh A. (2003) 'Application of Bayesian decision networks to life cycle engineering in Green design and manufacturing', Engineering *Applications of Artificial Intelligence*, Vol.16, pp.91–103.

[9] Xu Y. T. and Zhang Y. (2009) 'A online credit evaluation method based on AHP and SPA', *Communications in Nonlinear Science and Numerical Simulation*, Vol.14, pp.3031–3036.

[10] Li T. S. and Huang H. H.(2009) 'Applying TRIZ and Fuzzy AHP to develop innovative design for automated manufacturing systems', *Expert Systems with Applications*, Vol.36, pp.8302–8312.

[11] Tan X. C., Liu F., Cao H. J. and Zhang H. (2002) 'A decision-making framework model of cutting fluid selection for green manufacturing and a case study', *Journal of Materials Processing Technology*, Vol.129, pp.467–470.

[12] Saaty T. L. (1990) 'How to mark a decision: The analytic hierarchy process', *European Journal of Operational Research*, Vol.48, pp.9–26.

[13] Wittmaier M., Langer S. and Sawilla B.(2009) 'Possibilities and limitations of life cycle assessment (LCA) in the development of waste utilization systems-Applied examples for a region in Northern Germany', *Waste Manage*, Vol.29, pp.1732–1738.

[14] Weinzettel J., Reenaas M., Solli C. and Hertwich E G. (2009) 'Life cycle assessment of a floating offshore wind turbine', *Renew Energy*, Vol.34, pp.742–747.

[15] Chang C. W., Wu, C. R. and Lin C. T. (2007) 'Evaluating digital video recorder systems using analytic hierarchy and analytic network processes', *Information Sciences*, Vol.177, pp.3383–3396.

[16] Khoo H. H. (2009) 'Life cycle impact assessment of various waste conversion technologies', *Waste Manage*, Vol.29, pp.1892–1900.

[17] Renou S., Thomas J. S., Aoustin E. and Pons M. N. (2008) 'Influence of impact assessment methods in wastewater treatment LCA', *Journal of Cleaner Production*, Vol.16, pp.1098–1105.

[18] Ardente F., Beccali M., Cellura M. and Mistretta M. (2008) 'Building energy performance: A LCA case study of kenaf-fibres insulation board', *Energy and Buildings*, Vol.40, pp.1–10.

[19] Chang C. W., Wu, C. R. and Lin H. L. (2009) 'Applying fuzzy hierarchy multiple attributes to construct an expert decision making process', Expert Systems with Applications, Vol.36, pp.7363–7368.

Drying, Roasting, and Calcining of Minerals
Edited by: Thomas P. Battle, Jerome P. Downey, Lawrence D. May, Boyd Davis, Neale R. Neelameggham, Sergio Sanchez-Segado, and P. Chris Pistorius
TMS (The Minerals, Metals & Materials Society), 2015

BEHAVIOR OF PHOSPHORUS DURING REDUCTION ROASTING OF OOLITIC HEMATITE ORE WITH HIGH-PHOSPHORUS CONTENT

Guanghui Li, Chongzhong Ouyang, Mingjun Rao*, Yuanbo Zhang, Shuhui Zhang

(School of Minerals Processing & Bioengineering, Central South University, Changsha, Hunan 410083, China)

Keywords: Phosphorus, Direct reduction, Oolitic hematite, Metallic iron

Abstract

Oolitic hematite ore with high-phosphorus content is currently recognized as one of the most refractory iron ore resources in the world. The high-phosphorus content and unique structure of the oolitic hematite ore make it difficult to be beneficiated. The process of direct reduction followed by magnetic separation has significance to the utilization of oolitic hematite ore, and sodium salts can improve the dephosphorization. However, the behavior of phosphorus during reduction roasting, and the intensified dephosphorization mechanism of sodium salts are not clear. In this study, by using chemical analysis, chemical phase analysis, XRD analysis, optical microscopy, and scanning electron microscopy analyses, the behavior of phosphorus during reduction roasting of high-phosphorus content oolitic hematite ore and the effect of sodium salts on behavior of phosphorus were studied.

Introduction

Since the oolitic hematite is found with huge reserves in China, the efficient utilization has been a challenge for Chinese scholars [1]. For that, various processes have been raised to resolve this issue in recent years, such as reverse flotation, chemical leaching, and magnetized roasting. As this ore is characterized by low iron content, complex structure, fine mineral dissemination and high phosphorus content, these processes are proved to be inefficient [2, 3, 4].

The process of direct reduction followed by magnetic separation was proposed for oolitic hematite, and sodium salts were used as the dephosphorization agent during roasting [5, 6]. In a

* Corresponding authors: Dr. Mingjun Rao, Email: raomingjun2003@126.com, Tel./fax: +86 731 88830542

previous study, the Fe-P separation results of the oolitic hematite reduced at 1050 °C for 120 min were obviously improved in the presence of 7.5% sodium sulfate and 1.5% borax. The P content decreased to 0.09% in the magnetic concentrate, and dephosphorization ratio reached 96.6% [1]. Therefore, the results showed that the sodium salts could dramatically reduce the phosphorus content of the magnetic concentrate.

In this study, the behavior of phosphorus at different reduction temperatures, and effects of Na_2CO_3 and Na_2SO_4 on the distribution of the phosphorus were studied by chemical analysis, XRD analysis, optical microscopy, and scanning electron microscopy analysis.

Experimental

Materials

Oolitic Hematite Ore The oolitic hematite used in this research was sampled from the west of Hunan Province. The chemical compositions are given in Table I. The main mineral composition is shown in Figure 1, and fluorapatite is the main phosphorus mineral in the ore.

Table I. Main Chemical Compositions of Oolitic Hematite Ore /%

TFe	P	SiO_2	Al_2O_3	CaO	MgO	Mn	S	LOI*
51.28	1.30	9.53	5.66	3.57	0.18	0.05	0.05	3.06

LOI*—loss on ignition

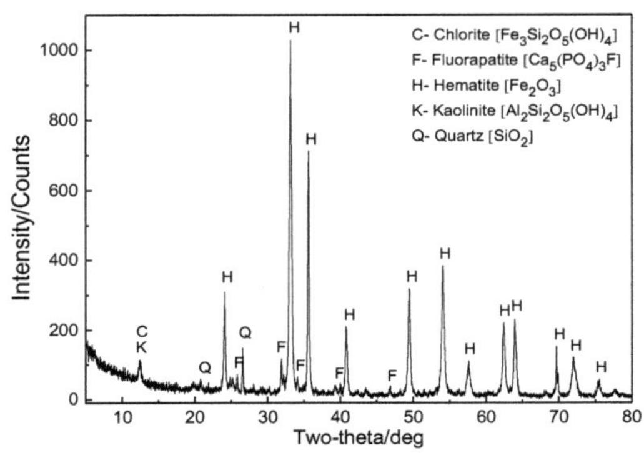

Figure 1. XRD Pattern of the Oolitic Hematite Sample

Methods

Experimental Flowsheet The oolitic hematite ore was ground to 100 wt. % less than 0.074 mm in a ball mill. The ground sample was mixed with a certain amount of Na_2SO_4 and agglomerated into green pellets with 12-16 mm in diameter. The green pellets were dried in a baking box at 105 °C for 4 h. The lignite was crushed and screened to a size range of 1-5 mm and used as the external reductant.

Reduction experiments were performed in a vertical resistance furnace with an inner diameter of 70 mm. Excess coal was used to ensure the sufficient reducing atmosphere. Then, the pot was placed into a fixed high temperature zone of the vertical resistance furnace for reduction. The reduced pellets were taken from the pot and cooled down in oxygen-isolated atmosphere. Some reduced pellets were prepared for SEM-EDS analysis, and others were ground to 100 wt. % passing 0.074 mm for XRD analysis.

Results and Discussion

Effects of Reduction Temperature on the Behavior of Phosphorus

The results in Figure 2 evaluate the effect of the temperature on the P content in the magnetic concentrate. The experiments were performed under the condition of reduction roasting at different temperatures (900 °C—1200 °C) for 90 min. The products from the experiments were ground for 20 min and magnetic separation was performed at 1000 Gs. XRD and SEM-EDS analyses of the samples are shown in Figure 3 and Figure 4, respectively.

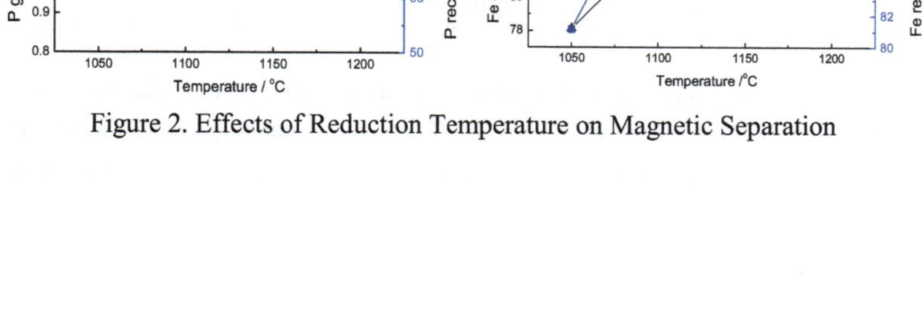

Figure 2. Effects of Reduction Temperature on Magnetic Separation

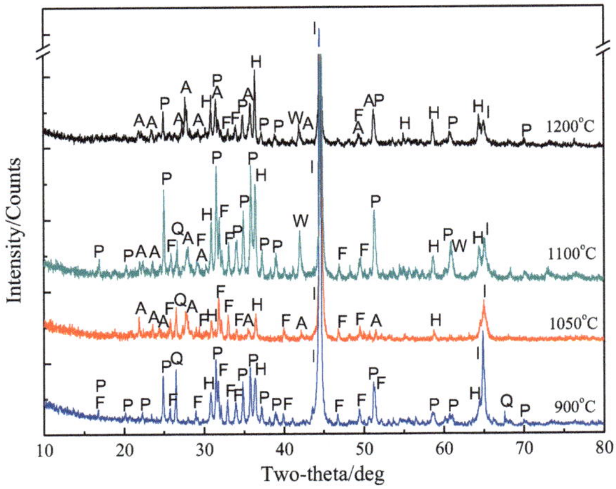

(A-$CaAl_2Si_2O_8$; F-$Ca_5(PO_4)_3F$; H-$FeAl_2O_4$; I-Fe; P-Fe_2SiO_4; Q-SiO_2; W-FeO)

Figure 3. X-ray Patterns of Reduced Pellets (reducing time: 90min)

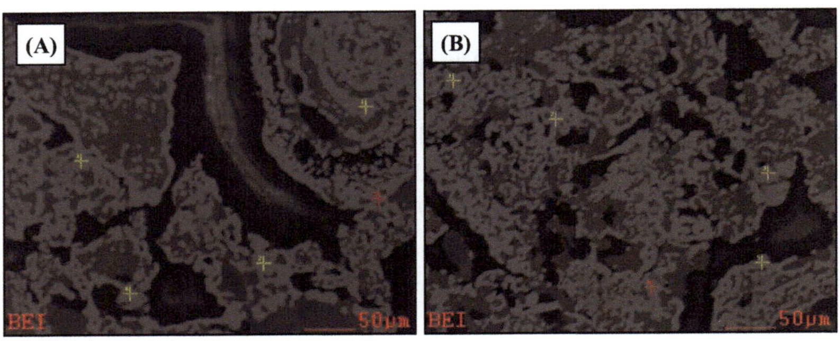

Figure 4. BSE Images of Reduced Pellet (reduced at 1050°C for 90 min)

Table II. Phosphorus Content in the Metallic Iron Grains

Sector	Serial number of metallic iron grain				
	1	2	3	4	5
A	0.64	1.5	1.31	1.47	1.21
B	0.83	0.90	1.15	1.56	1.24

The roasting temperature had significant effects on phosphorus grade of the magnetic product. The phosphorus grade increased from 1% to 1.26% with the temperature increasing from 1050°C to 1150°C. And the grade of iron and iron recovery increased obviously with reduction temperature increasing from 1050°C to 1200°C.

The X-ray diffraction patterns of reduced pellets showed that $CaAl_2Si_2O_8$ appeared when the temperature increased from 900 °C to 1050 °C. Combined with the ESM-EDS results of the reduced pellet at 1050 °C, we found that most iron particles contained some amount of P. And the content of P in particle is about 1% as the Table II showed. This means that the fluorapatite ($Ca_5(PO_4)_3F$) was reduced during the reduction roasting and the produced P entered into the iron phase during the process. That is the main reason of the high P content of the magnetic concentrate.

Effects of Na_2SO_4 on the Behavior of Phosphorus

The pellets with Na_2SO_4 (5%, 10%, 20%) were reduced at 1050 °C for 90 min followed by grinding-magnetic separation. The grade and recovery of the magnetic concentrate are shown in Figure 5, and the SEM-EDS result of the reduced pellet with 10% Na_2SO_4 is shown in Figure 6.

The additive Na_2SO_4 had significant effects on dephosphorization as shown in Figure 5, and the phosphorus content decreased from 1% to 0.56% with the dosage of the Na_2SO_4 increased to 20%. The Fe grade and recovery was improved at the same time.

The P content in metallic iron particles of the 10% Na_2SO_4 reduced pellet is about 0.6% as the SEM-EDS results showed in Table III. It was lower than results of the no additive pellet. This means that the Na_2SO_4 prevented the P from entering into the metallic iron particle. The less P content in the metallic iron particle indicates that the separation of the P and Fe would be more effective by the magnetic separation.

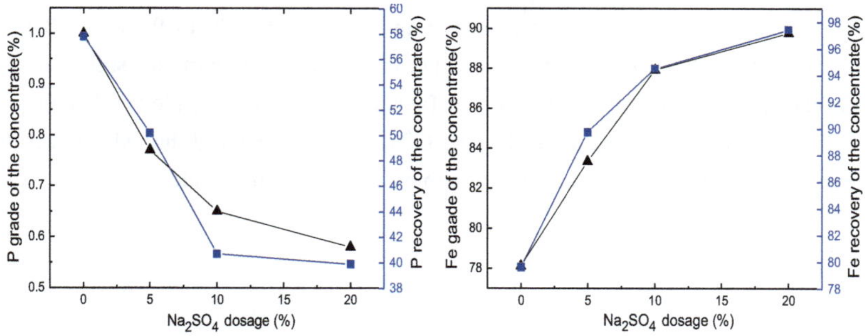

Figure 5. Effects of Na_2SO_4 Dosage on Magnetic Separation

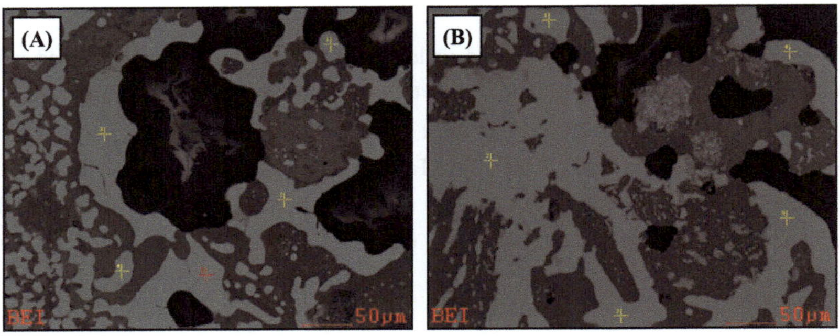

Figure 6. BSE Images of Reduced Pellet with 10% Na_2SO_4

Table III. Phosphorus Content in the Metallic Iron Grains

Sector	Serial number of metallic iron grain				
	1	2	3	4	5
A	0.73	0.74	0.62	0.55	0.61
B	1.02	0.44	0.41	0.00	0.00

Effects of Na_2CO_3 on the Behavior of Phosphorus

The pellet with different dosage of the Na_2CO_3 (5% 10% 20%) were reduced at 1050°C for 60min, ground for 20min and magnetic separation was performed at 1000Gs. The results of magnetic product are given in Figure 7, and the SEM-EDS analysis of the reduced pellet with 10% Na_2CO_3 is given in Figure 8 and Table IV.

It can be observed that the P content and recovery of the magnetic product decreased from 1%, 57.8% to 0.31%, 21.4% with the dosage of the Na_2CO_3 increased to 20%. And the Fe content and recovery of the magnetic product increased from 78.13%, 79.63% to 90.84%, 98.45%. The results from SEM-EDS of the pellet showed that the metallic iron particles still contains some mount of the P, it was about 0.4% as shown in Table IV which was much lower than the particles in the reduced pellet without additive. The Na_2CO_3 not only improved the reduction of the iron oxide but also prevented the P from entering into the metallic iron particles.

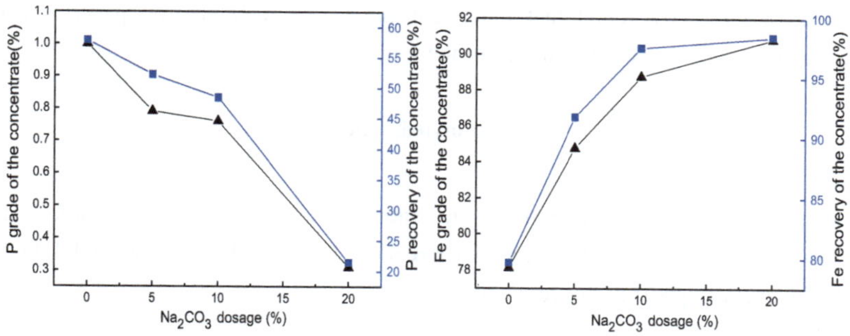

Figure 7. Effects of Na_2CO_3 Dosage on Magnetic Separation

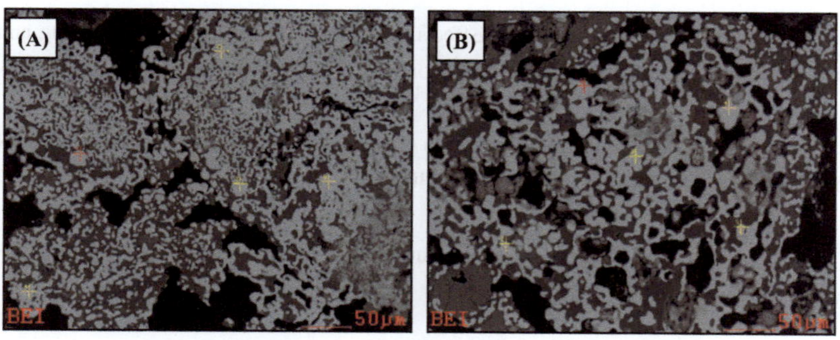

Figure 8. BSE images of reduced pellet with 10% Na_2CO_3

Table IV. Phosphorus Content in the Metallic Iron Grains

Sector	Serial number of metallic iron grain				
	1	2	3	4	5
A	0.44	0.32	0.31	0.51	0.24
B	0.46	0.34	0.27	0.37	0.43

Conclusions

[1]. During the reduction roasting, the apatite was reduced, and the produced P entered into the metallic iron particles. It made the separation of the P and Fe difficultly by magnetic separation.

[2]. Both of the sodium salts have the effects on the dephosphorization and improving the reduction of the iron oxide. Comparing the SEM-EDS results with experiments performed

without additive, we found that the P content of the metallic iron particles decreased significantly with increased additions of the sodium salts.

Acknowledgement

The authors wish to express their thanks to National Natural Science Foundation of China (51174230 and 51234008), and New Century Excellent Talents in University (NCET-11-515) for the partial financial support of this research.

References

[1] G. Li, S, Zhang, and M. Rao et al., “Effects of sodium salts on reduction roasting and Fe–P separation of high-phosphorus oolitic hematite ore,” *Int. J. Miner. Process.* 2013, 124: 26-34.

[2] S. Weissberger and Y. Zimmels, “Studies on concentration and direct reduction of the Ramim iron ore,” *Int. J. Miner. Process.* 1983, 11: 115-30.

[3] J. Wang, Y. Shang and Q. Zhang, “Preliminary study on the Flotation Test of oolitic hematite flotation,” *Mining engineering*, 2004, (3).

[4] S. Lu, “Investigation on the Dephosphorization of Ningxiang Type Colloidal Phosphorous Iron Ore by Decolloidization Leaching,” *Metal Mines*. 1994 (8): 30-33.

[5] G. Li, T. Lei and M. Rao et al., “Effects of Sodium Salts - Modified Paigeite on Dephosphorization of High - Phosphorus Oolitic Hematite during Reduction,” *3rd International Symposium on High-Temperature Metallurgical Processing: Wiley Online Library,* 2012 480-90.

[6] G. Li, C. Xie and Y. Zhang et al., “Influence of additives on dephosphorization of oolitic hematite by direct reduction process,” *TMS 2010 - 139th Annual Meeting and Exhibition, Seattle, WA, United States: Minerals, Metals and Materials Society*, 2010 505-13.

Drying, Roasting, and Calcining of Minerals
Edited by: Thomas P. Battle, Jerome P. Downey, Lawrence D. May, Boyd Davis, Neale R. Neelameggham, Sergio Sanchez-Segado, and P. Chris Pistorius
TMS (The Minerals, Metals & Materials Society), 2015

EFFECT OF ADDITIVES ON PHASE TRANSFORMATION OF NICKEL LATERITE ORE DURING LOW-TEMPERATURE REDUCTION ROASTING PROCESS USING CARBON MONOXIDE

Shiwei Zhou [1], Bo Li * [1], Yonggang Wei [1], Hua Wang [1], Chengyan Wang [2], Baozhong Ma [2]

1 State Key Laboratory of Complex Nonferrous Metal Resources Clean Utilization, Faculty of Metallurgical and Energy Engineering, Kunming University of Science and Technology, Kunming 650093, China

2 Beijing General Research Institute of Mining and Metallurgy, Beijing 100160, China

Keywords: Nickel laterite ore; Additives; Phase transformation; Gas reductant

Abstract

The effects of additives with Na_2S, Na_2SO_4 and $CaSO_4$ on phase transformation of nickel laterite ore in low-temperature reduction roasting process using carbon monoxide were investigated. The results showed that diffraction peaks of forsterite increased with increasing reduction temperature for 30min with the addition of 10wt% Na_2SO_4 and 50vol% CO, and diffraction peaks of forsterite increased with increasing the content of Na_2SO_4 at 900°C for 30min. Comparing the above three additives, phase transformation of nickel laterite ore were strongly influenced with the addition of 10 wt% Na_2SO_4 and 10 wt% Na_2S. Calcium sulphide from the thermal decomposition of $CaSO_4$ was carried out in the reaction product. SEM-EDS analysis result deduced that increasing the dosage of Na_2SO_4 from 10wt% to 15wt% could promote the polymerization of ferronickel and improve the grade of nickel in ferronickel.

Introduction

In recent years, there has been an increased focus on the utilisation of low-grade nickel laterite ore, along with a growing demand for stainless steel and a declining supply of sulphide ores [1]. Among the world's nickel resources, nickel laterite ore comprises 73% and will be the dominant source of nickel in the future. With the continuous depletion of high-grade nickel ores such as millerite and niccolite, nickel laterite ores have become the major source for the production of nickel metal. However, currently, only 42% of the world's production of nickel comes from nickel laterite ore because the concentration of nickel in the ore is low [2]. Therefore, in the long term, it will be necessary to optimise valuable metals extraction from nickel laterite ore.

Corresponding author: Bo Li; Tel:+86-15987127468; E-mail: libokmust@163.com

Several investigations of the additives effects and phase transformation on nickel laterite ore have been performed. For example, Jie Luet al. studied the effect of sodium sulphate on the hydrogen reduction process of nickel laterite ore [3]. The presence of sulfur has been found to markedly improve the Ni and Co recoveries from laterite ores by Valix, M. et al. (2002) [4]. Man Jiang et al completed a simulation study on mechanism of sodium sulfate in promoting selective reduction of nickel laterite ore during reduction roasting process [5]. The mineralogical composition and phase transformation of Greek nickeliferous laterites was investigated by Emmanuel N. Zevgoliset al (2010) during preheating and reduction with carbon monoxide [6]. Yongfeng Chang et al investigated the phase transformation in reduction process with microwave heating by XRD and the reduction degree of iron by chemical method [7].

In this study, the author conducted carbon monoxide reduction of low-grade nickel laterite ore with the addition of Na_2S, Na_2SO_4 and $CaSO_4$ to determine the effects of temperature and additives content on phase transformation. The phase transformation accompanying reduction and the catalytic mechanism of additives were analysed using instrumental analysis techniques such as XRD, SEM–EDS and thermodynamic calculations.

Materials and methods

Raw material

The chemical composition of low-grade nickel laterite ore was analysed by chemical analysis, as shown in Table I.

Table I. Chemical analysis of nickel laterite ore (mass fraction, %).

TFe	Ni	Co	Al_2O_3	MgO	CaO	SiO_2	Cr_2O_3
9.12	1.09	0.023	2.47	29.08	0.03	36.48	0.34

The additives of sodium sulphide (Na_2S), sodium sulphate (Na_2SO_4) and calcium sulphate ($CaSO_4$) used in this study were chemical grade. Reduction was achieved by mixing reducing gases, which included CO and CO_2.

Experimental methods

Nickel laterite ores were ground to 95wt% passing 100 mesh-sizes using a laboratory scale ball mill, and then mixed with different additives, respectively. The sample was then granulated by tablet machine according to the experimental requirements. The experiments were carried out in a vertical tube furnace. The reducing gas, which total rate of 50 ml/min (CO: 25 ml/min, CO_2: 25ml/min), were metered into the tube while it was heated to the required temperature. After 30 minutes of reduction at temperature, the reducing gas metered to the reactor was replaced with nitrogen.

XRD and SEM-EDS analysis

The XRD experiments were performed on a Japan Science D/max-R diffractometer apparatus with Cu Kα radiation (k = 1.5406 Å), an operating voltage of 40 kV and a current of 40 mA. The diffraction angle (2θ) was scanned from 5° to 90°. The morphology and particle size were determined by Scanning Electron Microscopy (Philip XL30) equipped with an Energy Dispersive X-ray Spectroscopy.

Results and discussion

Effect of reduction temperature

The laterite ores with the additive of 10 wt% Na_2SO_4 were reduced at temperatures from 600 to 1000°C.

The XRD patterns of the roasted ore in different conditions are shown in Figure 1. The peaks intensity of forsterite increased with increased reduction temperature from 600 to 900°C, but it had no obvious increase when the reduction temperature was above 900°C. The theoretical decomposition temperature of lizardite was 585°C [3]. But, when the reduction temperature was 600°C the diffraction peaks of forsterite was so weak, it was deduced that the lizardite was not decomposed and the actual decomposition temperature was higher than theoretical decomposition temperature above 600°C. In addition, the quartz can be observed in XRD pattern under 800°C, and the reasons may be attributed to the presence of quartz in raw laterite ores and the thermal decomposition of lizardite [8]. However, it disappeared above 800°C. According to the thermodynamic calculation, quartz could react with Na_2SO_4 at 684°C in the CO atmosphere and the equation was an endothermic reaction (enthalpy $\Delta H > 0$). Therefore, increasing the reduction temperature could accelerate the reaction and resulted in the amount of quartz decreased gradually. As revealed from the XRD, the peak intensity of quartz decreased gradually, and finally disappeared above 800°C. Thus, the reaction equation could be represented by [9]

$$Na_2SO_4 + SiO_2 + CO(g) - Na_2SiO_3 + SO_2(g) + CO_2(g) \qquad \Delta G_T^{\theta} = 137582.64\text{-}143.63T \qquad (1)$$

Effect of additive dosage

The laterite ores with different dosage of Na_2SO_4 from 0 to 15wt% were reduced at the reduction temperature of 900°C. The XRD pattern of the roasted ore in different conditions is shown in Figure 2. With the increase of Na_2SO_4 dosages from 0 to 10wt%, the peak intensity of forsterite increased. However, when the dosage of Na_2SO_4 was increased to 15wt%, the diffraction peaks of forsterite had no obvious change.

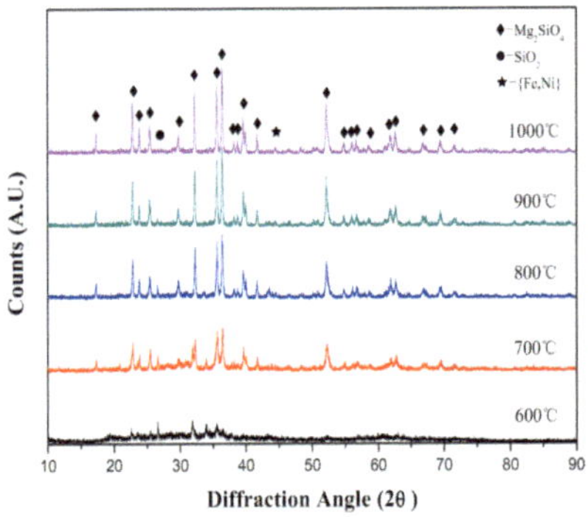

Figure 1. XRD patterns of roasted ore at varying temperatures and 10wt% Na_2SO_4.

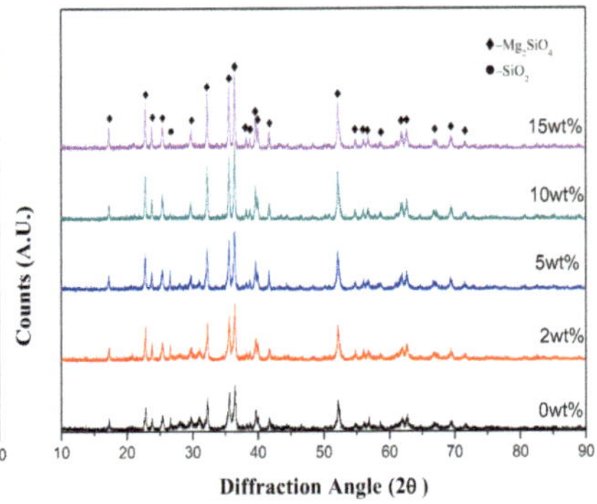

Figure 2. XRD patterns of roasted ore at 900°C and varying dosages of Na_2SO_4.

Effect of different additives

The laterite ores with different additives of 10wt% were reduced at the reduction temperature of 900°C and the XRD pattern of the roasted ore is shown in Figure 3. As revealed from the XRD, the peaks intensity of forsterite varies with the kinds of additive. The diffraction peaks of forsterite with additive of Na_2S had an increase, in contrast to the XRD pattern of the roasted ore with the additives of Na_2SO_4 and $CaSO_4$. The result illustrated that Na_2S had more effect on decomposition of lizardite. Moreover, the XRD pattern of roasted ore with the additive of $CaSO_4$ included calcium sulphide, and the reason may be attributed to the thermal decomposition of $CaSO_4$. The reaction equation is shown below [9]:

$$1/4CaSO_4 + CO(g) - 1/4CaS + CO_2(g) \qquad \Delta G_T^{\theta} = -10435.98+0.17T \qquad (2)$$

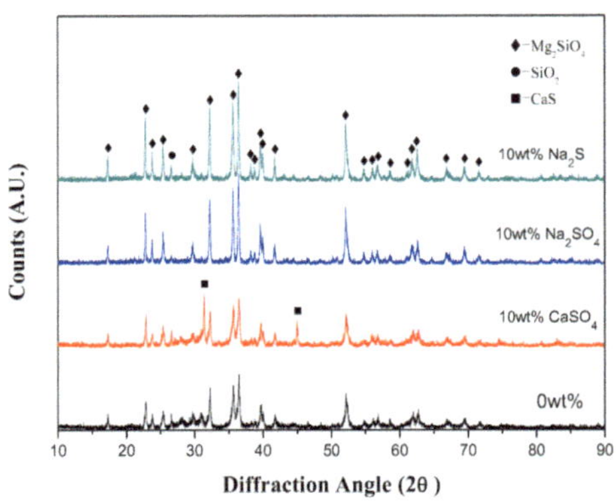

Figure 3. XRD patterns of roasted ore with different additions at 900°C.

SEM-EDS analysis

The microstructure of the roasted ores with different additives and dosages was analyzed by SEM and EDS, as shown in Figure 4. These figures from (a) to (d) were the microstructures of the roasted ore and other figures from (a') to (d') were the magnified view of the marked areas in the figures from (a) to (d), respectively.

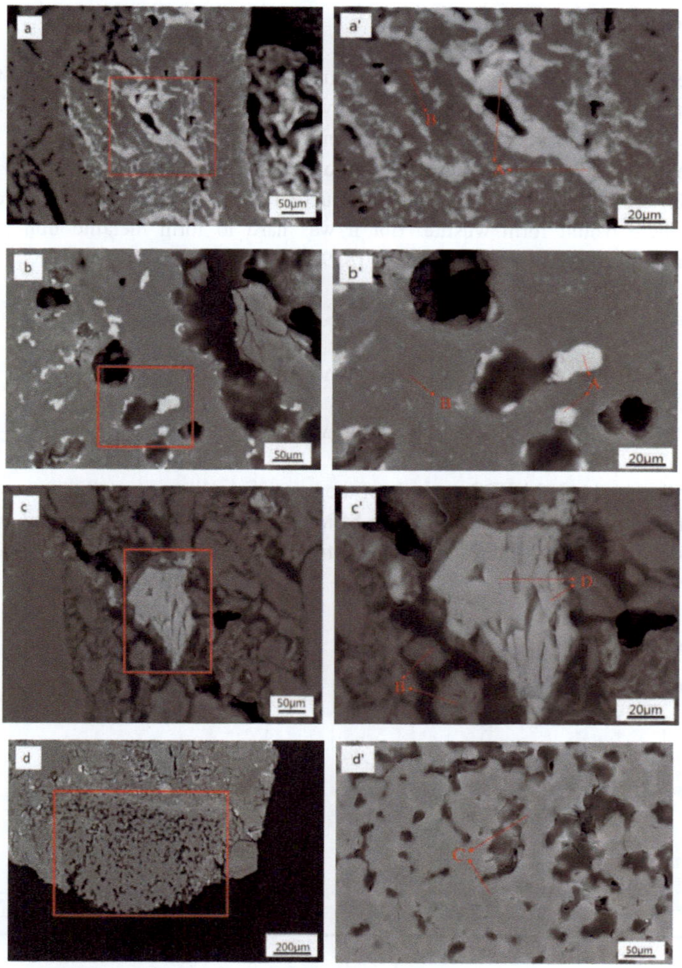

Figure 4. SEM images of the roasted ores with different additives at 900°C: (a): 10wt% of Na_2SO_4; (b): 15wt% of Na_2SO_4; (c) and (d): 10wt% of Na_2S. A-FeS, B-forterite, C-quartz, D-wustite.

Figure 4 (a) and (b) showed the microstructure of the roasted ores with 10wt% and 15wt% Na_2SO_4, respectively. As revealed from the figures, forsterite and ferronickel polymer were observed. The compositions of ferronickel polymer were determined by EDS, and the contents of Ni were 1.58wt% and 6.28wt%, as shown in Figure 5 and 6 respectively. This result illustrated that increasing the dosage of Na_2SO_4 could promote the polymerization of metal nickel, and the reason was that the thermal decomposition of Na_2SO_4 generated sulfur, which could reduce the surface tension and make the metal particle size grow [10]. Figure 6 shows that ferronickel polymer interlinked with the forsterite, and the reason was the most of Ni and Fe existed in silicates in the form of isomorphism.

Figure 4 (c) and (d) show the microstructure of the roasted ores with 10wt% Na_2S. Wustite, quartz and forsterite were observed in these figures. Removal of crystal water for goethite formed ferric oxide during the drying and roasting process [11-14] and the ferric oxide was reduced by gaseous reductant (CO/CO_2), and then formed ferroferric oxide. Nevertheless, further reduction of ferroferric oxide in the reducing atmosphere would form wustite. But it was hard to form metallic iron [8]. The aforementioned equations are as follow [8, 9]:

$$3Fe_2O_3 + CO(g) - 2Fe_3O_4 + CO_2 \qquad \Delta G_T^\theta = -52131-41.0T \qquad (3)$$

$$Fe_3O_4 + CO(g) - 3FeO + CO_2(g) \qquad \Delta G_T^\theta = 35380-40.16T \qquad (4)$$

$$FeO + CO(g) - Fe + CO_2(g) \qquad \Delta G_T^\theta = -22800+24.26T \qquad (5)$$

According to the SEM-EDS analysis of the roasted ore, it was noted that wustite and ferronickel polymer were detected by SEM-EDS, but were not identified by XRD analysis, as shown in figures. This was attributed to the fact that those phases were present in a low content.

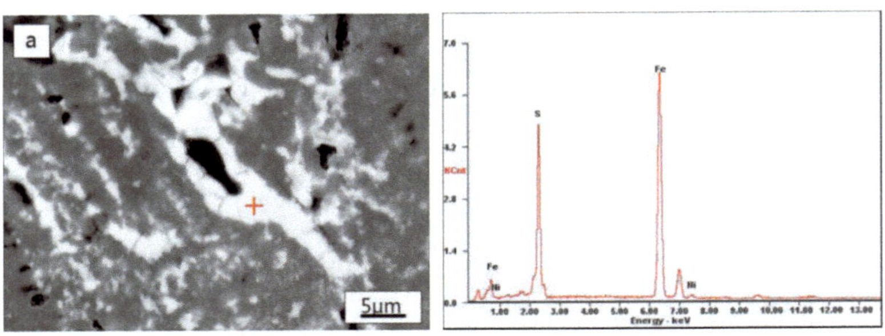

Element	FeK	SK	NiK
Wt%	64.85	33.57	1.58
At%	51.96	46.84	1.20

Figure 5. EDS result of roasted ore grain with 10wt% Na_2SO_4 at 900°C.

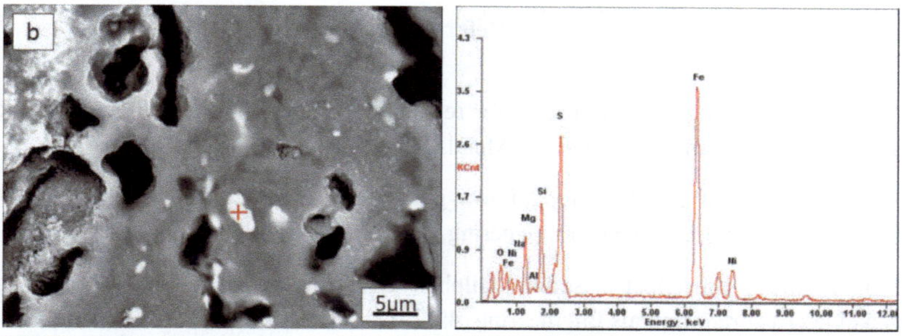

Element	FeK	SK	OK	NaK	MgK	AlK	SiK	NiK
Wt%	29.69	18.65	15.03	4.02	12.76	0.79	12.78	6.28
At%	15.90	17.40	28.09	5.22	15.69	0.88	13.61	3.20

Figure 6. EDS result of roasted ore grain with 15wt% Na_2SO_4 at 900°C.

Conclusions

This article was focused on the phase transformation of nickel laterite ore, which added different additives, during low-temperature reduction roasting process with a gaseous reducing mixture (50vol.% CO and 50vol.% CO_2) for 30min.

The results of XRD analysis in this study demonstrate that use of selected additives promoted phase transformation in nickel laterite ore. Diffraction peaks of forsterite increased with increasing reduction temperature with the addition of 10wt% Na_2SO_4, and it was also increased with higher amounts of Na_2SO_4 at 900°C. Comparing the three additives (Na_2SO_4, Na_2S and $CaSO_4$), Na_2S had more promoting effect on the phase transformation of nickel laterite ore. Meanwhile, the $CaSO_4$ would decompose to form CaS in the reduction atmosphere. As it was verified by SEM-EDS, increasing the dosage of Na_2SO_4 could promote the polymerization of ferronickel and improve the grade of nickel in ferronickel.

Acknowledgments

Financial support for this study was supplied from the National Natural Science Foundation of China (Project Nos. U1302274 and 51304091), the Applied Basic Research Program of Yunnan Province (No. 2013FZ007 and 2013FD009), the Candidate Talents Training Fund of Yunnan Province (2012HB009) and the Analysis and Testing Foundation of Kunming University of Science and Technology.

References

1. Y.C. Zhai et al., A green process for recovering nickel from nickeliferous laterite ores. Transactions of the Nonferrous Metals Society of China. 20(2010), 65–70.

2. K. Jungah et al., Calcination of low-grade laterite for concentration of Ni by magnetic separation. Minerals Engineering, 23(2010), 282–288.

3. J. Lu et al., The effect of sodium sulphate on the hydrogen reduction process of nickel laterite ore, Minerals Engineering. 49(2013), 154-164

4. M. Valix , W.H. Cheung. Effect of sulfur on the mineral phases of laterite ores at high temperature reduction. Minerals Engineering. 15(2002), 523-530

5. J. Man et al., Mechanism of sodium sulfate in promoting selective reduction of nickel laterite ore during reduction roasting process. International Journal of Mineral Processing. 123(2013), 32-38

6. Emmanuel N. Zevgolis et al., Phase transformations of nickeliferous laterites during preheating and reduction with carbon monoxide. J Therm Anal Calorim. 100(2010), 133-139

7. Y.F. Chang et al., Phase transformation in reductive roasting of laterite ore with microwave heating. Transactions of Nonferrous Metals Society of China. 18(2008), 969-973

8. Li Bo, Wang Hua, Wei Yonggang, The reduction of nickel from low-grade nickel laterite ore using a solid-state deoxidization method. Minerals Engineering. 24(2011), 1556–1562.

9. Shi Tangming, Research on the Solid·State Reduction Roasting of Nickeliferous Laterite Ore Intensified by Sulfur-bearing Additives.

10. Nagata, K., Bolsaitis, P., 1986. Selective removal of iron oxide from laterite by sulfurization and chlorination. International Journal of Mineral Processing. 19(1987), 157–172.

11. O'Connor F, Cheung WH, Valix M. Reduction roasting of limonite ores: effect of dehydroxylation. International Journal of Mineral Processing. 80(2006), 88–99.

12. Swamy YV, Kar BB, Mohanty JK. Physico-chemical characterization and sulphatization roasting of low-grade nickeliferous laterites. Hydrometallurgy. 69(2003), 89–98.

13. Mackenzie R. Differential thermal analysis, London: Academic Press; 1973.

14. Walter D, Buxbaum G, Laqua WJ. The mechanism of the thermal transformation from goethithe to hematite. J Therm Anal Calorim. 63(2001), 733–48.

Drying, Roasting, and Calcining of Minerals
Edited by: Thomas P. Battle, Jerome P. Downey, Lawrence D. May, Boyd Davis, Neale R. Neelameggham, Sergio Sanchez-Segado, and P. Chris Pistorius
TMS (The Minerals, Metals & Materials Society), 2015

EFFECT OF TEMPERATURE ON REDUCTION ROASTING OF LOW-GRADE IRON ORE AFTER GRANULATING WITH COAL

Zhucheng HUANG [1, 2], Ronghai ZHONG [1], Jun ZOU [3], Tao JIANG [1]

[1]School of Minerals Processing and Bioengineering, Central South University, Changsha, Hunan, 410083, China. [2]zchuangcsu@126.com

[3]Qidong County Shunda Mining Limited Company, Hengyan, Hunan, 421600, China.

Keywords: low-grade iron ore; granulating with coal; reduction roasting

Abstract

The method of reduction roasting and magnetic separation of low-grade, micro-fine disseminated refractory iron ore following granulation with coal produced unexpected success. Pellets of iron ore with diameter between 3 and 8mm were granulated with coal, bentonite and water. The raw pellets and reduction products were analyzed by scanning electron microscopy and x-ray diffraction. Results show uniformly distributed coal powders inside the pellets improved the kinetics of the reduction process and accelerated the reduction rate. Reduction time was higher at temperatures below 890℃, which resulted in the formation of substantial amounts of fayalite. When temperature was increased, micro-fine hematite was then quickly reduced to iron metal and the amount of fayalite decreased sharply. When pellets were reduced at 950℃ for 25min, reduction products with 56.73% metallization rate were obtained. After grinding and magnetic separation, a concentrate with 72.18% Fe_{tot} was obtained.

Introduction

Low-grade, micro-fine disseminated hematite ore is an abundant source of refractory iron ore in China. This type of iron ore is characterized by its low iron grade, high silicate content and micro-fine disseminated grains of iron oxides. It also contains iron silicate gangue minerals with similar physical and chemical properties. As a result of these characteristics, it is difficult to concentrate the ore by conventional processing methods [1, 2, 3]. If these iron ore resources could be used effectively, it would help to satisfy the demand for iron ore in China. Direct reduction-magnetic separation is an effective way to deal with this type of iron ore. When the ore is reduced at high temperature (1050℃-1200℃) for 80-180mins, nearly 90% metallization rate can be achieved. After grinding and magnetic separation, concentrate with 70%-90% iron grade is obtained [4, 5, 6, 7]. However, solid-phase reaction between SiO_2 (high content) and FeO occur during the reduction process, causing substantial amounts of fayalite to form [8]. Fayalite is difficult to reduce, and its low magnetism can result in loss of iron during magnetic separation process. Fayalite will melt at 1177℃ and form liquid phase [9], which hinders further reduction of iron oxide. The reduction of Fe_2O_3 follows the order: Fe_2O_3, Fe_3O_4, FeO and Fe. Research indicates that Fe_2O_3 can be reduced to Fe_3O_4 easily, but reducing Fe_3O_4 to FeO and FeO to Fe are more difficult, especially reducing FeO to Fe [10, 11]. So, accelerating reduction rate of iron oxide and avoiding forming large numbers of fayalite are important. The method of granulating iron ore with coal powder can ameliorate reducing atmosphere inside the pellets, improve reduction rate of the iron oxide, and decrease the generation of fayalite. In this study, a method of reduction roasting and magnetic separation of iron ore after granulating with coal is adapted to deal with a low-grade, micro-fine disseminated refractory iron ore. XRD and SEM were employed to analyze the effectiveness of the process.

Experimental

Materials

Materials used in this study include a low-grade, micro-fine disseminated refractory iron ore and coal powder used as reducing agent. The chemical analyses of iron ore and coal are shown in Table I and Table II respectively. The particle size of iron ore powder used in this study was less than 0.037mm and that of coal was less than 0.074mm. The chemical analysis of iron ore shows that the grade of iron ore was 33.86%, the main gangue mineral was SiO_2 which accounted for 40.61%. Also, the content of harmful elements phosphorus and sulfur was low.

Table I. Chemical analysis of low-grade iron ore (*wt.* %)

Fe_{tot}	FeO	SiO_2	Al_2O_3	CaO	MgO	P	S	K_2O	Na_2O	LOI
33.86	1.67	40.61	4.36	1.57	1.05	0.38	0.082	1.58	0.03	1.49

Table II. Industrial analysis of coal powder (*wt.* %)

Mad	Fcad	Ad	Vdaf
8.16	58.53	5.44	36.03

The microstructures of iron ore and composite pellet are shown in Figure 1.

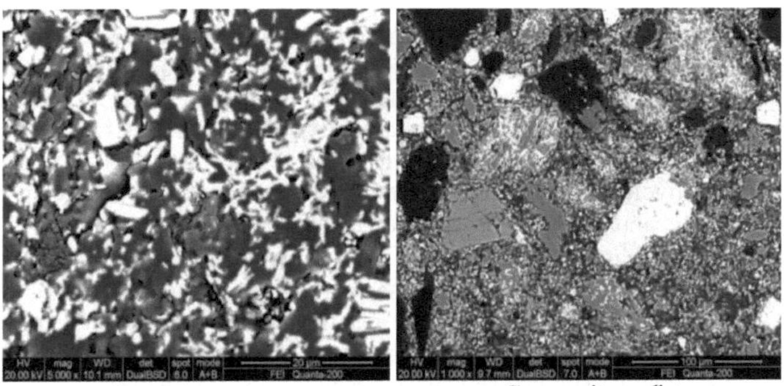

Iron ore Composite pellet

Figure 1. The microstructure of the iron ore and the composite pellet

The main minerals presented in the iron ore were hematite (white) and quartz (black). Hematite crystal was widely dispersed within the gangue minerals. The average hematite crystal size was less than 0.005mm, with a few particles that were about 0.03mm in size. The relationship between hematite crystal and gangue minerals was extremely complex. The structure was compact which hinders the diffusion of reduction gas. Pores were uniformly distributed throughout the pellets, which helped improve the diffusion of reducing atmosphere in the pellets, thus accelerating iron oxide reduction.

Experimental Method

The iron ore and coal were crushed and milled before experiments began. Coal powder and iron ore were mixed at a specific ratio of fixed carbon C_{Fix} and T_{Fe} in iron ore particles, and 1% bentonite was used as binder. Pellets with diameters between 3 and 8mm were made in a disc pelletizer (diameter 1000 mm, rim height 300 mm, angle of inclination 45°and rotating 30 rpm) and dried at 110 ℃ for 2 hours. The schematic diagram of the experimental device for reduction roasting experiments is shown in in Figure 2.

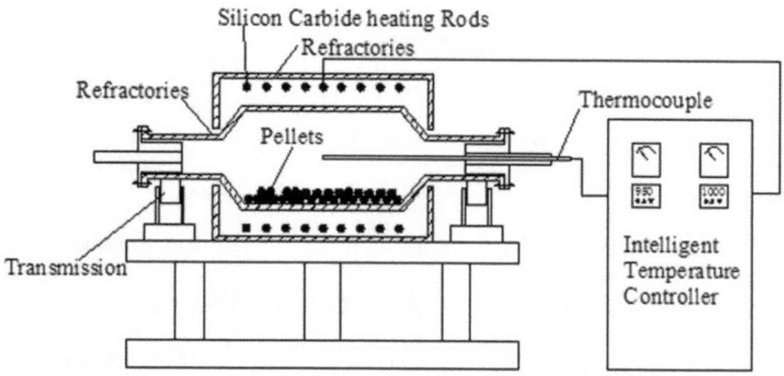

Figure 2. Schematic diagram for reduction roasting apparatus

The furnace was heated to the desired temperature, and then the samples (3 kg charges of pellets each experiment) were placed into the furnace to reduce for the desired time, while a specific amount of coal with diameter between 3 and 8mm was put into the furnace to provide the proper reducing atmosphere. At the end of the reduction roasting, the pellets were removed from the furnace and put into coal. After cooling to room temperature, several samples were selected for milling and magnetic separation. Before milling, the pellets were crushed to less than 1mm. The grinding process milled 90% of the material to a particle size less than 0.074mm. The magnetic field strength was 180KA/m for magnetic separation.

Results and Discussion

Effect of Coal Dosage on the Results of Reduction Roasting

When reduced for 25 min at 950℃, the ratio change of C_{fix} /T_{Fe} had a large effect on results as shown in Figure 3. When the ratio of C_{fix}/T_{Fe} was 0 (no coal addition to pellets), concentrate yield and iron recovery were 11.26% and 23.86% respectively. The recovery increased as the C_{fix}/T_{Fe} ratio increased. The reduction kinetics in the iron ore-coal composite pellets were expected to be enhanced due to the presence of reductant, which shorten of diffusion distances of reductant and availability of large number of reaction sites. The reducing atmosphere was improved by increasing the coal dosage inside pellets, accelerating the reduction rate of iron oxide. The concentrate yield and iron recovery were highest when the C_{fix}/T_{Fe} was 0.4.

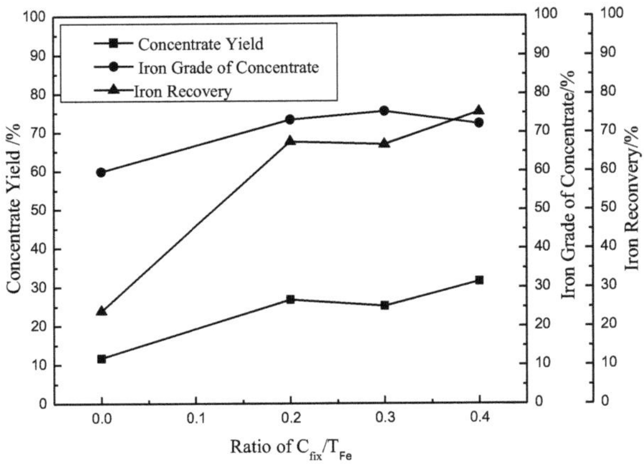

Figure 3. Effect of coal dosage on results of reduction roasting

Effect of Reduction Roasting Time on the Results of Reduction Roasting

At a ratio of 0.4 C_{fix}/T_{Fe} and a hold temperature of 950℃, the effects of varied roasting time can be seen in Figure 4.

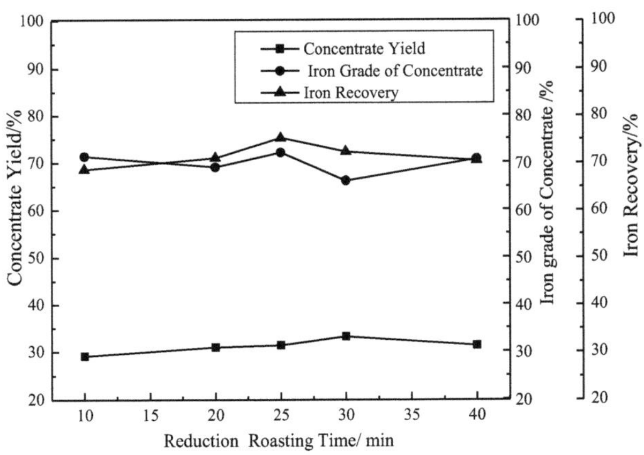

Figure 4. Effect of time on results of reduction roasting

The reduction roasting time had an insignificant effect on iron recovery. When reduced for 10min, iron recovery was lower than recovery of reduced for a longer period, which was attributed to incomplete reduction reaction of iron oxide. Reduction reaction was thoroughly when the reduction time was increased. However, when roasting time was increased to 40min, the recovery was decreased by a small amount because of the consumption of the reducing agent. The best recovery was achieved at a roasting time of 25min.

Effect of Reduction Roasting Temperature on the Results of Reduction Roasting

At a ratio of 0.4 C_{fix}/T_{Fe} and a reduction roasting time of 25 min, the reduction roasting temperature was varied, the effect can be seen in Figure 5. Table III shows the metallization rate of reduction roasting products.

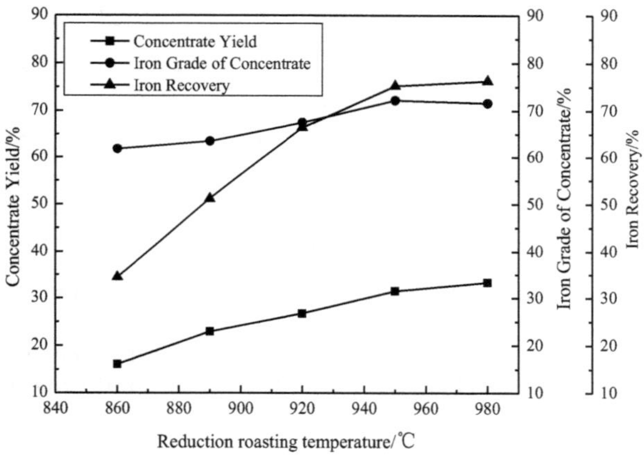

Figure 5. Effect of temperature on results of reduction roasting

Table III. Metallization rate of reduction roasting products

Reduction roasting temperature(℃)	860	890	920	950	980
Metallization ratio (%)	20.14	32.43	42.08	56.74	58.07

One of the most significant factors influencing reduction roasting results was temperature. It can be seen that the concentrate yield, iron grade of concentrate, iron recovery, and metallization ratio increased as the roasting temperature increased. In order to find the degree of the reduction reaction, the reduction products from the 860℃ to 980℃ were crushed and further analyzed. X-ray diffraction (XRD) was used to identify the formed phases. The results are shown in Figure 6. At 860℃, the chemical composition of the reducing product was complex. The composition was mainly SiO_2 and Fe_2SiO_4, and the weight of metallic iron was low. The weight of Fe_2SiO_4 decreased greatly as the reduction temperature increased, and disappeared at 950℃. At the same time, the weight of metallic iron increased. There was no Fe_2O_3. The weight of Fe_3O_4 was low, and decreased as the temperature increased, which indicated that both Fe_2O_3 and Fe_3O_4 could be reduced quickly.

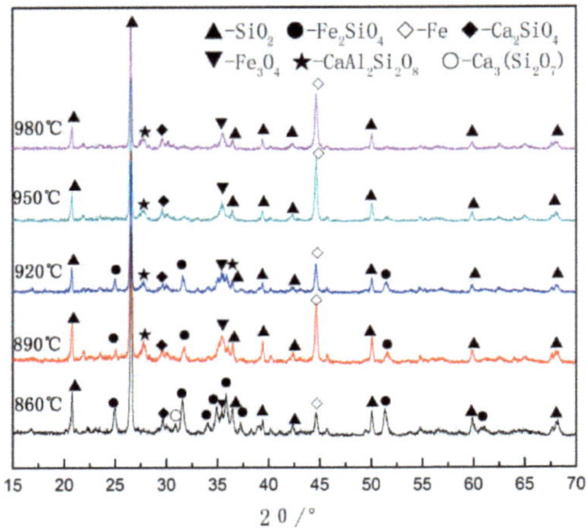

Figure 6. XRD analysis of samples at different temperatures

When the reduction temperature was higher than 570℃, reduction of Fe_2O_3 follows this order: Fe_2O_3, Fe_3O_4, FeO, Fe. The intermediate products FeO and Fe_3O_4 can react with SiO_2, forming Fe_2SiO_4

$$2FeO + SiO_2 = Fe_2SiO_4$$
$$\Delta G^O = -36233 - 21.09T \quad (1)$$

Reduce reactions of Fe_3O_4 and FeO by CO are as follow:

$$Fe_3O_4 + CO = 3FeO + CO_2$$
$$\Delta G^O = 31500 - 37.06T \quad (2)$$

$$FeO + CO_2 = Fe + CO_2$$
$$\Delta G^O = -13175 + 17.24T \quad (3)$$

Fe_2SiO_4 could be reduced by CO:

$$Fe_2SiO_4 + 2CO = Fe + SiO_2 + 2CO_2$$
$$\Delta G^O = 12510 + 21.09T \quad (4)$$

Table IV shows the concentration of CO required at beginning of the reduction reactions of Fe_3O_4, FeO and Fe_2SiO_4 at different temperatures, according to (2), (3) and (4).

According to Table IV, it can be seen that concentration of CO required at beginning of the reduction reactions of FeO and Fe_2SiO_4 by CO is high, so the reduction reaction of Fe_2SiO_4 can be neglected. When the roasting temperature was 860℃, Fe_2O_3 can be reduced to FeO quickly.

However, the reduction of FeO requires a higher CO concentration. Also, the reduction reaction of FeO has higher activation energy than that of Fe_3O_4 [12, 13]. In addition, the carbon gasification reaction was not spontaneous in the temperature conditions, leading to a low reducing atmosphere in the furnace. This caused the reduction rate of FeO to slow down, which led to the formation of large amounts of fayalite. The carbon gasification reaction was facilitated as temperature increased. The more strongly reducing atmosphere caused an increase in reduction rate of FeO and Fe_3O_4, and a decrease the amount of fayalite generated.

Table IV. Concentration of CO required at beginning of the reduction reactions for Fe_3O_4, FeO and Fe_2SiO_4 by CO as a function of temperature

Temperature(℃)	860	890	920	950	980
Reduce Fe_3O_4 (%)	24.76	23.16	21.74	20.44	19.26
Reduce FeO (%)	66.25	67.05	67.80	68.51	69.17
Reduce Fe_2SiO_4 (%)	78.53	78.24	77.96	77.69	77.44

Conclusions

(1) This iron ore is characterized by its low iron grade, high silicate content and micro-fine disseminated grains of iron oxides. It also contains iron silicate gangue minerals with similar physical and chemical properties. As a result of these characteristics, it is difficult to effectively process the ore by conventional methods.

(2) Accelerating the reduction rate of iron oxide and avoiding large amounts of fayalite generation are important during the reduction roasting process. When the roasting temperature is low, the reduction rate of Fe_3O_4 and FeO and carbon gasification reaction of carbon are low, which leads to the formation of large amounts of fayalite.

(3) The best results of 56.73% metallization rate and 72.18% concentrate iron grade were achieved with a C_{fix}/T_{Fe} ratio of 0.4, a roasting temperature of 950℃, and a roasting time of 25min.

References

1. J.S. Zhang ,"Status and trend of exploitation and utilization of iron Ore resources in China," *China Metallurgy*, 17(1) (2007),1-6.

2. Z.T. Yuan et al., "Status Quo and Development Orientation of China's Refractory Ore Resource Utilization," *Metal Mine*, 1(2007), 1-6.

3. Z.L. Hou, "Current situation and potential of iron ore resources in China," *The Exposition of Geology Prospecting*, 20(4) (2005), 242-247.

4. H.E. Yang, "The function and mechanism of an aidant reducer in the deep reduction process of iron ore," *Journal of Harbin Engineering University*, 32(12) (2011), 1630-1636.

5. D.Q. Zhu et al., "Research of Superfine Low Grade Hematite by Coal-based Direct Reduction-low Intensity Magnetic Separation," *Metal Mine*, 2(2012), 60-62, 66.

6. B. Xu et al., "Study of low-grade iron ore with coal based direct reduction," *Multipurpose Utilization of Mineral Resources*, 06 (2001), 20-24.

7. X.M. Qin et al., "Study on the direct reduction magnetic separation of a refractory iron ore," *Metal Mine*, 6(2010), 73.

8. J.Y. Fu et al., *Sintering and Pelletizing*, (Changsha: Central South Univ. Press, 1996), 127-131.

9. J.Y. Fu et al., *Sintering and Pelletizing*, (Changsha: Central South Univ. Press, 1996), 78-84.

10. Y. M, "Influence of temperature and time on reduction behavior in iron ore–coal composite pellets," *Powder Technology*, 256(2014), 361-366

11. R.J. Fruehan, "The Rate of Reduction of Iron Oxides by Carbon," *Metallurgical Transactions B*, 8(1) (1977), 279-286.

12. K.L. Trushenski et al., "Non-topochemical reduction of iron oxides," *Metallurgical Transactions*,(5) (1974),1149-1158.

13. Q.T. Tsay et al., "The modeling of hematite reduction with hydrogen plus carbon monoxide: Part I The behavior of single pellets," *American Institute of Chemical Engineers Journal*, 22(6)(1976),1064-1072.

Drying, Roasting, and Calcining of Minerals
Edited by: Thomas P. Battle, Jerome P. Downey, Lawrence D. May, Boyd Davis, Neale R. Neelameggham, Sergio Sanchez-Segado, and P. Chris Pistorius
TMS (The Minerals, Metals & Materials Society), 2015

Thermo-Physical Properties of Petroleum Coke during Calcining Graphitization Process

Mujun Long*[1)], Junhao Sheng[1)], Tao Liu[1)], Dengfu Chen[1)], Yi Yang[2)], Shikai Gong[2)], Chunmei Chen[1)]

[1)] College of Materials Science and Engineering, Chongqing University, Chongqing 400044, P.R. China. *Email: longmujun@cqu.edu.cn
[2)] Guiyang Aluminium Magnesium Design & Research Institute Co., Ltd, Guiyang 50004, Gui Zhou, P.R. China.

Keywords: Thermo-physical properties; Calcining; Graphitization; Petroleum coke

Abstract

The relationship between thermo-physical properties of petroleum coke and temperature is of importance for determining of the operating parameters during the calcining graphitization process. In this paper, the variations of thermo-physical properties of petroleum coke versus temperature were investigated, such as the DSC-Tg curve, the specific heat capacity, the thermal diffusion coefficient, the thermal conductivity, and so on. The thermal analysis curve of petroleum coke showed synchronous thermal hysteresis at higher heating rates, indicating that a larger heating rate will cause delay of thermal effects. The thermal diffusivity of petroleum coke decreased during the calcining process. With increasing temperature, coefficient of heat conductivity and specific volume heat of the petroleum coke increased with a decreasing variation rate. The research provides a database for the calcining graphitization process of petroleum coke.

Introduction

Petroleum coke is a by-product from oil refining, the residue after the cracking of heavy oil in a delayed coking unit [1]. Petroleum coke is widely used for prebaked anodes, graphite electrodes, metallurgical carburant, and so on [2-10]. The petroleum coke needs to be calcined and graphitized before using. The thermo-physical properties of petroleum coke, such as the specific heat, thermal diffusion coefficient, thermal conductivity and the thermal behavior during heating, were important base parameters for the simulation and control of calcining graphitization process. Generally, the physical properties of petroleum coke depend on the composition and characteristics of the feedstock [11]. Works have been performed to obtain the thermal properties of petroleum coke [7,11,12]. The thermo-physical properties of petroleum coke vary in different calcining temperature, especially at different heating rates. Simulating the calcining graphitization process, the specific heat, thermal diffusion coefficient, thermal conductivity and the thermal behavior of petroleum coke were experimentally studied.

Experimental

Petroleum coke (A and B) obtained from two petroleum coke production enterprise were used as raw materials in this research. The two petroleum cokes had been calcined at approximately 1100°C. Chemical compositions of the two petroleum cokes are presented in Table 1.

Table 1 Chemical compositions of the petroleum cokes

Cokes	S (%)	N (%)	Si (%)	Fe (%)
Coke A	0.44	0.48	0.0065	0.028
Coke B	0.89	0.47	0.067	0.085

Thermal analysis experiments were conducted using a simultaneous thermal analyzer STA (DSC/DTA-TG, NETZSCH, Germany) to monitor both the weight change and ex-endothermal behavior of the samples during calcining process. The samples were heated from room temperature up to 1400 °C at constant heating rates of 10 °C/min and 20 °C/min in a nitrogen atmosphere, respectively.

Thermal conductivity analysis (LFA-457, NETZSCH, Germany) was used to investigate the specific heat, thermal diffusivity, and thermal conductivity. The testing was performed from room temperature to 1000 °C.

Results and Discussion

Thermal behavior of the petroleum cokes

Generally, the calcining process can be monitored by the change of heat and weight. DSC-TG curves characterized the absorbing or releasing of heat and weight changes during the calcining process. By studying the DSC-TG curves, the ex-endothermal behavior and volatile matter removal of the petroleum cokes could be obtained.

Simulating the calcining process of petroleum coke, the ex-endothermal behavior and weight loss of the two petroleum cokes at heating rates of 10 °C/min and 20 °C/min were studied with the simultaneous thermal analyzer. Results of the petroleum cokes A and B are illustrated in Figure 1 and Figure 2, respectively.

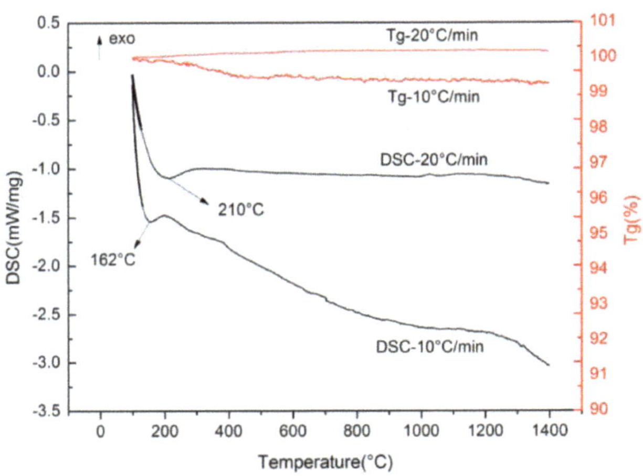

Fig. 1 DSC-TG curves of the petroleum coke A

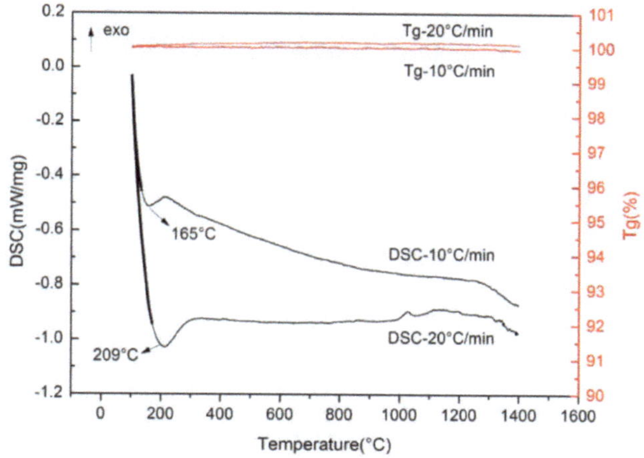

Fig. 2 DSC-TG curves of the petroleum coke B

As shown in Figure 1 and Figure 2, the mass loss of the two petroleum cokes changed little during the heating process, indicating that there were little emission of volatile matter and moisture during the process. Since the petroleum cokes had been calcined at temperatures around 1100 °C before the experiments, most of the volatile matter and moisture had already been removed. There was a small weight loss for petroleum coke A at the heating rate of 10 °C/min, it may owe to the absorption of moisture of coke A before the DSC experiment.

The DSC curves in Figure 1 and Figure 2 were smooth. Only some small exothermic peaks and endothermic peaks were found at around 200 °C and around 1000 °C, which indicated again that most of the volatile matter and moisture had already escaped at the temperature below 1100 °C. The temperature of the exothermic peaks and endothermic peaks were higher when the heating rates were increased, indicating that a faster heating rate would delay the thermal effects[13], such as the emission of moisture and volatile matters. As the heating rate increased from 10 °C/min to 20 °C/min, the endothermic peaks of coke A and B moved from 162 °C to 210 °C, and 165 °C to 209 °C, respectively.

Heat capacities of the petroleum cokes

Specific heat (*Cp*) represents thermal properties of materials and relates to the status and the types of materials. Specific heat is one of the important parameters for the heat transfer simulation and temperature control during the graphitization process.

The variations of specific heat versus temperature for petroleum cokes A and B are presented in Figure 3, showing higher specific heats with increasing temperature for both petroleum cokes. The increasing rate of both specific heats were lower at higher temperature. From room temperature to 1000 °C, the *Cp* increased from 0.396 J/(g·K) to 1.314 J/(g·K) for petroleum coke A, and from 0.416 J/(g·K) to 1.45 J/(g·K) for petroleum coke B. Therefore, compared with petroleum coke A, more energy is needed for petroleum coke B to be calcined to the same temperature. The difference of *Cp* between petroleum cokes A and B is bigger at higher temperature.

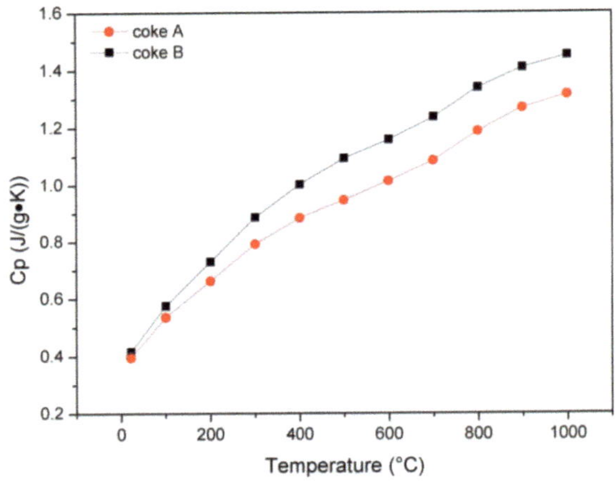

Fig. 3 Variations of specific heat with temperature for petroleum cokes A and B

Thermal conductivities of the petroleum cokes

Figure 4 shows the thermal diffusion coefficients α (mm^2/s) of petroleum coke A and B tested from room temperature to 1000 °C. As Figure 4 showed, the lower temperature, the higher the thermal diffusion coefficients. The thermal diffusivity of petroleum coke A is higher than petroleum coke B, especially at low temperature.

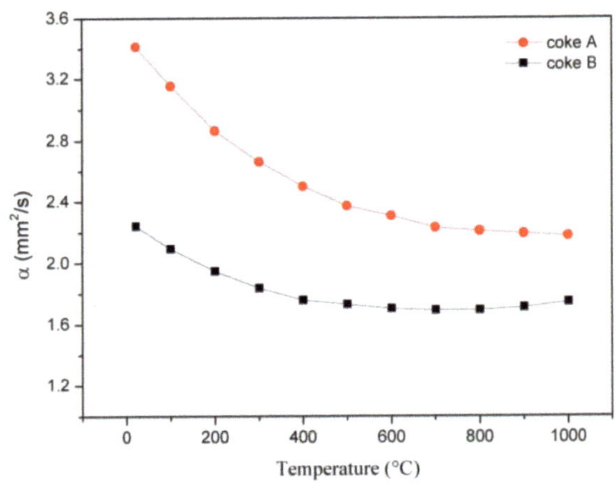

Fig. 4 The thermal diffusion coefficients versus temperature for petroleum cokes A and B

As one of the important parameters for heat transfer, the thermal conductivity coefficient λ (w/(m·K)) can be calculated by

$$\lambda = \alpha \cdot Cp \cdot \rho \tag{1}$$

where Cp is specific heat in J/(kg·K); α is thermal diffusion coefficients in m^2/s; and ρ is density in kg/m^3.

With Eq.(1), the thermal conductivity coefficients for petroleum coke A and B were obtained, as shown in Figure 5.

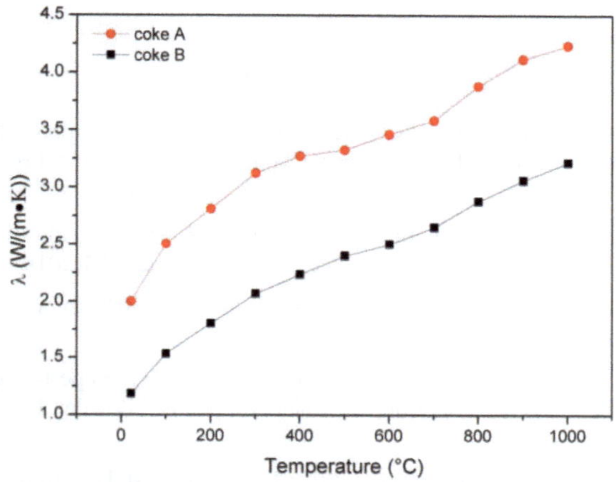

Fig. 5 The thermal conductivity coefficients versus temperature for petroleum cokes A and B

When the temperature increased, the number of phonons was larger, the specific heat capacity was higher, hence the thermal conductivity coefficient was larger. Furthermore, the thermal conductivity coefficient was larger at higher graphitization degree [14]. As shown in Figure 5, the heat conductivity coefficients of the both petroleum cokes increased with a decreasing variation rate. From room temperature to 1000 °C, the heat conductivity coefficient increased from 2.002 w/(m·K) to 4.236 w/(m·K) for petroleum coke A, while from 1.184 w/(m·K) to 3.215 w/(m·K) for petroleum coke B. The heat conductivity coefficient of petroleum cokes A was larger than that of petroleum cokes B.

Summary

The thermo-physical properties of petroleum cokes, such as the specific heat, thermal diffusion coefficient, thermal conductivity, and the DSC-TG curve, were experimentally investigated. Results indicated: (1) Higher heating rates will cause delay of thermal effects, such as the emission of moisture and volatile matter. Most of the volatile matter and moisture had already escaped at the temperature below 1100 °C because the petroleum cokes had been pre-calcined. (2) The thermal diffusivity of petroleum cokes decreased during calcining process. With increasing temperature, the heat conductivity coefficient and specific heat of both petroleum coke increased with a decreasing variation rate.

Acknowledgements

Authors would like to acknowledge the financial support of Guiyang Aluminum Magnesium Design & Research Institute Co., Ltd. The technical assistant from the Analytical and Testing Center of Chongqing University is gratefully acknowledged.

References

1 E.M.A. Edreis, G.Q. Luo, et al.,"CO_2 co-gasification of lower sulphur petroleum coke and sugar cane bagasse via TG-FTIR analysis technique," Bioresource Technology, 136(2013), 595-603.

2 C. Gao, F. Vejahati, H. Katalambula and R. Gupta, "Co-gasification of Biomass with Coal and Oil Sand Coke in a Drop Tube Furnace", Energy & Fuels, 24(2010), 232-240.

3 A. Sarkar, D. Kocaefe, Y. Kocaefe, et al., "Coke-pitch interactions during anode preparation," Fuel, 117(2014), 598-607.

4 K. Azari, H. Alamdari, G. Aryanpour et al., "Compaction properties of carbon materials used for prebaked anodes in aluminum production plants," Powder Technology, 246(2013), 650-657.

5 Z.I. Syunyaev, "Petroleum Electrode Coke," Chemistry and Technology of Fuels and Oils, 13 (9-10) (1977), 749-760.

6 M.S. Gorpinenko, V.V. Zelenina and V.A. Smolentseva, "Effect of Petroleum Coke Properties and Structure on Quality of Graphitized Electrodes," Chemistry and Technology of Fuels and Oils, 17 (3-4) (1981), 114-17.

7 J.A. Menendez, M.A. Diez, G. de la Puente, et al., "Thermal behaviour and reactivity of green petroleum cokes used as additives in metallurgical cokemaking," Journal of Analytical and Applied Pyrolysis, 45 (1) (1988), 75-87.

8 P. Sedov, N. Pokhodenko, et al., "Effect of conditions of calcining and cooling petroleum coke on its physicomechanical properties," Chemistry and Technology of Fuels and Oils, 9 (4) (1973), 268-271.

9 Y.Q. Wu, S.Y. Wu, et al., "Differences in physical properties and CO2 gasification reactivity between coal char and petroleum coke," Process Safety and Environmental Protection, 87 (5) (2009), 323-30.

10 E.M.A. Edreis, "Synergistic effects and kinetics thermal behaviour of petroleum coke/biomass blends during H2O co-gasification," Energy Conversion and Management, 79 (2014), 355-66.

11 M. Kakuta, M. Kohriki and Y. Sanada, "Relationships between the Characteristics of Petroleum Feedstocks and the Graphitizability of the Petroleum Cokes," Journal of Materials Science, 15 (7) (1980), 1671-79.

12 S. Birghila, I.C. Popovici and A. Dumitru, "Study on Physical-Chemical Properties of Petroleum Cokes", Romanian Journal of Physics, 56 (7-8) (2011), 976-82.

13 Bin Xu, Xuexin Zhang, "Research of petroleum coke calcining process," Carbon Techniques, (1) (1990)14-19. (in Chinese)

14 V.I. Kasatochkin, V.K. Zamoluev and A.T. Kaverov, "RELATION BETWEEN THE THERMAL PROPERTIES AND THE ATOMIC AND MOLECULAR STRUCTURE OF CARBON IN HOMOGENEOUS GRPHITIZATION," Atomnaya Energiya, 7 (3) (1959), 272-75.

DRYING, ROASTING, AND CALCINING OF MINERALS

INDURATION AND SINTERING

Session Chairs:
Thomas P. Battle
P. Chris Pistorius

Drying, Roasting, and Calcining of Minerals
Edited by: Thomas P. Battle, Jerome P. Downey, Lawrence D. May, Boyd Davis, Neale R. Neelameggham, Sergio Sanchez-Segado, and P. Chris Pistorius
TMS (The Minerals, Metals & Materials Society), 2015

PREDICTION OF NON-ISOTHERMAL OXIDATION OF MAGNETITE PELLETS

Petrus Christiaan Pistorius[1] and Ming Tang[1]

[1]Center for Iron and Steelmaking Research, Carnegie Mellon University
Dept. of Materials Science & Eng., 5000 Forbes Avenue, Pittsburgh, PA, 15213, USA

Keywords: Pellets, magnetite, oxidation, size distribution

Abstract

Magnetite concentrate is pelletized and then hardened by elevated-temperature oxidation to hematite, to produce pellets for ironmaking. Previous work showed that the rate of oxidation is under mixed control of gaseous mass transfer of molecular oxygen through the pellet pores, and solid-state diffusion into the oxidizing grains. Because of the partially limiting role of solid-state diffusion, concentrate size distribution strongly affects pellet oxidation. Previously, a simple mixed-control model was formulated to describe isothermal oxidation. In that model, the main effect of concentrate size distribution was on the plateau level of oxidation which is reached at a given temperature. However, the extent of plateau oxidation increases with increased temperature, requiring a more complex model for non-isothermal oxidation. In this paper, predictions of the non-isothermal model confirm the strong effects of concentrate size distribution and oxygen enrichment. In particular, a greater fraction of coarse magnetite particles would be beneficial, decreasing the size of the macroscopic unoxidized pellet core.

Introduction

Pellets prepared from magnetite concentrate are commonly hardened by oxidative firing, typically using grate-kiln or straight-grate processes. During firing, pellets are strengthened by the oxidation of magnetite to hematite. The oxidation process has been studied for several decades [1,2]; it is well-established that pellet oxidation is under mixed control, with both gaseous mass transfer of molecular oxygen (into the pellets) and solid-state diffusion (within the oxidizing magnetite/hematite particles) limiting the reaction rate. The result is that pellets often display shrinking-core morphology on two very different length scales (see Figure 1): limited pore mass transfer can cause a macroscopic reaction front to develop within the pellet (length scale of millimeters), and limited solid-state diffusion can lead to a microscopic reaction front within each oxidizing particle (length scale of microns to tens of microns).

Limited gaseous mass transfer into pellets can lead to persistence of a macroscopic core of unoxidized magnetite at the end of the oxidation step; such an unoxidized core would tend to lower pellet strength. It has been shown that oxygen enrichment can ensure full oxidation [3]. Given that oxidation kinetics is under mixed control, not only gaseous oxygen enrichment, but also the solid-state oxidation behavior of microscopic magnetite particles, affects the macroscopic reaction front. Specifically, more sluggish solid-state kinetics (caused by larger concentrate particles, lower oxidation temperatures, or both) would promote better gaseous oxygen penetration throughout the pellet volume. The effect of concentrate particle size on oxidation kinetics has been demonstrated; it was shown that the size effect could be quantified by assuming shrinking-core solid-state kinetics and a Rosin-Rammler particle size distribution [4].

A mixed-control model for isothermal oxidation of pellets, including comparison with experimental results, was presented before [3]. However, practical pellet oxidation is not at all isothermal, with initial heating rates of 2-4 K/s. In this work, a preliminary non-isothermal mixed-control model is presented.

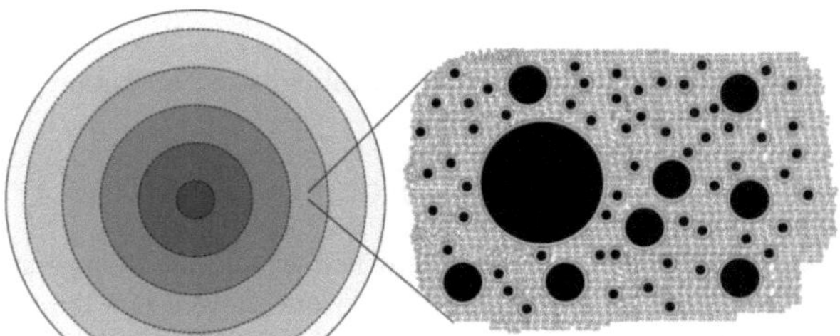

Figure 1. Schematic illustration of the grain model for oxidation of polydisperse magnetite concentrate. The pellet is divided into concentric shells. Each shell contains a wide distribution of oxide grains.

Modeling approach

The model is a type of grain model, describing both the macroscopic pellet reaction front (due to limited gaseous oxygen mass transfer) and the microscopic reaction fronts in each of the oxidizing particles (caused by limited solid-state diffusion). As indicated in Figure 1, the pellet was divided radially into six shells. Within each shell, the oxidation of particles of four different sizes was calculated. All particles of a particular size in a specific shell were taken to have the same degree of oxidation.

The four particle sizes were based on the Rosin-Rammler size distribution (see Figure 2); each of the four particle sizes was chosen to represent one quarter of the concentrate mass. A baseline ("standard") size distribution (as determined previously [4]) was used in the calculations, as were two additional variants – one with smaller fine particles, and one with larger coarse particles; see Figure 2.

Approximating the oxidizing magnetite particles as spheres with uniform reaction fronts, the rate of oxidation under solid-state diffusion control is given by [4]

$$\frac{df}{dt} = \frac{12k_p}{d_p^2\left[(1-f)^{-1/3}-1\right]} \tag{1}$$

where f is the degree of oxidation (amount of oxygen taken up by particle, relative to the amount required for full conversion of magnetite to hematite), k_p is the parabolic rate constant for oxidation of magnetite to hematite, and d_p is the particle diameter. (Equation [1] only applies while $f<1$; for $f=1$, $df/dt=0$.)

The parabolic rate constant in principle depends on the local partial pressure of oxygen, through the effect of the oxygen activity on concentration of oxygen anion vacancies in hematite. For hematite, the concentration of oxygen anion vacancies is proportional to $p_{O_2}^{-0.5}$ [5], leading to the following relationship between the parabolic rate constant and oxygen activity:

$$k_p = k_1\left[1-\sqrt{\frac{p_{O_2}^{\text{equilib}}}{p_{O_2}}}\right] \tag{2}$$

where k_1 is the parabolic rate constant for p_{O_2}=1 bar; $p_{O_2}^{\text{equilib}}$ is the partial pressure of oxygen at magnetite-hematite equilibrium, and p_{O_2} is the actual partial pressure of oxygen. Figure 3 shows this relationship graphically, and illustrates that the parabolic rate constant can be taken to be constant, except at very low oxygen activities close to the equilibrium activity.

Equation (1) gives the oxidation rate under solid-state diffusion control; oxidation is in general also limited by oxygen mass transfer. Transfer of molecular oxygen to the pellet exterior by diffusion in the surrounding gas atmosphere (calculated by assuming stagnant conditions) and within the pellet by pore diffusion was quantified, as was oxygen transfer by bulk flow. Bulk flow occurs during pellet oxidation because molecular oxygen is consumed when magnetite oxidizes to hematite. In principle, bulk flow also occurs because of thermal expansion of the gas in the pellet. (While bulk flow due to thermal expansion was included in the calculations, it was found to be negligible compared with reaction-driven bulk flow.)

In general oxidation within a particular shell of the pellet was limited by both molecular oxygen transfer into the shell of the pellet, and solid-state diffusion within the particles in the shell. The approach was to first calculate the amount of molecular oxygen available in that shell (calculated from diffusion and bulk flow, for the particular time step). This available oxygen was then allocated to particles of the four different particle sizes in proportion to what the oxidation rate of that particle size would have been under full solid-state control (Equation [1]).

Numerical integration was used to find the degree of oxidation of each particle size fraction in each pellet shell after every time step; the time step size was 1 s. Numerical values and expressions for the main numerical inputs into the model are summarized in Table I.

Table I. Main inputs used in model calculations

Size distribution of magnetite concentrate: Cumulative mass fraction = $1-\exp[-(d/d^*)^n]$; d=magnetite particle diameter; d^* and n are the parameters of Rosin-Rammler distribution; for baseline distribution d^*=17.5μm and n=0.73 [4]	**Pellet dimensions:** Diameter=12 mm Porosity: ε=0.3; tortuosity: τ =2.6 (estimated from ε [6]) **Standard heating profile:** Initial temperature 673K; heated to 1123K at 2K/s, then held at 1123K; total time 600 s.
Gaseous diffusion of oxygen: Mass transfer outside pellet: $Sh = 2$ Binary diffusivity in gas: $D = 4.9\times10^{-8}\,T + 1.1\times10^{-10}\,T^2$ m²/s; T in kelvin (fitted to the calculated binary diffusivity in O_2-N_2 [7]) Pore diffusivity $D_{pore} = D(\varepsilon/\tau)$	**Oxidation of magnetite to hematite:** Equilibrium partial pressure of oxygen: $\log(p_{O_2}^{\text{equilib}}/\text{bar})$ $= -24660/T + 5.854 + 2.592 \log T$; T in kelvin [5] Parabolic rate constant: k_p= 220 exp(-65100/RT) μm²/min; T in kelvin [4]

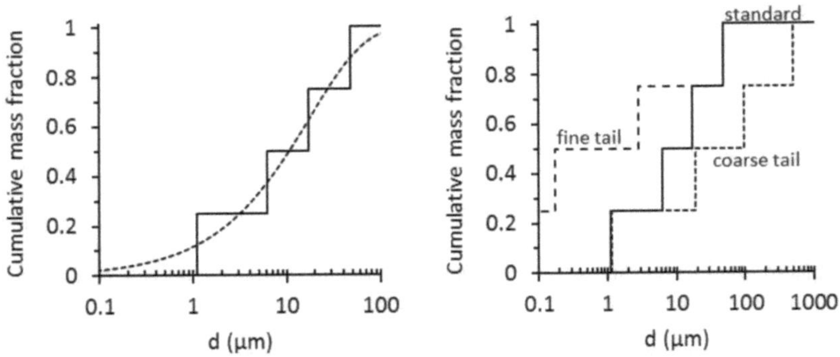

Figure 2. Assumed size distribution of magnetite particles, showing (left) continuous Rosin-Rammler distribution (d^*=17.5μm, n=0.73), and approximation by four particle sizes. Variants of the size distribution considered in the calculations are shown at right.

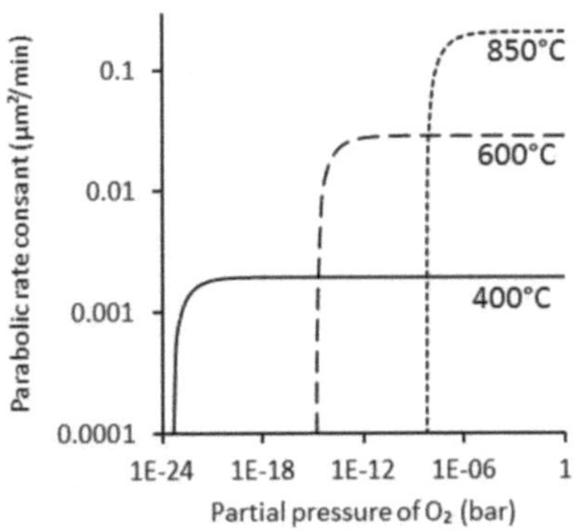

Figure 3. Calculated dependence of parabolic rate constant, for oxidation of magnetite to hematite, on temperature and partial pressure of oxygen. The parabolic rate constant is essentially independent of oxygen activity at oxygen pressures slightly greater than at magnetite-hematite equilibrium.

Results and discussion

Examples of typical calculation results are given in Figure 4, showing the increase in extent of oxidation at different radial positions within a magnetite pellet, for gas atmospheres with lower (5% O_2) and higher (20% O_2) oxygen contents, and for isothermal oxidation and continuous heating. The results indicate that, as observed experimentally, oxidation in 5% O_2 leads to persistence of an unreacted core, whereas 20% O_2 eliminates the core (for oxidation times of around 10 minutes) [3,8].

Comparison of Figure 4 (a) (continuous heating) and (b) (isothermal oxidation) indicate that the difference in degree of oxidation within the pellet (core structure) is rather similar, whether the pellet is isothermally oxidized, or oxidizes during continuous heating (with the same total oxidation time). This model prediction also agrees with experimental observations [8].

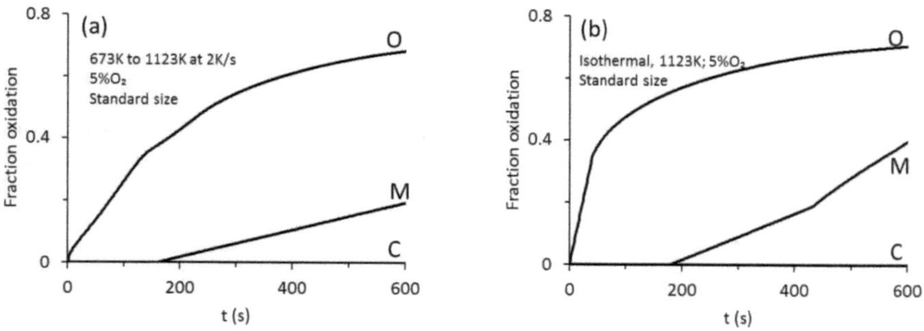

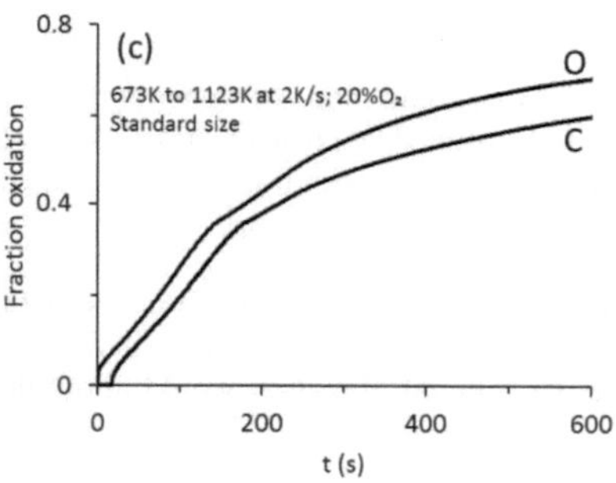

Figure 4. Examples of model calculations, showing the predicted change in degree of oxidation with time at the outer radius of the pellet ("O"), at the center of the pellet ("C") and at mid-radius ("M").
a) Continuous heating from 673K to 1123K at 2K/s, in 5% O_2 atmosphere;
b) isothermal holding at 1123K in 5% O_2 atmosphere;
c) continuous heating from 673K to 1123K at 2K/s, in 20% O_2 atmosphere

The model was used to test the effect of changes in the particle size distribution. It is expected that the unreacted core would be most sensitive to the proportion of larger particles, since small particles would oxidize rapidly (under solid-state diffusion control). Given the dependence of the parabolic rate constant on temperature (Table I), it is calculated that the diameter of particles which would just oxidize fully within ten minutes at 1123 K is $(24k_{p}t)^{0.5}$=7 μm. This means that the proportion of particles which are larger than approximately 7 μm would affect pellet oxidation, with little effect of the detail of the size distribution of smaller particles.

This prediction was tested by calculating the oxidation of pellets of concentrate with two modified size distributions (see Figure 5 for results, and Figure 2 for the assumed size distributions). One distribution had the same top size as the standard distribution, but with a longer tail of finer particles. The other distribution had the same size of fine particles as the standard size distribution, but the top size of particles was much larger.

The model calculations (Figure 5) support the idea that the detail of the size distribution of fine particles has little effect on the size of the unreacted core. The overall extent of oxidation is larger if the tail of the distribution extends to fine particles, simply following from the faster oxidation of smaller particles under solid-state control (Equation 1). However, the size of the unreacted core (the radial position where the degree of oxidation reaches zero) is the same as for the standard size distribution (to within the spatial resolution of the discretized shells). In contrast, adding more coarse particles (while keeping the same size and proportion of fines) is predicted to both decrease the fraction of oxidation at the pellet surface (because the larger particles oxidize more slowly under solid-state diffusion control), and decrease the size of the unreacted core.

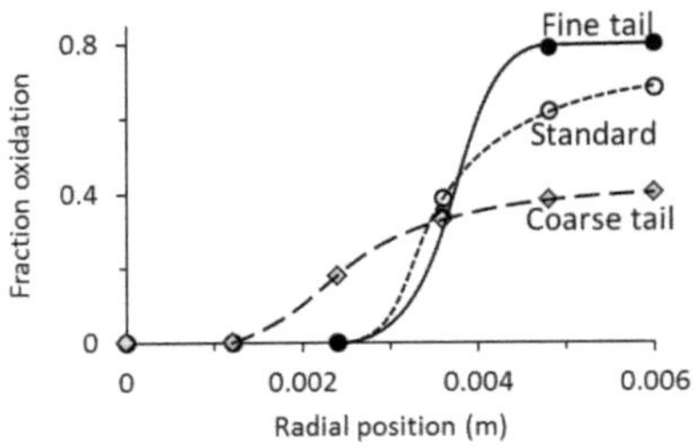

Figure 5. Effect of concentrate size distribution on the size of the macroscopic unreduced core, for heating from 673K to 1123K at 2K/s, followed by holding at 1123K (total time 600s), in 5% O_2 atmosphere.

Conclusion

A model has been constructed to predict non-isothermal oxidation of magnetite pellets containing a wide range of magnetite particle sizes, under mixed control by gaseous oxygen mass transfer and solid-state diffusion. The model predictions agree with previous experimental observations.

The practical implication for plants producing magnetite pellets is that unreacted cores in fired pellets can be eliminated by various combinations of close pellet size and porosity control, increasing the proportion of coarse magnetite particles (if mineral liberation would allow this), and adding oxygen enrichment.

References

1. J.D. Zetterstrom: *Oxidation of magnetite concentrates*, Bureau of Mines Report of Investigations 4728, United States Department of the Interior, 1950.
2. H. Kokal: *The oxidation mechanism of single magnetite pellets*, MSc Thesis, University of Minnesota, 1970.
3. M. Tang, H.J. Cho and P.C. Pistorius: "Early gaseous oxygen enrichment to enhance magnetite pellet oxidation." *Metallurgical and Materials Transactions B*, 45 (2014), 1304-1314.
4. H.J. Cho, M. Tang and P.C. Pistorius: "Magnetite Particle Size Distribution and Pellet Oxidation." *Metallurgical and Materials Transactions B*, 45 (2014), 1213-1220.
5. R. Dieckmann: "Point defects and transport in hematite ($Fe_2O_{3-\varepsilon}$)", *Philosophical Magazine A*, 68 (1993), 725-745.
6. J. Wu, B. Yu and M. Yun: "A resistance model for flow through porous media." *Transp. Porous Med.* 71 (2008), 331–343.
7. J.O. Hirschfelder, R.B. Bird and E.L. Spotz: "The transport properties of gases and gaseous mixtures", *Chemical Reviews*, 44 (1949), 205-231.
8. M. Tang and P.C. Pistorius: "Magnetite Pellet Oxidation: Effects of Water Vapor and Non-Isothermal Reaction." *AISTech 2014 Proceedings* (Warrendale, PA: Association for Iron & Steel Technology, 2014), 635-641.

Drying, Roasting, and Calcining of Minerals
Edited by: Thomas P. Battle, Jerome P. Downey, Lawrence D. May, Boyd Davis, Neale R. Neelameggham, Sergio Sanchez-Segado, and P. Chris Pistorius
TMS (The Minerals, Metals & Materials Society), 2015

MONITORING THE RING FORMATION IN ROTARY KILN FOR PELLET FIRING

Deqing Zhu[1], Xianlin Zhou[1*], Yanhong Luo[1], Jian Pan[1], Cailing Zhen[2], Guixiang Huang[2]
[1] School of Minerals Processing and Bioengineering, Central South University, Changsha 410083, Hunan, China
[2] WISCO Minerals, Wuhan 430080, Hubei, China

Keywords: monitoring, soft-sensing, ring formation, pellet rotary kiln

Abstract

The grate-kiln process has been widely applied for oxide pellet production with the quick development of steel industry in China, but ring formation takes place which adversely affects the productivity and properties of fired pellets. Rotary kiln at high temperature seems a black box, and ring formation is difficult to measure directly by current technology. It's of great importance to monitor ring formation during firing of pellets. Therefore, a method for estimating the thickness of ring formation in rotary kilns was studied, based on working out theoretically the relationship between ring thickness and the outside surface of the kiln, according to steady-state heat transfer across a cylinder, which provides a soft-sensing technique for ring formation at high temperature, and achieves stable pellet production .

Introduction

Rotary kilns are widely applied in chemical, metallurgical and pharmaceutical industries [1]. In steel industry, kilns are used for roasting pellets to produce oxidized or reduced pellets. In China, over 55% of fired pellets were produced by grate-kiln process because of its mature technology, high capacity, strong adaptability of raw materials and high uniformity of products [2]. However, ring formation takes place easily during firing in the kiln, which leads to higher energy consumption, heavier equipment load, and the destruction of refractory [3].

Analogue technology is widely used for modeling industrial production processes. Many models were established to simulate rotary kiln pellet production [4-9], but no models have focussed on ring formation. Currently inner real-time status of kiln is difficult to be detected directly because of high temperature over 1000°C during production, which makes kiln a black-box. Meanwhile, ring formation is difficult to be measured directly by current technology, but it causes the outer surface temperature decrease of the kiln due to the weak thermal conductivity of the ring. Therefore, a thickness model of ring formation in kiln is worked out based on the measurement of kiln's outer surface, and theoretical calculation of the ring thickness, which can reveal detailed

*Corresponding author: Xianlin Zhou, e-mail: xlzhou_csu@csu.edu.cn

information of real time ring formation, and provide guidance with the operation of the rotary kiln. Therefore, this method will help reduce and eventually eliminate ring formation, enhance the quality and output of product pellets.

Method

Relationship between ring formation and temperature of outer shell

Figure 1 shows that the temperature of kiln outside surface has a trend of wavy change, which was captured by infrared thermometer at midnight every day at a typical location of outer shell with a 7.39m distance from kiln head. The temperature in circumference has a similar trend. Combined with historical records of site operation listed in Table I, it can be concluded that the temperature of kiln outside surface drops with the ring forms inside the kiln, and it rises up while ring is removed by cooling down, because of different expansion shrinkage coefficients between ring and refractory. The peak value of the temperature corresponds to a clean inside of kiln, with a good heat conductivity of the kiln shell. A new thermal resistance forms while the ring takes place, leading to a drop of the temperature. There is direct relationship between the kiln outside surface temperature and thickness of ring formation inside the kiln. Therefore, it is possible to monitor the thickness of ring formation by determining the outer shell temperature.

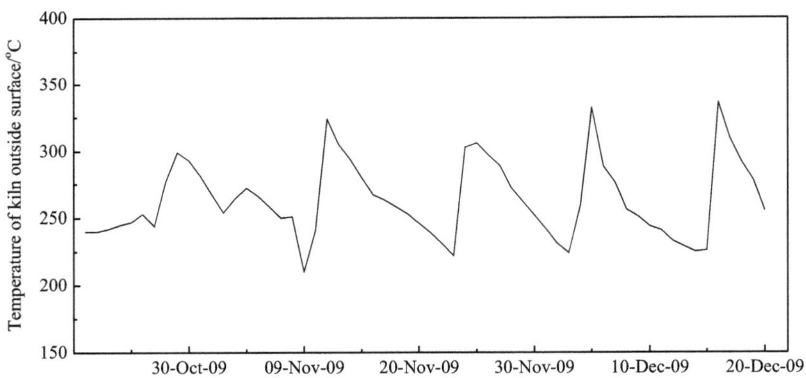

Figure 1. The change trend of temperature of kiln outside surface.

Table I. Historical records of site operation.

Date and time	Operation
01-Nov-09, 8PM	Removing ring for 1h
04-Nov-09, 8PM	Removing ring by cooling
09-Nov-09~10-Nov-09	Idling period
23-Nov-09, 9PM	Removing ring
01-Dec-09, 8PM~0AM	Removing ring by cooling
03-Dec-09, 11PM	Removing ring by cooling

Theoretical calculation of the ring thickness in the kiln

The heat transfer mechanisms of a rotary kiln during firing iron ore oxide pellets in cross-section identifies a group of heat transfer phenomena. The phenomena consists of heat transfer between the gas and the exposed wall, between the gas and the exposed surface of the bed, between the exposed wall and the exposed bed, and between outer shell and environment, including radiative, convective, and conductive transfer.

Considering a radial small zone of length dz at any position along the kiln axis, the following assumptions are made for simplification in this work:

1) The process is at steady state;
2) Heat conduction is negligible in the axial direction of kiln;
3) The temperature outside the rotary kiln is constant during pellet production;

Perfect contact between kiln shell layers is exerted.

Fourier's law describes the rate of heat flux Q (W/m) from radius r of cylinder as:

$$Q = -2\pi \cdot r \cdot \lambda \frac{dT}{dr} \tag{1}$$

Where λ is thermal conductivity, and $\lambda=\lambda_0+\beta \cdot T$, λ_0 is the thermal conductivity when object temperature equals to zero, β is temperature coefficient determined by experiment; T is the temperature at radius of r. Figure 2 shows the temperature boundary conditions, where subscripts *ew*, *rt*, *st* and *sh* represent for inner exposed wall, inner refractory, steel plate and outer shell, respectively.

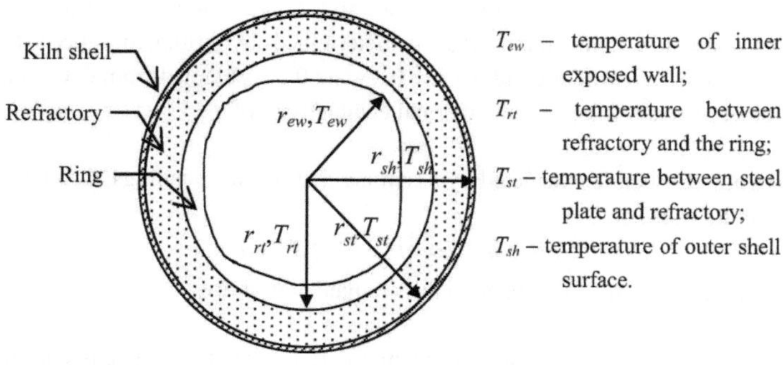

Figure 2. Distribution of rotary kiln shell's temperature in cross-section.

During pellet production, there is no energy accumulation in kiln shell at steady state. Therefore, the rate of heat transfer Q equals the heat lost from outer shell to ambient environment Q_{sh}, that is

$$Q=Q_{sh} \tag{2}$$

By solving equation (1), temperature T_{st} and T_{rt} can be modeled as follows:

$$T_{st} = T_{sh} + \frac{Q_{sh}}{2\pi\lambda_{st}} \ln\frac{r_{sh}}{r_{st}} \tag{3}$$

$$T_{rt} = -\frac{\lambda_{rt0}}{\beta_{rt}} + \sqrt{\left(T_{st} + \frac{\lambda_{rt0}}{\beta_{rt}}\right)^2 + \frac{Q_{sh}}{\pi\beta_{rt}} \ln\frac{r_{st}}{r_{rt}}} \tag{4}$$

Where λ_{st} is thermal conductivity of steel plate, λ_{rt0} is the thermal conductivity of refractory when its temperature is zero, β_{rt} is temperature coefficient of refractory. T_{ew} equals to T_{rt} if no ring formation occurs in the kiln.

The heat loss of the outer shell to ambient environment Q_{sh} includes two heat transfer processes: the convective resistance between the outer wall and the environment (h_{sh}^C) and the radiant resistance (h_{sh}^R). So the rate of heat loss is presented by:

$$Q_{sh} = \left(h_{sh}^C + h_{sh}^R\right) \cdot A_{sh} \cdot \left(T_{sh} - T_a\right) \tag{5}$$

Where A_{sh} is heat radiating area of kiln outer shell per unit length; T_a and T_{sh} are the temperature of air and shell surface, respectively; the heat convective coefficient h_{sh}^C between the outer shell and environment can be obtained from the correlation recommended by Nusselt [10] for natural convection from a rotating cylinder

$$Nu_m = C \cdot \left(Gr \cdot \Pr\right)_m^n = \frac{h_{sh}^C \cdot l}{\lambda_a} \tag{6}$$

Parameters Nu, Gr and Pr are Nusselt, Grashof and Prandtl numbers, respectively; subscript m means the qualitative temperature that is determined by arithmetic mean temperature as $T_m=(T_{sh}+T_a)/2$; coefficient C and index n are 0.48 and 0.25, obtained from document [11]; l is the outer diameter of rotary kiln; λ_a is the thermal conductivity of air.

Radiation from the shell is calculated by assuming that the surroundings absorbed all the energy emitted from it. Thus,

$$h_{sh}^R = \sigma\varepsilon\left(T_{sh}^4 - T_a^4\right)/\left(T_{sh} - T_a\right) \tag{7}$$

Where σ is Stefan-Boltzmann number and ε is radiant emissivity.

The outer shell temperature of kiln will decline if ring formation happens due to poor thermal conductive performance of the ring. As a result, under the condition of steady state, the thickness of ring is solved as an inverse heat conduction problem based on temperature variations:

$$Th_{rf} = r_{rt} - r_{ew} = r_{rt} - r_{rt} \cdot \exp\left\{\left[\left(T_{ew} + \frac{\lambda_{rf0}}{\beta_{rf}}\right)^2 - \left(T_{rt} + \frac{\lambda_{rf0}}{\beta_{rf}}\right)^2\right] \cdot \frac{\pi\beta_{rf}}{Q}\right\} \tag{8}$$

Where λ_{rf0} is the thermal conductivity of the ring when temperature equals to zero, and β_{rf} is the temperature coefficient of the ring.

Determination of boundary conditions: T_{sh} and T_{ew}

As shown above, the thickness of ring in the kiln is modeled based on the outer shell temperature T_{sh}, the inner exposed temperature T_{ew} and the thermal conductivity of shell materials.

Outer shell temperature T_{sh} can be directly measured by infrared thermometer to predict the conditions in kiln, i.e. whether ring formation happens or not, which has been applied in cement rotary kilns for temperature measuring instrument [12]. In Figure 3 different color refers to different temperatures, and three bars with dark color refer to the locations of supporting and driving wheels of kiln. X axis represents the length of kiln from head to tail, with measurement range of 40m, Y axis stands for temperature from 100°C to 650°C. Figure 3 was captured from site operating computer connected with infrared thermometer.

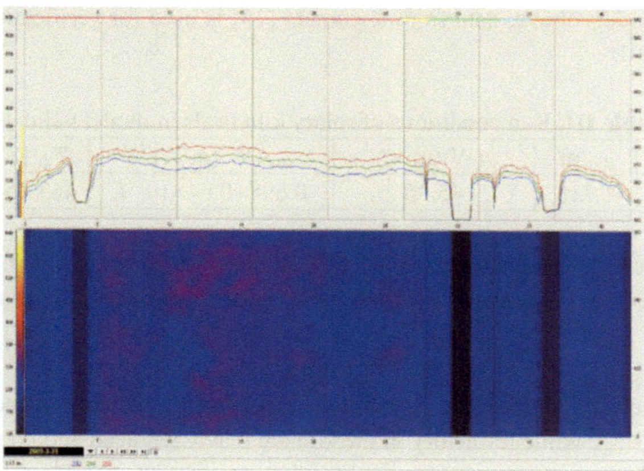

Figure 3. Temperature of outer shell determined by infrared thermometer along the length of rotary kiln.

During pellet production, the highest inner temperature will reach over 1300°C, and no techniques can be used to directly measure the temperautre distribution on the whole surface of inner wall in the rotary kiln. To keep the steady state of production, manual operation is always involved to guarantee good quality of fired pellets. It's assumed that the temperature of the inner exposed wall T_{ew} keeps constant whether the ring forms or not. That is, T_{ew} can be obtained based on T_{sh} during initial production stage while refractory in the kiln are newly replaced and no ring formation takes place yet.

Dimension of kiln and physical parameters

Rotary kiln used in WISCO E'zhou pellet plant is designed by Metso Minerals, with 6.858m of inner diameter without refractory, and 45.72m of length. The detailed parameters of the kiln are tabulated in Table II, where the temperature is stated in Kelvin.

Table II. Parameters of rotary kiln of E'zhou pellet plant.

r_{sh}/m	r_{st}/m	r_{rt}/m	λ_{st}/(W·(m·K)$^{-1}$)	λ_{rt}/(W·(m·K)$^{-1}$)	λ_{rf}/(W·(m·K)$^{-1}$)	T_a/K
3.469~3.559	3.429	3.179	45.36	1.75+0.00045T	3.431-1.901×10^{-3}×T	293.15

The radius of outer shell of kiln r_{sh} changes along with the axial length because of varying thicknesses of the steel plate. The thermal conductivity of the ring is measured by experiment based on GB11108-89 [13] and ISO11357-4-2005 [14] ranged from 100°C to 1310°C. It can be seen that the ring possesses weak thermal conductivity of 3.431-1.901×10^{-3}T W·(m·K)$^{-1}$, which will cause a decrease in outer shell temperature of rotary kiln.

To validate the models above, data from document [15] is used for verification in present work, as listed in Table III:

Table III. Run conditions of rotary kiln trials in model validation.

r_{sh}/m	r_{st}/m	r_{rt}/m	λ_{st}/(W·(m·K)$^{-1}$)	λ_{rt}/(W·(m·K)$^{-1}$)	T_{sh}/K	T_a/K	T_{ew}/K
1.500	1.478	1.318	45.36	0.698+0.64×10^{-3}T	463.15	293.15	1243.15

By solving equation (4) using data listed in Table III, T_{ew} is achieved as 1237.75K, while the real value is 1243.15K as presented above, which shows good reliability of the above model.

Results

Figure 4 shows the axial temperature distributions of rotary kiln inner wall and outer shell during initial production stage while refractory in the kiln are newly replaced and no ring is formed.

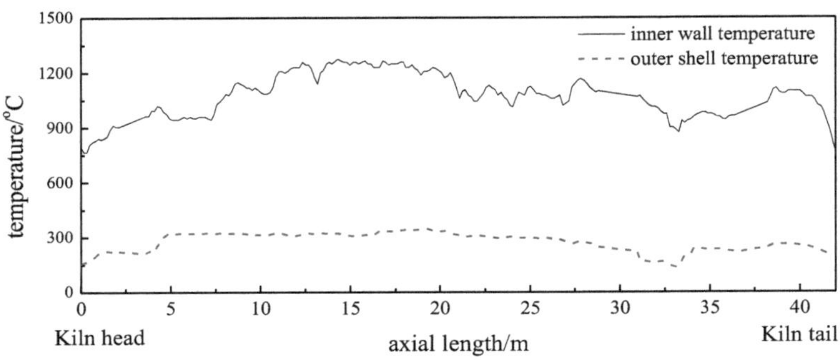

Figure 4. The surface temperature distribution of rotary kiln.

The pellet kiln is heated by a burner with a length around of 3m into the kiln from head. Some 1100°C and 950°C of gas temperature at kiln head and tail are achieved, and preheated pellets are roasted from 750°C when entering to 1300°C in firing zone, and then cooled to 1200°C while unloading from kiln. The inner temperature of the kiln increases and reaches the peak value of about 1300°C at axial length of 18m from kiln head, and then drops. In contrast, the outer shell temperature keeps steady when there is no ring forms inside the rotary kiln. The results of heat transfer coefficients of outer shell are worked out from Eqs.(6) and (7) and given in Table IV.

Table IV. Heat transfer coefficients obtained from Eqs.(6) and (7).

h_{sh}	h_{sh}^{C}	h_{sh}^{R}
Simulation results/W·(m^2·K)$^{-1}$	6.70~8.52	6.77~15.34

The kiln shell can't be directly measured during pellet production, and the whole inside conditions is unavailable to be visualized, which makes the kiln a black box. Therefore, a visualization of the ring inside the kiln was developed based the above model, and programed by MATLAB, as seen in Figure 5. X axis is the length of rotary kiln from its head to the tail, Y axis means circumferential position of kiln, and Z axis presents the thickness of the ring formed. Combined with Figure 4, it can be concluded that the ring forms more easily around the area with higher temperature in general. By simulating and displaying of the models above, it can provide a visual reference for the site operation in real time.

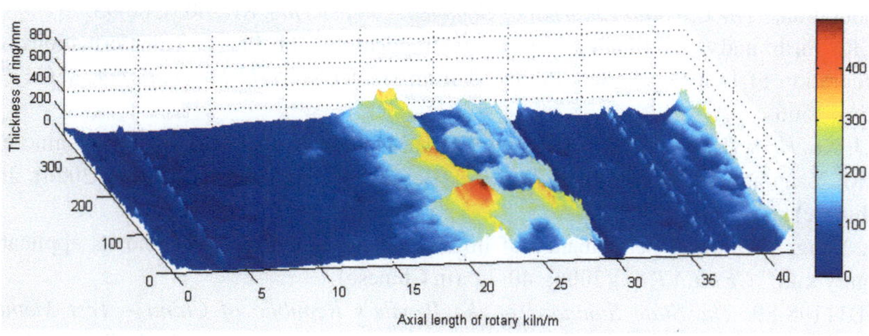

Figure 5. 3D show of thickness of ring formation in rotary kiln.

The method of estimating the thickness of the ring formed has been applied in WISCO Minerals, China. A real time monitor was developed by MATLAB based on T_{sh} which was determined every 15min.

Conclusions

1) Ring formation caused some decrease of rotary kiln outer shell temperature, so that a method for estimating the thickness of ring formation in oxidized pellet kiln was developed based a steady-state heat transfer across cylinder with the surface temperature of ring being constant.

2) The model worked well using T_{sh} measured by infra-red thermometer every 15min, and a 3D show of thickness of ring formation in rotary kiln was developed by MATLAB.
3) The method was applied in WISCO, offering a real time monitor of ring formation, and contributing to stabilizing pellet production consequently.

References

1. S.Q. Li et al., "A Mathematical Model of Heat Transfer in a Rotary Kiln Thermo-Reactor," *Chemical Engineering Technology*, 28(12) (2005), 1480-1489.
2. W.Q. Liu, and X. Hao, "Development Status and Trend of Sintering and Pelletizing in China," *China Steel*, 10(2009), 25-27. (in Chinese)
3. J.Y. Fu and D.Q. Zhu, *Theory, Techniques and Equipment of Iron Concentrate Oxidized Pellet* (Changsha, Central South University Press, 2005), 314. (in Chinese)
4. R.W. Young and M.Cross, "Mathematical Model of Grate-Kiln-Cooler Process Used for Induration of Ire Ore Pellets," *Ironmaking and Steelmaking*, 1(1979), 1-13.
5. J.A.Thurlby, "A Dynamic Mathematical Model of the Complete Grate/Kiln Iron - Ore Pellet Induration Process," *Metallurgical Transactions B*, 19(B)(1988), 103-112.
6. J.A.Thurlby, "Gas Flow and Pressue Balancing in Modelling Grate/Kiln Induration," *Metallurgical Transactions B*, 19(B)(1988), 113-121.
7. R.A.Davis, "Mathematical Model of Magnetite Oxidation in a Rotary Kiln furnace," *The Canadian Journal of Chemical Engineering*, 12(74)(1996), 1004-1009.
8. R.A.Davis and D.J. Englund, "Model and Simulation of a Ported Kiln for Iron Oxide Pellet Induration," *The Canadian Journal of Chemical Engineering*, 81(2003), 86-93.
9. A.R. Firth and J.R. Manuel, "Thermal Implications of Phase Transformations during Induration of Iron Ore Pellets Produced from Hematite," *ISIJ International*, 45(11)(2005), 1561-1566.
10. Z.J. Xu, *Heat Transfer Theory* (Beijing: China Machine Press, 1980), 8-11. (in Chinese)
11. S.M. Yang and W.Q. Tao, *Heat Transfer* (Beijing: Higher Education Press, 2006), 269. (in Chinese)
12. B. Yang, "Technical performance of infrared scanning thermometer and its application in rotary kiln," *CEMENT*, (7)(2006), 40-43. (in Chinese)
13. GB11108-89, *The State Standard of the People's Republic of China – Test Method for Thermal Diffusivity of Hardmetal (Carbide)* (Beijing: China Standards Press, 1990). (in Chinese)
14. ISO 11357-4:2005(E). *Plastics - Differential Scanning Calorimetry (DSC) - Part 4: Determination of Specific Heat Capacity*. (Switzerland: ISO copyright office, 2005).
15. J.W. Wang and Z.S. Li, *Handbook of Cement Producing Process Calculations* (Beijing: China Building Materials Press, 2001), 317-318. (in Chinese)

Drying, Roasting, and Calcining of Minerals
Edited by: Thomas P. Battle, Jerome P. Downey, Lawrence D. May, Boyd Davis, Neale R. Neelameggham, Sergio Sanchez-Segado, and P. Chris Pistorius
TMS (The Minerals, Metals & Materials Society), 2015

STUDY ON THE IMPROVEMENT OF PREHEATING AND ROASTING CHARACTERISTICS OF PELLET MADE BY ORGANIC-BENTONITE COMPOUND BINDER

Yufeng Guo[1],Ting Duan[1],Andrew Yakovlevich Travyanov[2], Tao Jiang[1], Shuai Wang[1] , Fuqiang Zheng[1]

[1]School of Minerals Processing and Bioengineering;Central South University; Changsha, 410083, China
[2]National University of Science and Technology "MISiS";Leninsky Prospect 4; Moscow,119049,Russia

Keywords: Iron ore pellet, Organic compound binder, Additives, Compressive strength

Abstract

Organic compound binder can effectively enhance the strength of green pellet and reduce the addition amount of binder during pelletizing. However, compared with traditional bentonite binder pellet, the compressive strength of the preheated and roasted pellets significantly declined. It is of significance to resolve this problem for the application of organic compound binder. In this paper, magnetite concentrate was used as raw material, carboxymethyl cellulose (CMC)-bentonite compound was used as binder, characteristics of preheated and roasted pellets and the effects of additives on that were systematically studied, The results show that sodium acetate can effectively improve the characteristics of the finished compound binder pellets. The compressive strength of the preheated and roasted pellets were increased by $52/N{\cdot}p^{-1}$ and $342/N{\cdot}p^{-1}$ respectively when the addition amount of sodium acetate is 2.2%wt of compound binder. Moreover, sodium acetate has no adverse effect on pelletizing process and drying characteristics of green balls.

Introduction

With the gradual depletion of iron ore resources, low-grade iron ore resources must be upgraded by the means of mineral processing. Then, the iron concentrate must be pelletized to be used in industrial production [1,2]. During pelletizing, binders must be added to improve the strength of pellets, of which bentonite was firstly introduced and widely used in the world. The pellets which used bentonite（calcium bentonite or sodium bentonite）as binder have higher compression strength and explosion temperature－an important technological parameter of green pellets. however, due to the high contents of Al and Si, this kind of pellets usually contain more impurities, resulting in lower iron grade and poorer reducibility[3,4]. This problem is particularly prominent in China, there more than 90% of bentonite is calcium bentonite [5], which has poorer physicochemical property and application effect than sodium bentonite. Thus the addition amount in pellets is much higher[6,7]. For these reasons, several substitutes have been developed to replace bentonite and a few have been commercially applied including organic binders. The green pellets made by organic binder usually have great strength, however, it is difficult to completely mix with iron ore concentrate. Furthermore, the balling dynamics, drying rate of green pellets and compressive strength of preheated and roasted pellets are unsatisfactory. Therefore, for a long period, bentonite will still remain as the dominant binder for pelletizing process. Recently, binders composed of mixtures of bentonite and other organic compounds were investigated to minimize the dosage of bentonite. However, the results showed that the compressive strength of preheated and roasted pellets made by composite binders were still too low causing by adding of organic binder.

Many studies have been carried out to improve the compressive strength of the pellets made with organic binders as follows [8,9]:

1) Wet-grinding: This approach can increase the content of iron ore fines, enhance the surface activity of iron ore concentrate and reduce the porosity of pellets, thus improve the particle migration and crystallization during preheating and roasting process.

2) Additives: Additives, such as limestone, contain substance that can promote the generation of low-melting compounds (LMC). LMC plays an important role for liquid consolidation; Moreover, it can accelerate the particles migration and promote crystal growth.

3) Microwave drying: This method selectively evaporates moisture and keeps the binder from decomposing at low temperatures, thus improves the efficiency of preheating, which is conducive to the magnetite oxidation and grain growth.

A great deal of effort has been done to eliminate or weaken the adverse effects causing by organic binders. However, most of these efforts have not been commercially applied because they are heavy consumers of additives and energy, have high equipment requirements. For instance, wet-grinding will increase energy consumption and the generated iron ore fines may worsen drying and preheating process. In addition, additives may change the alkalinity of pellets and reduce the strength of preheated pellets; while microwave drying increases costs and makes uneven heating.

In this paper, a magnetite concentrate was used as raw material, carboxymethyl cellulose (CMC)-bentonite compound was used as binder. The primary purpose of this work was to find performance-superior additives, which can improve the property of preheated and roasted pellets, and promote the application of organic compound binder in pelletizing.

Raw Materials and Methods

Raw Materials

A magnetite concentrate was used as raw material; Bentonite and carboxymethyl cellulose (CMC)-bentonite compound were used as binders; Sodium acetate, ferric sulfate, and sodium oxalate were used as additives.

The chemical compositions of magnetite concentrate are given in Table I, the grain size distribution and specific surface area are shown in Table II, and the physical properties of bentonite are shown in Table III[10].

Table I. Chemical Compositions of Raw Materials (mass percent, %)

TFe	FeO	SiO_2	Al_2O_3	CaO	MgO	K_2O	Na_2O	P	S	Mn	TiO_2
63.49	27.53	10.96	0.15	0.69	0.24	0.07	0.11	0.02	0.06	0.06	0.03

Table II. Size Distribution and Specific Surface Area of Iron-bearing Material

Grain Size Distribution /%				Specific Surface Area/$cm^2 \cdot g^{-1}$
+0.147mm	0.147～0.075mm	0.075～0.045mm	-0.045mm	
2.98	9.97	28.01	56.02	1305.77

Table III. Physical Properties of Bentonite

Colloid Percentage/ml·(15g) $^{-1}$	Expanding Volume/ ml.g^{-1}	Water Absorbency/%	Methylene Blue Uptake /g·$(100g)^{-1}$	Montmorillonite Content /%
576	30.0	415.7	35.75	80.88

From Table I and Table II, it can be seen that the magnetite concentrate is acidic gangue based with low contents of sulfur and phosphorus. The -0.045mm content and specific surface areas of magnetite concentrate are lower than the required specification for pelletizing process.

Methods

In this experiment, Measures of the quality of pellets are the compressive strength, drop number and drying rate of green pellets , as well as the compressive strength of preheated and roasted pellets. Pelletizing was conducted in a rolling disc. During pelletizing, the moisture of mixture was controlled at 8.0% and the pelletizing time was controlled at 12min. Parts of the qualified green pellets were submitted for drop number, compressive strength and drying speed test. Drop number was measured by vertically dropping down the samples from 0.5m height to a steel plate (thickness 10mm); compressive strength of green balls was tested successively on a KQ-2 compressive strength tester; drying speed was carried out in a vertical tube furnace at 450℃with hot gas stream flow through the furnace at a rate of 1.8m/s. The rest of qualified green pellets were dried at 105±5 ℃ for 24h. Preheating and roasting experiments were carried out in a horizontal tube furnace with an automatic temperature control system (diameter 50mm). The corundum porcelain boat filled with pellet samples would be placed into the furnace to preheat and roast for a period of time and then removed, cooled to the room temperature in the air. The compressive strength of the sample obtained would be tested on a ZQYC-compression gauge.

Results and Discussion

In each pellet batch, 20 randomly selected pellets were tested to determine pellet quality(drop number and compressive strength), indicated by the mean value and calculated 95% confidence level. The standard deviation is relatively small for the following reasons: 1) Screen and use the 13-15mm pellets to eliminate the influence of pellet particle-size ; 2) Only 4 pellets were preheated or roasted once to ensure they are basically at the same temperature field. Results of the pelletizing experiment tests (drop numbers ,compressive strength and drying rate of green pellets, compressive strength of preheated and roasted pellets) are given below.

The Characteristics of Green, Preheated and Roasted Pellets Made by Bentonite

Effects of the dosage of bentonite on the characteristics of green pellets were studied. The results are shown in Figure1.

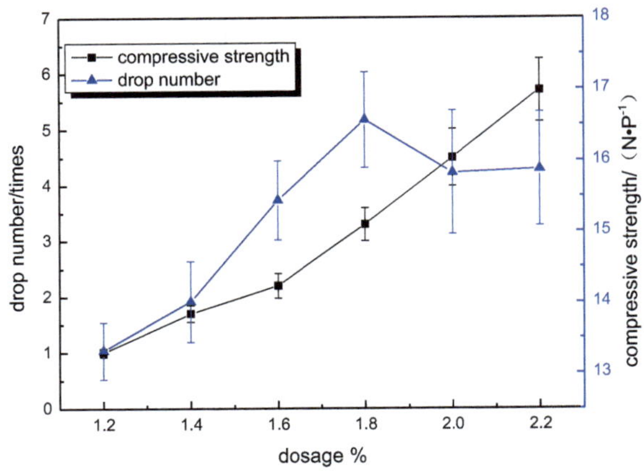

Fig.1. Effects of bentonite on the characteristics of green pellets

As can be seen in Fig.1, with the dosage of bentonite binder increasing from1.2% to 1.8%, the drop number and compressive strength of the green pellets effectively increased, but the compressive strength descends when the dosage was over 1.8%. When the dosage of the binder was 2.0%, the drop number and compressive strength of the green pellets reached 4.5/times, 15.83/N·p^{-1} respectively, which meet the requirements of production. Table IV shows the characteristics of the green, preheated and roasted pellets with 2.0% bentonite.

Table IV. The Characteristics of 2.0%Bentonite Pellet

Drying Rate / Kg·(m^2·s)$^{-1}$	Compressive Strength /N·P^{-1}	
	Preheated Pellet	Roasted Pellet
0.0109	700±40.6	3861±162.16

The results in Table IV show that the preheated and roasted pellets are qualified at the predetermined preheating and roasting conditions.

The Characteristics of Green, Preheated and Roasted Pellets Made by Compound Binder

CMC was added to bentonite to form a CMC-bentonite compound binder (bentonite dosage 2.0%). The dosage of CMC ranged from 0 to 1.5%wt of bentonite. The effects of CMC on the characteristics of green pellets were studied. The results are shown in Fig.2.

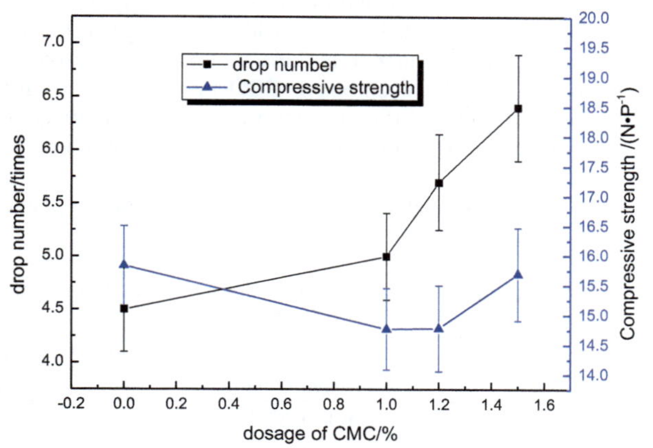

Fig.2. Effects of the dosage of CMC on the characteristics of green pellets

From Fig.2, it can be seen that with the increasing dosage of CMC, the drop number of the green pellets effectively increase, and it is higher than that of 2.0% bentonite addition, the compressive strength descends a little, but it still meet the requirements of production. When the dosage of CMC is 1.5%, the drop number increases by a factor of 1.9 compared with a 2.0% bentonite addition . When the CMC-bentonite compound binder was chosen as the binder, in which CMC is 1.5wt% of the bentonite addition ,The effect of the dosage of CMC-bentonite compound on the characteristics of pellets was studied, the results are shown in Fig.3

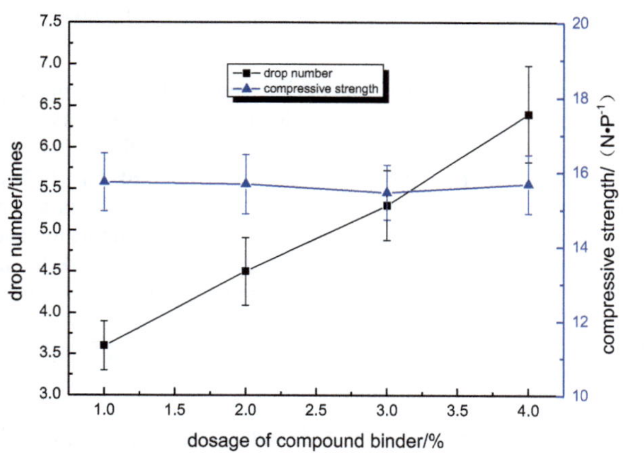

Fig.3. Effects of dosage of CMC-bentonite compound binder on the characteristics of green pellets

The results in Fig.3 show that the drop number and compressive strength of the green pellets made with1.5% CMC-bentonite compound binder were basically at the same level with 2.0% bentonite pellets. The results indicated that the addition of organic binder in bentonite could decrease the addition amount of binder from 2.0% to 1.5%. The reason is that CMC contains carboxyl(COO-) and hydroxy (-OH). Carboxyl(COO-) adsorbs on the surface of the iron ore concentrate and then reacts with Fe^{3+}(Fe^{2+}) to form Complexes. Hydroxy (-OH) is hydrophilic. Thus organic binder makes the iron ore concentrate hydrophilic resulting in increased capillary force between the particles.
Table V shows the drying rate and compressive strength of pellets made with 1.5% CMC-bentonite compound binder.

Table V. Characteristics of the Drying Rate and Compressive Strength of Pellets Made by Compound Binder

Drying Rate / $Kg{\cdot}(m^2{\cdot}s)^{-1}$	Compressive Strength /$N{\cdot}P^{-1}$	
	Preheated Pellet	Roasted Pellet
0.098	604±30.41	3375±114.75

Comparison of Pellets Made by CMC-bentonite Compound Binder and Bentonite Binder

The characteristics of pellets made with 1.5% CMC-bentonite compound binder and 2.0% bentonite are shown in Fig.6.

Table VI. Characteristics of Pellets Made by CMC-bentonite Compound Binder and Bentonite

Kinds of Binders	Drop Number /times	Drying Rate /$Kg{\cdot}(m^2{\cdot}s)^{-1}$	Compressive Strength/ $N{\cdot}P^{-1}$		
			Green Pellet	Preheated Pellet	Roasted Pellet
bentonite	4.5±0.51	0.110	15.83±0.87	700±40.60	3861±162.16
Compound binder	4.5±0.41	0.098	15.27±0.79	604±30.41	3375±114.75

The results in Table VI show that the drying speed of pellets made by CMC-bentonite compound binder decreases slightly. It may be due to the high viscosity of the binder, which is harmful to the migration of moisture. The compressive strength of the preheated and roasted pellets made by CMC-bentonite compound binder decreased by 96/$N{\cdot}p^{-1}$ and 486/$N{\cdot}p^{-1}$, respectively. This phenomenon may be due to the increasing porosity in green pellets causing by the high viscosity of organic binder, or the holes left for the decomposition and disappearance of the organic binder when the pellets were preheated, which would be harmful to particle migration and crystal growth, resulting in the low strength of preheated and roasted pellets.

Effect of Additives on The Characteristics of Green ,Preheated and Roasted Pellets Made by Compound Binder

Since the compressive strength of preheated and roasted pellets made by CMC-bentonite compound binder are obviously lower than that of bentonite under the same level of green pellets, additives were used to improve the property of pellets. Fixed the dosage of additives to 1.5wt% of the binder, effects of additives on the characteristics of compound binder pellets were studied. The results are shown in Table VII.

Table VII. Effects of Additives on the Characteristics of Green, Preheated and Roasted Pellets Made by Compound Binder

Kinds of Additives	Drop Number/ times	Drying Rate/ $Kg\cdot(m^2\cdot s)^{-1}$	Compressive Strength/ $N\cdot P^{-1}$		
			Green Pellet	Preheated Pellet	Roasted Pellet
No	4.5±0.43	0.098	15.27±0.79	604±30.41	3375±114.75
Diiron Trisulfate	4.6±0.40	0.087	14.73±0.71	620±24.83	3369±120.89
Sodium acetate	4.5±0.40	0.094	15.00±0.78	641±31.15	3578±119.67
Sodium oxalate	4.5±0.43	0.096	14.82±0.89	627±29.80	3417±123.03

The results in Table VII indicates that the effect of additives on the drop number of the pellets was not obvious; while diiron trisulfate and sodium oxalate decrease the compressive strength of pellets slightly. Also, diiron trisulfate reduced drying speed reduced from 0.098 $Kg\cdot(m^2\cdot s)^{-1}$ to 0.087 $Kg\cdot(m^2\cdot s)^{-1}$. Most of the additives were conductive to the strength of preheated and roasted pellets, especially sodium acetate which increases by 37/ $N\cdot P^{-1}$ and 203/ $N\cdot P^{-1}$, respectively.

The obtained results indicated sodium acetate showed the most obvious improvements. When the dosage of sodium acetate was 2.2%, under the optimal condition, the characteristics of the obtained compound binder pellets is shown in Table VIII.

Table VIII. Effect of Sodium Acetate on the Characteristics of Compound Binder Pellets

Additive Amount/%	Drop Number/ times	Drying Rate/ $Kg\cdot(m^2\cdot s)^{-1}$	Compressive Strength/ $N\cdot P^{-1}$		
			Green Pellet	Preheated Pellet	Roasted Pellet
0	4.5±0.43	0.098	15.27±0.79	604±30.42	3375±114.75
2.2	4.6±0.49	0.092	15.69±0.80	656±30.67	3717±110.71

The results in Table VIII shows that when the dosage of Sodium acetate is 2.2%, the compressive strength of preheated and roasted pellets are 656 /$N\cdot P^{-1}$ and 3717 /$N\cdot P^{-1}$, increasing by 52 /$N\cdot P^{-1}$ and 342/ $N\cdot P^{-1}$, respectively. Furthermore, sodium acetate has no adverse effect on the pelletizing as well as the compressive strength and drying characteristics of green pellets. These results can be explained as follows. Firstly, Sodium acetate in the pellets decomposes and produces active sodium ions when heated, and the active sodium ions can promote the particle migration and the link between chemical bonds. Secondly, sodium acetate can adjust the system's pH value to decrease the viscosity of CMC, resulting in most closely arrangement between iron ore concentrate, which can be good to particle migration and crystallization growing.

Conclusions

1) During the iron ore pelletizing process, the dosage of bentonite binder can be effectively reduced from 2.0% to 1.5% by adding CMC, thus reduces the impurities content brought into the pellets by bentonite. However, the strength of preheated and roasted pellets made by CMC-

bentonite compound binder are weaker than that of bentonite binder under the same level of green pellets, which decrease by 96 /N·P^{-1} and 86 /N·P^{-1}, respectively.
2) Sodium acetate can improve the compressive strength of the pellets made by CMC-bentonite compound binder. When the additive amount is 2.2%, the compressive strength of preheated and roasted pellets are 656 /N·P^{-1} and 3717/N·P^{-1}, increasing by 52/N·P^{-1} and 342/N·P^{-1}, respectively. Furthermore, sodium acetate has no adverse effect on the pelletizing as well as the compressive strength and drying characteristics of green pellets.

Acknowledgements

This investigation was supported by the Program for New Century Excellent Talents in University (NCET-10-0834), China.

References

1. Y.H. Luo, Basic research on oxide pellets of magnetite concentrate:[Master paper] . Changsha : Central South University , 2000,33-39

2. M.S. Cassola and A.P. Chaves, "Study of Organic and Inorganic Binders on Strength of Iron Oxide Pellets," *Powder and Particle Journal*,(1998),136-141

3. Z.T. Huang, "Research and Development of Additives of Pellets," *Sintering and Pelletizing*, 22(3) (1997),1-7

4. X.B. Zhang and M.W. Zhu, "The Impact of Bentonite Binder on The Pellets Production," *Sintering and Pelletizing*,28(6)(2003),3-7

5. L. Hou, F.C. Yi, and J.R. Zhou, "Research on Sodium Modification of Calcium bentonite," *China mining magazine*,15(12)(2006),77-79

6. J.Y. Zhou et al., "Study on the Influence Factors of Sodium Modification Procession of Calcium Bentonite," *RESOURCES SURVEY & ENVIRONMENT*, 27(3)(2006), 195-199

7. P.Hu et al., "Experimental Study on pelletizing with addition of sodium bentonite in Jinan Iron and Steel Group," *Multipurpose Utilization of Mineral Resources*,(1)(2010),15-18

8. Z.J. Wang, " The State of Application and Research of Pellet Binders," *China Chemical Trade*,10(10)(2012),148

9. X.H. Fan, et al., "Measures to Improve the Strength of Oxidized Pellets Made By Organic Binder," *Journal of Iron and Steel Research*, 20(5)(2008),6-19

10. GB/T 20973—2007, Bentonite

Drying, Roasting, and Calcining of Minerals
Edited by: Thomas P. Battle, Jerome P. Downey, Lawrence D. May, Boyd Davis, Neale R. Neelameggham, Sergio Sanchez-Segado, and P. Chris Pistorius
TMS (The Minerals, Metals & Materials Society), 2015

EFFECTS OF ANTHRACITE ON PELLETIZATION OF HEMATITE ORE

Zhaokun Tang, Mingjun Rao, Yuanbo Zhang, Guanghui Li*

(School of Minerals Processing & Bioengineering, Central South University, Changsha, Hunan 410083, China)

Keywords: Hematite ore, Pellet, Anthracite, Roasting

Abstract

The ballability of hematite ore with addition of a certain amount of pulverized anthracite and the induration features of anthracite–containing hematite pellets were studied. The results indicate that adding a certain proportion of anthracite can improve the quality of green pellets as well as the roasting performance of roasted pellets. The suitable mass dosage of anthracite is 1.0%–1.5%. Under the conditions of roasting temperature 1280 °C–1350 °C and roasting time 15 min–20 min, the compressive strength of roasted pellets is found to be more than 2500 N/P. Compared with hematite pellets in the absence of anthracite, the proper induration temperature is diminished by 20 °C, whereas optimal induration temperature range expands by 50 °C, and induration time is shortened by 5 min–10 min. Adding anthracite is an effective measure to improve the pelletization of hematite.

Introduction

The pelletization of fine powder plays an important role in industry, particularly in the iron and steel industry [1]. The pellets of iron ore concentrates have been found to be an excellent feed for a blast furnace and markedly improve its efficiency [2]. Magnetite and hematite concentrates are two main raw materials for oxidized pellet production. Presently, magnetite concentrate has more predominance in China. However, with the rapid development of the iron and steel industry, magnetite resources are becoming scarce [3]. It is imperative to make good use of hematite resources to manufacture oxidized pellets.

In China, most hematite ores are imported, with a high iron grade, low gangue content and fine granularity. Compared with the magnetite, hematite concentrate has poor ballability, and the hematite pellet requires a higher roasting temperature and narrower firing temperature range (1300–1350 °C), which makes the operation of firing equipment difficult [4]. So, it is imperative to solve the problems of oxidized pellet preparation from hematite concentrates.

*Supported by the New Century Excellent Talents in University (NCET-11-0515)
Corresponding author: Guanghui Li, Professor, Ph.D; Tel: +86–731–88830542; E-mail: liguangh@csu.edu.cn

In the early stage of pelletizing industry, the practical results for adding a certain proportion of solid fuel to hematite concentrates in travelling grate route show that the oxidized pellet strength is obviously increased, and the energy consumption is reduced markedly, and the metallurgical properties are improved [5, 6]. At present the primary technology for the preparation of oxidized pellets is grate–kiln process in China. In the last 20 years, considerable research has been conducted on carbon–containing hematite pellets in China [7]. However, most of these researchers were mainly focused on the preparation of reduced pellets for non–blast–furnace ironmaking [8]. With the development of blast furnace ironmaking and as the increasing requirement of the fuel economy, adding solid fuel in pellets as an energy–saving measure has been greatly developed. In this paper, pellets were made from hematite concentrate with a certain percentage of pulverized anthracite. Effects of anthracite on properties of green hematite balls and the roasting features of hematite pellets were studied in order to better utilize hematite resources by grate–kiln process.

Experimental

Raw Materials

The hematite concentrate was taken from Brazil, characterized by high total iron grade (67.22% Fe_{total}) and less impurities (Table I), and the specific surface area measured by a permeability method (Blaine index) was 1630 cm^2/g. Pulverized anthracite was used as the carbon–containing material, of which the BET surface area was 6599 cm^2/g, and the proximate analysis tested according to GB/T 212–2008 is listed in Table II.

Table I. Chemical Composition of Materials /%

Materials	Fe_{total}	FeO	SiO_2	CaO	MgO	Al_2O_3	LOI*
Iron concentrate	67.22	0.55	2.17	0.01	0.05	0.55	0.59
Bentonite	7.07	/	60.61	0.94	2.2	17.98	10.41

*LOI–Loss on ignition

Table II. Proximate Analysis of the Anthracite /%

FC_d	V_d	A_d	Main chemical composition of ash				
			Fe_{total}	CaO	MgO	Al_2O_3	SiO_2
77.6	6.48	15.6	9.26	9.45	1.50	36.82	42.95

Note: FC_d= Fixed carbon, V_d= Volatile, A_d=Ash

Methods

Ball Preparation and Drying For each trial, 5 kg hematite concentrate was blended with pulverized anthracite and 1.25% bentonite. Green balls were prepared in a disc pelletizer with a diameter of 1000 mm, and the qualified green balls with 9 mm–15 mm in diameter were statically dried at 105 °C in a drying oven for 4 h.

Ball Measurement To measure the compressive strength of a ball, it is subjected to uniform loading between two parallel plates until the ball is ruptured. The average of compressive values at which the balls break down is measured in Newton unit as the compressive strength. The drop strength was measured by means of letting a green ball hit a steel plate repeatedly from a height of 0.5 m. The number of drops was able to withstand without damage was measured as the drop strength of the ball. A total of 10 balls were tested. The average value of the drop number was used as the drop strength.

Induration Tests and Strength Measurement Firing tests were carried out in an electrically heated shaft furnace. Roasting atmosphere was adjusted by mixed gas of N_2/O_2 at 20% oxygen partial pressure with a flowrate of 6 L/min. The dry balls were charged into a heat–resistant pot, and then the pot was pushed downward from the furnace mouth into the high temperature zone in the furnace every 1 min in five steps. The pellets were fired at the given temperature for a certain period. Afterwards, the fired pellets were taken out and naturally cooled to ambient temperature. The compressive strength of cooled pellets was measured with an LJ–1000 material experimental machine. An average value of 10 pellets was used as the compressive strength for each test.

Microscopic Analysis Microstructure of fired pellets at various roasting conditions was characterized by using Leica DMRXE microscope with an automatic image analyzer.

Results and Discussion

Effects of Anthracite on the Ballability of Hematite Pellets

Balling is the initial and a key step of pelletization for any powder material [9]. The ballability of powder material is directly related to the product quality. So, the investigation on ballability of hematite ore adding anthracite is of great importance.

Effects of Moisture Effects of moisture of raw material on compressive strength and drop strength of green balls are shown in Fig. 1. With increasing moisture content the drop strength of green balls increases gradually and the compressive strength first increases and then decreases after 8%. Therefore, the optimal moisture should be 7.5%–8.0%.

Effects of Balling Time Fig. 2 illustrates the effects of balling time on compressive strength and drop strength of green balls. As shown in Fig. 2, both drop strength and compressive strength of green balls increase gradually with the prolonging of time. The compressive strength almost reaches a maximum at the balling time of 12 min.

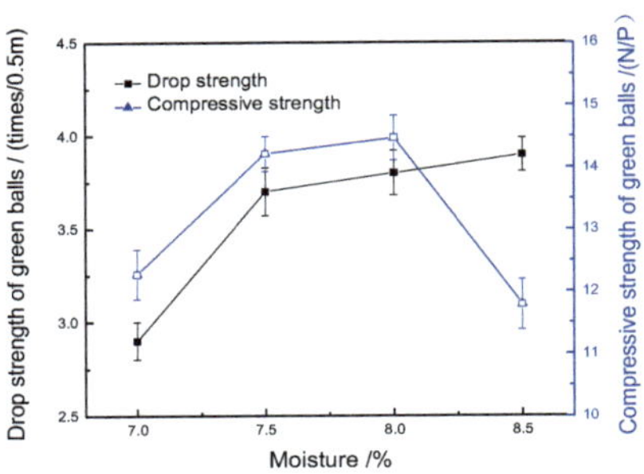

Fig. 1. Effects of moisture of raw material on the strength of green balls (anthracite: 1.0%, balling time:12min)

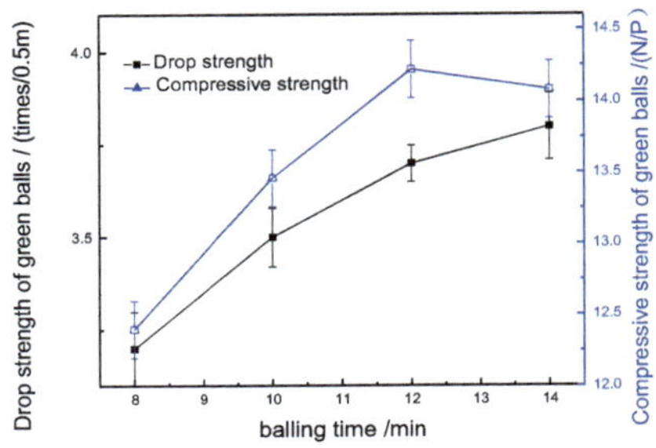

Fig. 2. Effects of balling time on the strength of green balls (anthracite: 1.0%, moisture: 7.5%)

Effects of Anthracite Dosage Effects of anthracite dosage on compressive strength and drop strength of green balls are shown in Fig. 3 and Fig. 4 respectively. Fig.3 shows that with the increase in anthracite dosage, the drop strength of green balls increased gradually, but the drop strength decreased adversely after the addition of anthracite reached 3%.

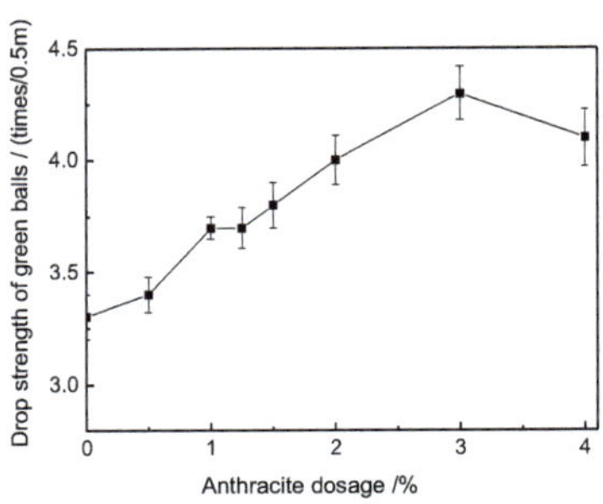

Fig. 3. Effects of anthracite dosage on drop strength of green balls (moisture:7.5%, balling time:12min)

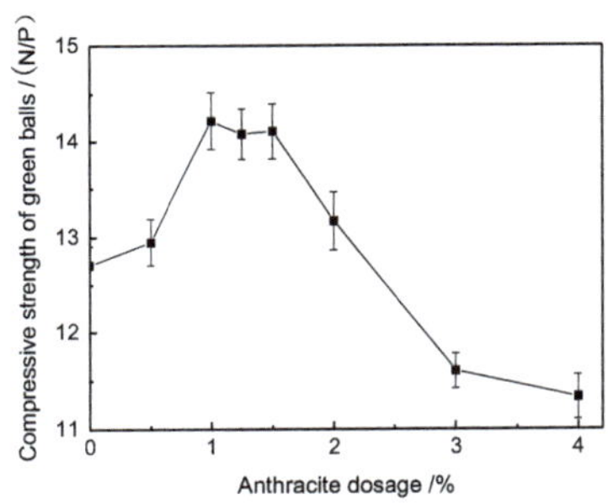

Fig. 4. Effects of anthracite dosage on compressive strength of green balls (moisture:7.5%, balling time:12min)

As shown in Fig. 4, when the anthracite dosage increased from 0 to 1%, the compressive strength increased from 12.71 N/P to 14.22 N/P. The compressive strength decreased greatly with further increase of anthracite dosage exceeding 1.5%. The ballability of materials is associated with particle size and hydrophilicity [10]. The smaller particle size and contact angle produce a better ballability. The particle of pulverized anthracite is fine, however it is hydrophobic. With the increase in anthracite dosage, the drop strength and compressive strength of green balls were both increased firstly and then decreased. It can be concluded that the proper addition amount of anthracite is about 1%–1.5 %.

The effects of anthracite dosage on the compressive strength of roasted pellets are shown in Fig. 5. The pellet strength with 0.5% anthracite was lower than that without anthracite. The compressive strength got a maximum with 1.25% anthracite, however, when the addition of anthracite exceeded 1.25%, the strength decreased greatly. The results indicate that the appropriate anthracite amount of 1.0–1.5% can improve the strength of the fired hematite pellets.

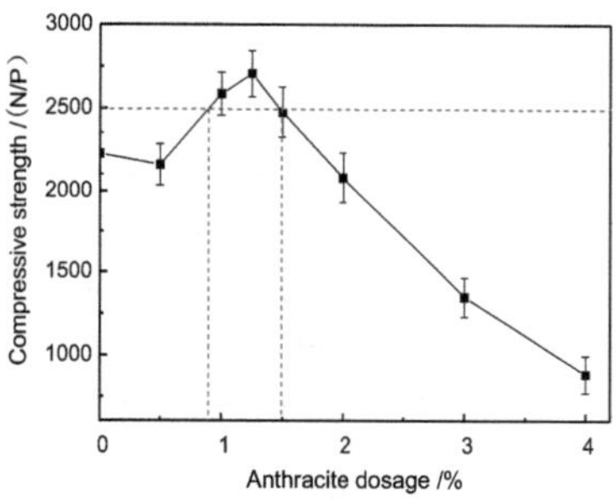

Fig. 5. Effects of anthracite dosage on the compressive strength of roasted pellet (temperature: 1280 °C, time: 20 min)

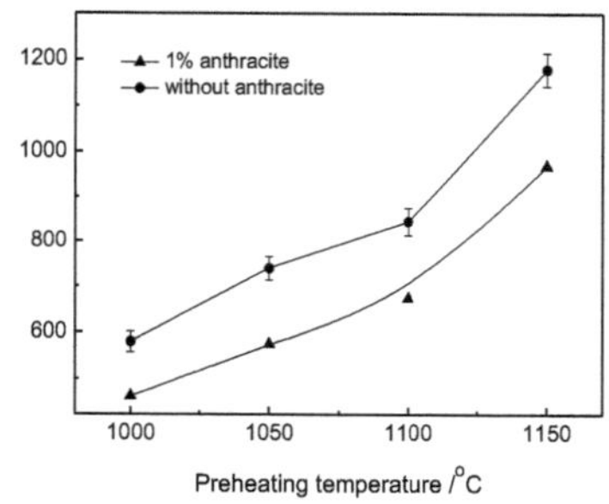

Fig. 6. Effects of preheating temperature on compressive strength of preheated pellets (preheating time: 20 min)

The compressive strength as functions of preheating and roasting temperature are shown in Fig. 6 and Fig. 7 respectively. Fig. 6 shows that the preheated pellet strength keeps increasing with the preheating temperature increasing from 1000 °C to 1150 °C. The addition of anthracite can improve the strength of preheated hematite pellets. The strength of the pellets with 1.0% anthracite reaches 579 N/P when the temperature is 1000 °C and it can satisfy the requirement of grate–kiln processes. As shown in Fig. 7, for the roasted pellets with 1.0% anthracite, the strength is always higher than that of the pellets without anthracite, and the compressive strength is more than 2500 N/P when the temperature reaches 1280 °C. However, regarding with pellets in the absence of anthracite, its compressive strength obviously decreases due to the decomposition of Fe_2O_3 at the temperature above 1320 °C. The results imply that the roasting temperature can be decreased and the suitable firing temperature range is enlarged by adding an appropriate amount of anthracite into hematite pellets.

Effect of the roasting time on the roasted pellet compressive strength is plotted in Fig. 8. It can be seen that the compressive strength of both pellets with 1.0% anthracite and pellets without anthracite increases gradually with the prolonging of roasting time, which almost reaches a maximum at the roasting time of 25 min. The strength of the pellets with 1.0% anthracite is always higher than that of the pellets without anthracite at the same roasting time. The results imply that the roasting time can be decreased by adding an appropriate amount of anthracite into hematite pellets.

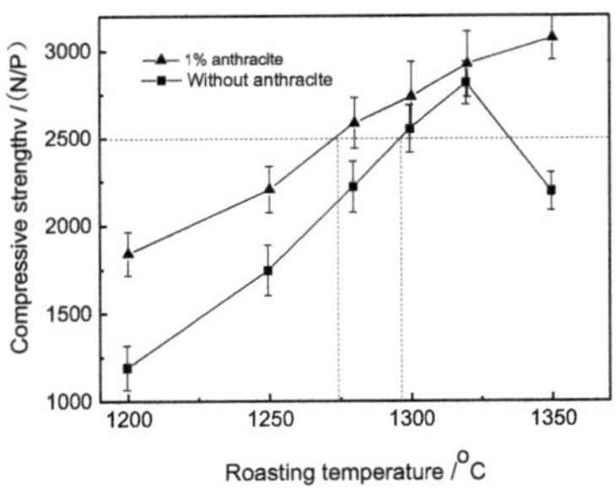

Fig. 7. Effects of roasting temperature on the compressive strength of pellets (roasting time: 20 min)

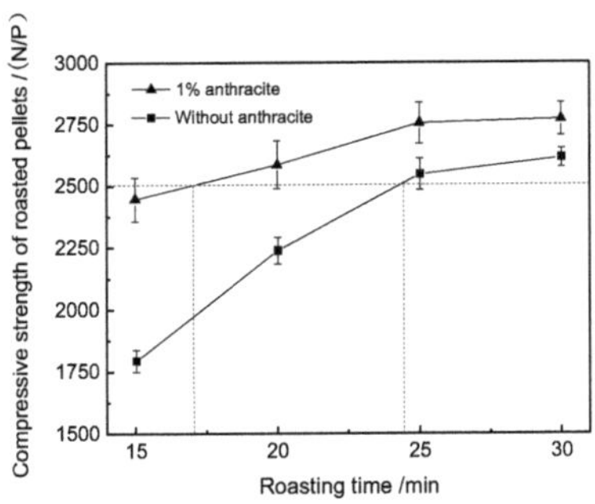

Fig. 8. Effect of roasting time on the compressive strength of roasted pellets (roasting temperature: 1280°C)

According to the above results, it can be concluded that the appropriate roasting conditions for the hematite pellet with 1.0% anthracite are 1280 °C–1350 °C roasting temperature, 15–20 min roasting time. The compressive strength of roasted pellet can reach 2500 N/P under these conditions. By adding 1% anthracite in hematite pellets, the proper roasting temperature is reduced by 20 °C, the roasting temperature range expands by 50 °C, and the roasting time is shortened by 5 min–10 min.

Effects of Anthracite on the Induration Procedure of Hematite Pellets

Microstructures of pellets with various dosage of anthracite are shown in Fig. 9. For the pellets roasted in the absence of anthracite, because of the high recrystallization temperature (>1300 °C) of original hematite (OH) grains, most of the original hematite grains kept their original shapes and discernible angularities (grain 1 in Fig. 9a). For the pellets with 0.5% anthracite, limited junctions between grains were formed (grain 2 in Fig. 9b). Due to the low anthracite dosage, a few secondary hematite (SH) grains were produced and the pores left by the combustion of anthracite particles decreased the compressive strength of roasted pellets. For the pellets with 1% anthracite, a large number of crystal bond junctions between grains were formed by recrystallization of SH and a compact microstructure was formed, accounting for the high strength.

There were lots of Fe_3O_4 existing in the roasted pellets with 2.0% anthracite (grain 3 in Fig. 9d). Because of the strong reducing atmosphere resulting from the excessive dosage of anthracite, lots of Fe_3O_4 grains remained within pellets until the end of roasting. Junctions between particles are mainly from the Fe_3O_4 recrystallization. Therefore, except that a recognized role in supplying a part of heat needed in roasting process via the combustion of carbon, anthracite plays a key role in the partial reduction of hematite grains in pellet, which changes the crystallization route of Fe_2O_3 during induration of pellets [7].

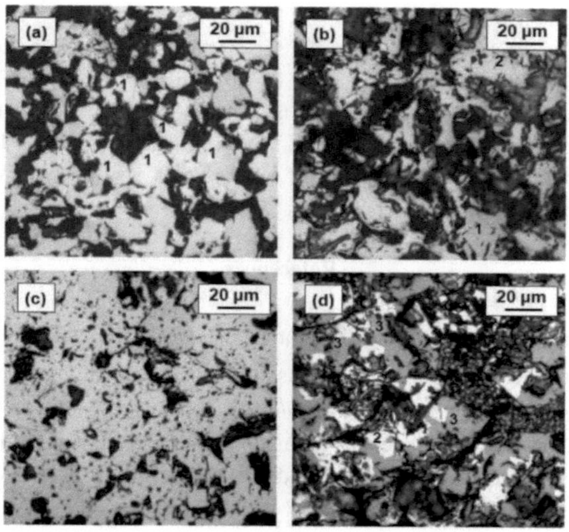

Fig. 9. Optical micro-images of pellet roasted with different amount of anthracite (a–0; b– 0.5%; c–1%; d–2.0%; 1–Original hematite; 2– Secondary hematite; 3–Magnetite)

Direct reduction reaction occurs when solid carbon contacts with iron oxides in the pellets. Therefore, a part of Fe_2O_3 grains in anthracite–containing hematite pellets would be reduced to Fe_3O_4 grains by the solid carbon of anthracite.

$$6Fe_2O_3 + C = 4Fe_3O_4 + CO_2 \quad (1)$$

Besides, CO and H_2 will be produced during the heating process of anthracite and the hematite inside pellets will also be reduced to magnetite [11]:

$$3Fe_2O_3 + H_2 = 2Fe_3O_4 + H_2O \quad (2)$$

$$3Fe_2O_3 + CO = 2Fe_3O_4 + CO_2 \quad (3)$$

The new–born magnetite grains can be subsequently oxidized into secondary hematite grains. The recrystallization between original hematite grains can be enhanced because the activity of the new–born secondary hematite grains is higher than that of original hematite grains, which is helpful to the recrystallization of Fe_2O_3 grains. Consequently, the compressive strength of roasted pellets is capable of being improved greatly.

Conclusions

[1]. The addition of a certain amount of pulverized anthracite to hematite pellets is able to improve both the compressive strength and drop strength of the green balls. The appropriate dosage of anthracite is 1.0–1.5 wt.%. The compressive strength of roasted pellets is greatly influenced by the dosage of anthracite. Based on microstructure analysis, anthracite changes the rule of Fe_2O_3 recrystallization.

[2]. The appropriate roasting parameters for the anthracite–containing hematite pellets are 1280–1350 °C roasting temperature, and 15–20 min roasting time. The addition of 1% anthracite in hematite pellets, can decrease the proper roasting temperature by 20 °C, expand the roasting temperature range by 50 °C, and shorten the roasting time by 5–10 min. Therefore, adding a certain proportion of anthracite is an effective way to enhance the roasting performance of hematite pellet.

References

[1] W. Pietsch, "Size enlargement by agglomeration," *Chichester: Wiley*, 1991.

[2] G. Qiu, T. Jiang, X. Fan, et al., "Effects of binders on balling behaviors of iron ore concentrate," *Scandinavian Journal of Metallurgy*, 2004, 33: 39–46.

[3] G.H. Li, Z.K. Tang, Y.B. Zhang, et al. "Reduction swelling behaviour of haematite/magnetite agglomerates with addition of MgO and CaO," *Ironmaking and Steelmaking*, 2010, 37(6):393–397.

[4] Y. Chen and J. Li, "Crystal rule of Fe_2O_3 in oxidized pellet," *Journal of Central South University: Science and Technology*, 2007, 38(1): 70–73. (in Chinese)

[5] C. Bryk, "Reduction phenomena in composites of iron ore concentrates and coals," *Ironmaking and Steelmaking*, 1986, 13(2): 70–75.

[6] V. Singh, V. Tathavadkar, M. S. Rao, et al. Estimating effect of chrome ore granulometry on sintered pellet properties. Ironmaking & Steelmaking, 2008, 35(1): 27-32.

[7] Q. Wang, Z. Yang, J. Tian, et al., "Mechanisms of reduction in iron ore–coal composite pellet," *Ironmaking and Steelmaking*, 1997, 24(6): 457–460.

[8] X. Chen, "Pyrometallurgical process physical chemistry," *Beijing: Metallurgy Industry Press*, 1984. (in Chinese)

[9] S.M. Iveson, J.D. Litster, K. Hapgood, et al., "Nucleation, growth and breakage phenomena in agitated wet granulation processes: a review," *Powder Technology*, 2001, 117. 3–39.

[10] S.M. Iweso and N.W. Page, "Tensile bond strength development between liquid bound pellets during compression," *Powder Technology*, 2001, 117, 113–122.

[11] X. He, "Coal chemistry," *Beijing: Metallurgy Industry Press*, 2010. (in Chinese)

Drying, Roasting, and Calcining of Minerals
Edited by: Thomas P. Battle, Jerome P. Downey, Lawrence D. May, Boyd Davis, Neale R. Neelameggham, Sergio Sanchez-Segado, and P. Chris Pistorius
TMS (The Minerals, Metals & Materials Society), 2015

Preparation of Straw Char by Preformation-Carbonization Process and Its Application in Iron Ore Sintering

Xiaohui Fan, Zhiyun Ji, Min Gan, Xuling Chen, Liang Yin, Tao Jiang

School of Minerals Processing & Bioengineering, Central South University, No.932, South Lushan Road, Changsha, Hunan, 410083, P.R.China

Keywords: Iron ore sintering, Straw char, Preforming technology, Pollutants emission reduction

Abstract

Straw char exhibited considerably faster burning speed than coke breeze for its porosity, which significantly degraded sinter quality when it replacing coke breeze as sintering fuel. In this investigation, preforming technology has been employed to densify raw straw. The properties of straw char preformed show that its density was increased, and the structure was also densified, which made the difference of combustibility between the straw char and coke breeze smaller. The proper preforming temperature, residence time and pressure were 200℃, 1min and 120MPa respectively. Sinter pot trials show that the product carbonized from preformed straw made the proper replacement ratio to coke breeze increase from 20% to 40%, and the emission reduction of SOx, NOx in flue gas increase from 11.50%, 14.15% to 37.85%, 23.60% respectively, compared with non-preformed straw char.

Introduction

Iron ore sintering is an energy-intensive and pollution-intensive stage in an integrated steelwork. About 9-12% of the total energy is consumed during sinter making, while 75-80% of the energy consumed is solid fuels, like coke breeze, anthracite etc.[1] The combustion of fossil fuels has been proved to be the main source of CO_2, SOx, NOx etc. [2] Therefore, substituting fossil fuels with clean and renewable bio-energy in sinter making seems to be an effective strategy to realize clean sinter production. Much work has been done to consider the application of biomass fuel and its effects on sintering indexes and pollutants emission.[3-7] Research gave positive results in terms of reducing CO_2, SOx, NOx etc. emissions, while showed adverse influences on sinter quality when the replacement ratio of biomass fuel to coke breeze exceeded a particular value.

China has abundant agricultural wastes like straw, which takes up 17% of the world total amount. However, the availability is less than 45% for open burning and being abandoned.[8, 9] Conversing straw resources into a substitute for fossil fuels serves as a win-win way for both environmental protection and comprehensive resource utilization. According to our investigation, straw char showed greater porosity and specific surface area than other biomass fuels, which resulted in its lower replacement ratio to coke breeze.[7] In this paper, a preformation-carbonisation process was researched for the purpose of preparing straw char of better adaptability for sinter making.

Materials and methods

Properties of materials

For making sinter, iron ores, fluxes (dolomite, limestone and quick-lime), fuels and return fines were used to produce sinter of TFe(Total iron content)=56.38%, SiO_2=4.92%, MgO=2% and basicity(CaO/SiO_2)=1.9 respectively. The chemical composition and proportions of raw materials are given in Table I.

Table I Chemical composition of raw materials and their proportion in the mixture (mass %)

Raw material	Chemical composition							Proportion
	TFe	SiO_2	CaO	MgO	Al_2O_3	FeO	LOI	
Mixed iron ores	60.74	4.65	1.9	1.54	1.92	9.24	2.64	64.93
Dolomite	0.21	0.87	31.57	19.68	0.22	0.13	46.91	0.53
Limestone	0.14	1.31	50.19	3.30	0.32	0.10	42.27	3.05
Quicklime	0.40	2.85	77.92	3.64	0.75	0.23	11.29	4.56
Return fines	56.38	4.92	9.36	2.00	2.07	8.58	0.00	23.08
Fuel	—	—	—	—	—	—	—	3.85

Table II gives the proximate and ultimate analyses of raw straw and coke breeze applied in this investigation, which showed that raw straw mainly consisted of volatiles, and its calorific value was lower than coke breeze. Since coke breeze came from coking plant, its major part was fixed carbon.

Table II Proximate and ultimate analyses of fuels (dry basis)

Fuel	Ultimate analyses (mass %)					Proximate analyses (mass %)			Calorific value(MJ/kg)
	C	H	O	N	S	Ash	Volatiles	Fixed carbon	
Coke breeze	81.84	2.46	1.03	0.72	0.224	19.54	5.88	74.68	27.06
Raw straw	45.44	5.33	46.96	0.25	0.078	5.10	76.15	18.75	18.80

Experimental methods

Preparation of straw char The preparation of straw char consisted of one preforming stage and the other carbonization stage. During the preforming stage, raw straw ground to 0-2mm is molded into pellets 10mm long and 10mm in diameter with a forming machine. Temperature, residence time and pressure were taken into consideration to determine the proper preforming conditions through single-factor trials. After that, preformed raw straw underwent a carbonization stage to prepare straw char. The designated factors and levels of preforming and carbonization process are described in Table III. An optical microscope, model DMRXP (Leica, Germany), and a simultaneous TG-DSC analyzer, model STA449C (Netzsch, Germany), were used to analyze the microstructure and combustibility of fuel samples.

Table III Factors and their levels for straw char preparation

Preforming factors	Levels				Carbonisation factors	Levels
Temperature/℃	150	200	250	300	Temperature/℃	700
Residence time/min	0.5	1	2	3	Residence time/min	30
Pressure/MPa	40	60	120	180	Heating rate/℃/min	25
					Atmosphere	N_2 with flow rate 0.1L/min

Sinter pot trials Sintering trials were carried out in a sinter pot Φ180×700mm. Specific experimental procedure included blending mixtures with prescribed water, and then charging blended materials into a drum with the size of Φ600×1400mm for granulating. Granulation time and revolving speed were 4min, 15r/min respectively. After granulation, mixtures were put into a sintering pot, before which 1kg sinter with size range of 10-16mm was charged to form hearth layer. Then natural gas was used to ignite the mixture from the top. Ignition temperature, ignition time, ignition negative pressure and heat-preservation time were 1150±50℃, 1min, 5kPa and 1min respectively. After ignition, adjust the negative pressure to 10kPa. When burning through arrived, the negative pressure was adjusted to -5kPa again and keeping cooling for 3min before unloading sinter cake. After that, sinter underwent crushing, screening to get finished product. For assessing the sinter quality, vertical sintering speed, yield, productivity and tumble index were selected.

During the whole sintering process, infra-red flue gas analyzers, model DELTA 65-3 and Vario Plus (MRU Corporation, Germany), were adopted to measure the concentration of O_2, CO, CO_2, NOx, and SO_2 in flue gas.

Results and discussion

Influences of preforming process on properties of straw char

Density of straw char Influences of preforming temperature, residence time and pressure on the density of straw char have been considered, of which results are given in Fig.1. Remarkable increase of the density of straw char was found after preforming raw straw. Improving preforming temperature, prolonging residence time and improving preforming pressure enabled to improve the density of straw char gradually. As can be seen in Figure 1(a), improving preforming temperature from 150 ℃ to 200℃ increased the density of raw straw from $0.995g/cm^3$ to $1.180g/cm^3$, with density of straw char increased from $0.577g/cm^3$ to $0.638g/cm^3$.

After further increasing preforming temperature to 250-300℃, densities of raw straw and straw char got further improvement, while the improvement was marginal. Therefore, 200℃ was selected as the proper preforming temperature. Figure 1(b) shows that when increasing residence time from 0.5min to 1min, densities of raw straw and straw char increased from 1.050g/cm^3, 0.605g/cm^3 to 1.180g/cm^3, 0.638g/cm^3 respectively. After prolonging residence time to 2-3min, only a slight improvement of both densities was observed. Consequently, 1min was selected as proper preforming residence time. Figure 1(c) indicates that improving preforming pressure from 40MPa to 60MPa merely increased densities of raw straw and straw char slightly, while increasing preforming pressure further to 120 MPa, densities of raw straw and straw char increased from 1.205g/cm^3, 0.647g/cm^3 to 1.320g/cm^3, 0.715g/cm^3 respectively. However, after increasing preforming pressure to 180MPa, the improvement of densities became rather small. Therefore, 120MPa was selected as the proper preforming pressure.

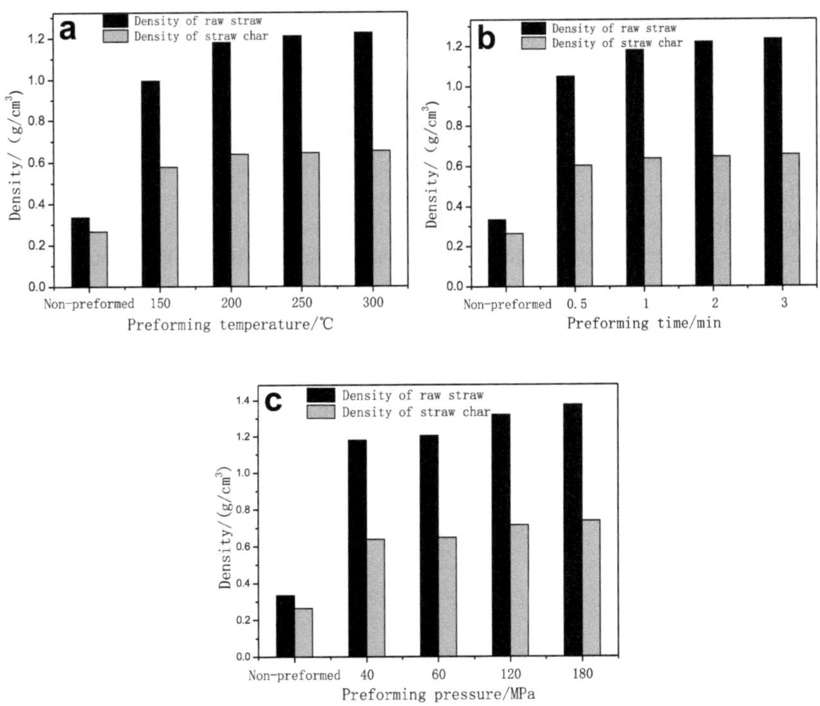

Figure 1. Effects of preforming factors on densities of straw and straw char: (a) Preforming temperature; (b) Preforming residence time; (c) Preforming pressure.

Structure of straw char Figure 2 gives the micro-structural differences of coke breeze, non-preformed straw char, and preformed straw char. Compared with non-preformed straw char, preforming facilitated the tight connection of carbon particles, which made the structure densified. Two possible reasons can be described to explain the positive effect of preformation. One is the function of mechanical pressure, which tightens the structure of raw straw and reduces inner voids. The other is that lignin in raw straw serves as an effective binder under given temperature, which builds bridges between straw particles to make the structure of raw straw even tighter. When moving the preparation process into the carbonization stage, tight connection structure showed great resistance to the damage generated by the removal of volatile matters. Therefore, preformation helped to reduce the structural differences between coke breeze and straw char.

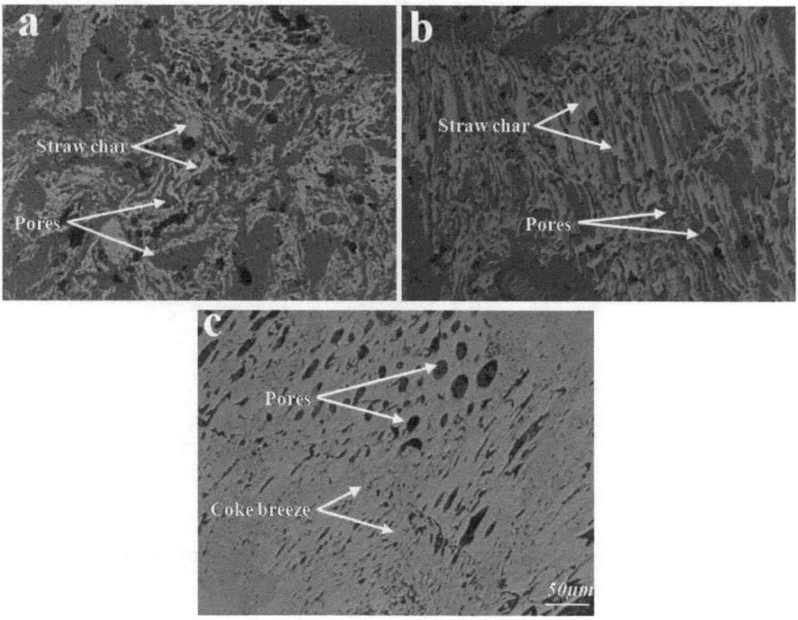

Figure 2. Microstructure of fuels: (a) Non-preformed straw char; (b) Preformed straw char; (c) Coke breeze.

Combustibility of straw char TG-DTG-DSC thermoanalysis was adopted to research the differences of coke breeze, non-preformed straw char and preformed straw char in combustibility. As can be observed in Figure 3, the intensive combustion phenomenon of non-preformed straw char was greatly improved during the combustion of preformed straw char since the heat release shown in DCS curve became more uniform and mild. Besides, the weight change shown in TG curve of preformed straw char became slower, which made the combustion duration of straw char longer. More direct proof of the change in combustibility can be observed in Table Ⅳ. The ignition temperature (Ti) of straw char was increased by 50℃, and the maximum weight loss rate (Vmax), maximum heat release value (Qmax) were reduced by 1.84%/min, 8.94 mW/mg respectively after preforming. Therefore, preformation benefited bridging the gap in terms of combustibility between straw char and coke breeze.

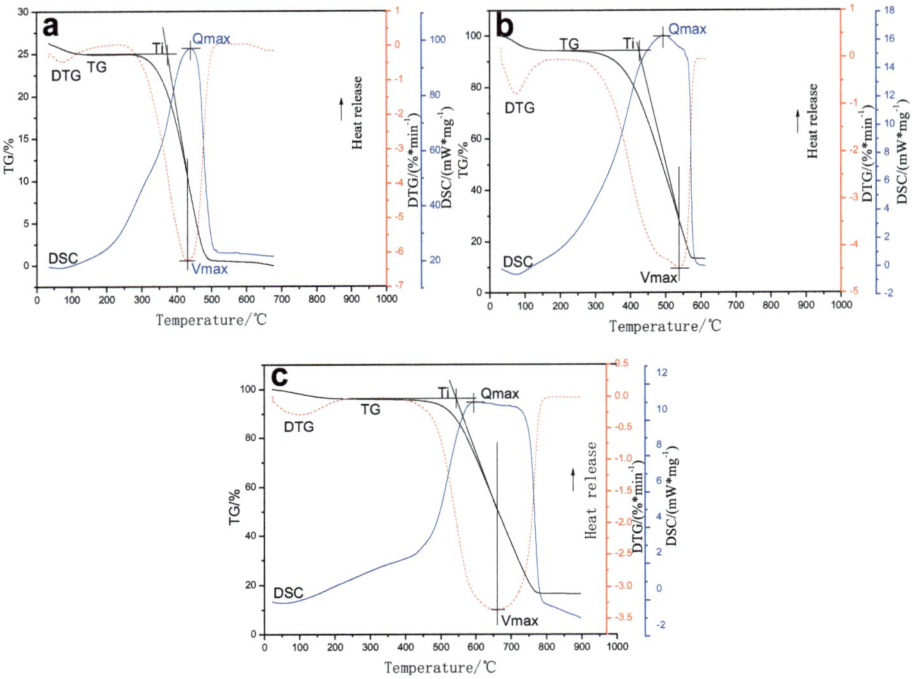

Figure 3. Combustibility of fuels: (a) Non-preformed straw char; (b) Preformed straw char; (c) Coke breeze.

Table Ⅳ Characteristic parameters of combusting curves

Fuel types	Preparation conditions	Ti/℃	Vmax/(%/min)	Qmax/(mW/mg)
Straw char	Non-preformed	376	6.31	25.27
	Preformed	426	4.47	16.33
Coke breeze	—	543	3.37	11.06

Application of preformed straw char in iron ore sintering

Influences on sintering indexes The application of preformed straw char in iron ore sintering process is given in Table V, where can be seen that when 20% coke breeze was replaced by both preformed straw char and non-preformed straw char, sintering indexes could match that of using coke breeze only. However, when 40% coke breeze was replaced, sintering indexes were still kept at a comparable level for the case using preformed straw char, while yield and tumble index dropped significantly for the case using non-preformed straw char. Increasing the replacement ratio of preformed straw char further to coke breeze by 60% led to remarkable damage to sinter quality as well. Therefore, the proper replacement ratio of straw char to coke breeze was increased from 20% to 40% after preformation.

Table V Effect of briquetting on sintering of straw char replacing coke breeze

Preparation conditions	Replacement ratio/%	Moisture /%	Sintering speed / $mm\cdot min^{-1}$	Yield /%	Tumble index /%	Productivity / $t\cdot m^{-2}\cdot h^{-1}$
—	0	7.25	22.01	73.30	65.32	1.51
Non-preformed	20	7.50	23.50	70.50	64.80	1.47
Preformed	20	7.50	23.02	71.85	65.04	1.48
Non-preformed	40	7.75	25.35	63.05	61.15	1.25
Preformed	40	7.75	23.98	68.76	64.20	1.45
Preformed	60	8.00	26.05	61.89	60.15	1.11

Influences on emissions of SOx and NOx The influences of non-preformed and preformed straw char replacing coke breeze on the emissions of SOx, NOx are described in Figure 4. Due to the lower levels of S, N in straw char, the emissions of SO_2, NO were reduced when employing straw char from the two processes to replace coke breeze. Since the proper replacement ratio of preformed straw char to coke breeze reached 40%, the peak concentration of SO_2 was reduced from 1407ppm, where 20% non-preformed straw char was used, to 815ppm, and the average concentrations of SO_2, NO were reduced from 350ppm, 303ppm to 251ppm, 272ppm respectively. Therefore, the emission capacities of SOx, NOx in flue gas were increased from 11.50%, 14.15% to 37.85%, 23.60% respectively.

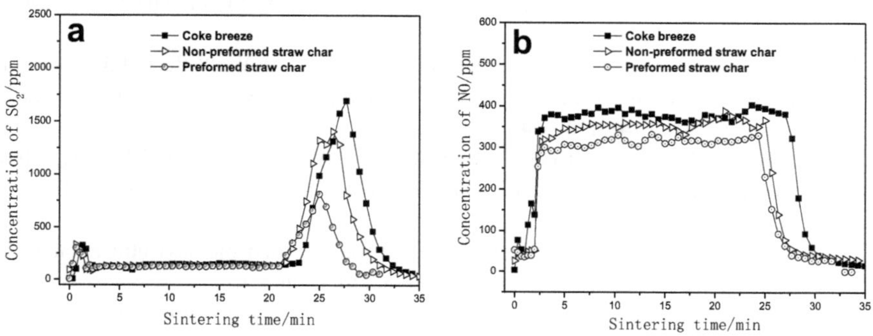

Figure 4. Influences on emissions of SO_2 and NO.

Conclusions

(1) Compared with non-preformed straw char, preformed straw char are significantly increased in density, and densified in structure, which makes the difference of combustibility smaller between the straw char and coke breeze. Recommended preforming temperature, residence time and pressure are 200℃, 1min and 120MPa respectively.

(2) Preforming process helps to improve the proper replacement ratio of straw char to coke breeze from 20% to 40%, with reducing emissions of of SOx, NOx in flue gas increase from 11.50%, 14.15% to 37.85%, 23.60% respectively.

Acknowledgements

The authors are grateful to programs supported by National Natural Science Fund (No.51174253, No.51304245), and also to the support of Chinese Postdoctoral Science Foundation (2013M540639).

References

[1] S.Q. Li, Z. J. Ji et al., "An analysis on the energy consumption of steel plants and energy-saving measures," *Journal of Industrial Heating*, 39(2010), 1–3.

[2] Y. G. Chen, Z. C. Guo, and Z. Wang, "Influence of CeO_2 on NO*x* emission during iron ore sintering," *Fuel Process. Technol.*, 90(2009), 933–938.

[3] R. Lovel, K. Vining, and M. Dell'Amico, "Iron ore sintering with charcoal," *Miner. Process. Extract. Metall.*, 116(2007), 85–92.

[4] T. C. Ooi, E. Aries, B. C. R. Ewan, D. Thompson Anderson, Fisher, R. D. R., T. Fray, and D. Tognarelli, "The study of sunflower seed husks as a fuel in the iron ore sintering process," *Miner. Eng.*, 21(2008), 167–177.

[5] M. Zandi, M. M. Pacheco, and T.A.T. Fray, "Biomass for iron ore sintering," *Miner. Eng.*, 23(2010), 1139–1145.

[6] M. Gan, X. H. Fan, Z.Y. Ji, X. L. Chen, T. Jiang, and Z. Y. Yu, "Effect of the distribution of biomass fuel in granules on iron ore sintering and NO*x* emission," *Ironmaking & Steelmaking*, 41(2014), 430–434.

[7] M. Gan, X. H. Fan, X. L. Chen, Z. Y. Ji, W. Lv, Y. Wang, Z. Y. Yu, and T. Jiang, "Reduction of pollutant emission in iron ore sintering process by applying biomass fuels," *ISIJ Int.*, 52(2012), 1580–1584.

[8] Y. M. Sun et al.,"The status and future of bioenergy and biomass utilization in China," *Journal of Renewable Energy*, 26(2006), 77–82. (in Chinese)

[9] L. Shi, Y. C. Zhao, and X. L. Chai, "Progress in comprehensive utilisation techniques of crop straws in China," *China Biogas*, 23(2005), 11–18. (in Chinese)

Drying, Roasting, and Calcining of Minerals
Edited by: Thomas P. Battle, Jerome P. Downey, Lawrence D. May, Boyd Davis, Neale R. Neelameggham, Sergio Sanchez-Segado, and P. Chris Pistorius
TMS (The Minerals, Metals & Materials Society), 2015

INFLUENCE OF MODIFIED BIOMASS FUEL ON IRON ORE SINTERING

Min Gan, Xiaohui Fan, Zhiyun Ji, Xuling Chen, Tao Jiang, Guanghui Li, Zhiyuan Yu

School of Minerals Processing & Bioengineering, Central South University, Changsha, Hunan, 410083, China

Keywords: Iron ore sintering, Biomass energy, Passivation, Energy saving and emission reduction

Abstract

With the increase of proportion of biochar(without modified) replacing coke breeze, the physical strength of sinter decreases due to the faster burning speed of biochar. An investigation upon passivating biochar with boric acid was then conducted to strengthen sintering in this paper. A layer of thin film of B_2O_3 distributed in the surface of biochar was observed, which could reduce the contact between reaction gas and biochar. So the burning speed of biochar was decreased properly after passivation, and the combustion efficiency increased due to the inhibition of carbon gasification. The comprehensive effects of above changes could increase the maximum temperature of sintering bed, which facilitated to the mineralization of sintering mixtures. Eventually sintering indices got reinforced. Thus the indices of sinter as applying 40% passivated biochar were comparable with that of 100% coke breeze.

Introduction

Sinter is the main iron-bearing burden in the blast furnace for ironmaking, which plays an important role in iron and steel production process. However, sinter process produces a large amount of flue gases that contain diverse pollutants, such as CO*x*, SO*x*, NO*x* and dioxins etc., and they are the main source of atmosphere pollution in iron&steel industry [1].

In sintering process, the combustion of coke breeze and anthracite is the main source of green house gas CO_2 and pollutants like gases SO*x*, NO*x*.[2] Therefore, it is of great importance to find clean and renewable energy sources to replace fossil fuels that make sinter production clean. Biomass is becoming an attractive alternative source of energy to traditional fossil fuels. Commercially available biomass materials, such as olive residues, sunflower husk, almond shells, bagasse pellets and so on, and their carbonization products, have been studied in a laboratory sinter pot.[3-5] The research results pointed out that biomass fuels were low in nitrogen and sulfur contents, and CO_2 liberated during combustion could be sequestered back into growing biomass.[6,7] Therefore, the emissions of CO*x*, SO*x* and NO*x* in sinter process were effectively reduced. However, since the properties of biomass and traditional fuels are different, the proportion of biomass replacing fossil fuels could not exceed 25%.[8] To achieve a higher replacement rate, sintering characteristics with biomass replacing fossil fuels were studied in the paper, and the strengthening measures were exploited to improve the sintering with bio-fuel.

Materials and methods

Properties of materials

Iron ore blend and three kinds of fluxes together with fuels were utilised to produce sinter. The chemical compositions of produced sinter are the typical characteristics of modern sintering plant, which the contents of TFe (total mass contents of Fe), SiO_2, MgO in sinter were 57.5%,4.82%,2.0%respectively and its basicity ((CaO/SiO_2)mass ratio) was 2.0. The chemical composition of raw materials and their percents are given in Table I.

Table I Chemical composition of raw materials and their percents in mixture

Types of raw materials	Chemical composition/%							Percent*/%
	TFe	FeO	SiO_2	CaO	MgO	Al_2O_3	LOI	
Iron ores blend	63.02	6.50	4.58	0.35	0.28	1.42	3.10	60.73
Dolomite	0.21	0.13	0.71	32.64	19.83	0.56	46.47	5.58
Limestone	0.14	0.10	1.49	50.66	2.28	0.43	40.72	2.16
Quicklime	0.4	0.23	2.86	76.69	1.18	1.20	12.36	4.62
Return fines	56.81	6.25	5.11	9.02	1.86	2.00	0.00	23.08

*The ratio is calculated on the conditions of coke being 3.85%.

Two types of solid fuels were applied in the experiments. One was coke breeze which came from sintering plant, and the other was biochar coming from carbonized acutissima(It has a wide distribution in China, which attains high economic value for hardwood product and can be managed as an energy tree species) at 700℃ for 30min in nitrogen gas. Ultimate and proximate analyses of fuels are illustrated in Table II. It can be seen that biochar is low in N, S content and ash content. Physical properties such as porosity rate and specific surface areas of two fuels are detected. The porosity rate of biochar is 58.22%, which is 12.47% higher than that of coke breeze. According to the results of detecting by the method of BET, the specific surface area of biochar is nearly 10 times bigger than coke breeze, which is $5.99m^2/g$, $54.76m^2/g$ respectively.

Table II Ultimate and proximate analyses of fuels

Fuel types	Ultimate analyses/%				Proximate analyses (dry base) /%			Calorific value $/MJ{\cdot}kg^{-1}$
	C	H	S	N	Fixed carbon	Ash	volatile	
Coke breeze	78.89	2.46	0.500	0.72	74.68	19.54	5.88	26.84
Biochar	94.64	2.77	0.037	0.19	87.34	5.10	7.55	30.77

Experimental methods

Sintering test Sintering process was simulated in a sinter pot of 180 diameter×700 mm deep. The procedure involved ore proportioning, blending, granulation, ignition, sintering and cooling. Raw materials were granulated in a drum of Φ600×1400mm for 4 min, and then charged into the sinter pot. After charging, the fuel in the surface layer was ignited at 1150 ± 50℃ for 1min by an ignition hood initially, and then the combustion front moved downwards with the support of downdraught system, which was mainly a draught fan used to enable sufficient air to be sucked into sinter pot from top at a negative pressure of 10kpa. In the sintering process, a infra-red analyzer was used to detect the CO, CO_2 contents of exhaust gas, and combustion efficiency of $CO_2/(CO+CO_2)$ was calculated to assess the burning degree of fuels. Besides, the temperature of sinter bed in the position of 200mm from the top of sinter pot was tested with type-B thermocouples. After sintering, the dropping test(2m×3times), screening and drum index were carried out to value the physical strength of sinter, yield and tumble index(TI).

The iron ore sintering was characterized as a relatively uniform process of solid material, coke combustion and various modes of heat transfer. Heat transfer in sintering bed contains very complicated modes of conduction, convection and radiation between gas and solid phases. Flame front moving is considered an important characterization of the bed combustion. It can describe the propagation speed of the combustion zone. The match of heat transfer and flame moving is very important for sintering. In the paper, heat front and flame front were tested in a sinter pot of Φ100mm×250mm. The iron balls used to simulate size distribution of sinter mixture were loaded into the sinter pot, and five thermocouples were placed into sinter bed evenly along height direct of sinter pot to test the change of temperature. The heat front testing starts with a thermal impulse of 1250℃ being given to charge surface for 1min without fuel in sintering mixture. Heat front achieved when temperature of sinter bed increases uniformly, which was subjected to the isothermal of 100℃. Unlikely, flame front experiment needed to add solid fuel with a ratio of 2.5% into iron ore pellets and ignite under 1150℃ for 1min, and the flame front achieved when temperature of sinter bed increased rapidly, which was subjected to the isothermal of 600℃.The speed of heat(flame) front was defined as the ratio of bed height to the transfer time.

Test of fuel's reaction properties Isothermal analysis method was employed to research the combustibility and reactivity (ability to react with CO_2) of fuels. The combustibility of fuel was tested in air, and the reactivity tested in CO_2 atmosphere. The experiment proceeded in an electrically heated furnace. 25g fuels with size range 3~5mm was loaded into a quartzose tube, and the reaction started at 1100℃ in simulated atmosphere, of which flow rate was 5L/min. During the whole heating process, the weight change of fuel was collected by computer. Fuel conversion rate x_c for corresponding time and instantaneous conversion velocity R were calculated. The value of R50% when x_c equals to 50% was used to reflect reaction speed.

$$x_c = (1 - \frac{m}{m_0}) \times 100\% \quad (1)$$

$$R = -\frac{1}{m_0}\frac{dm}{dt} \quad (2)$$

Where x_c is fuel conversion rate (%), R is instantaneous conversion velocity (%/ min), m_0 is initial mass (g), m is the mass at time t (g), dm/dt is the weight loss rate at time t (g/min).

Method of passivating biochar Boric acid was used to reduce the reaction capacity of biochar. At first, boric acid solution was prepared with a mass concentration of 15%. The additive proportion of solution was calculated according to the requirement of boric acid dosage. Specifically, if the proportion of boric acid accounts for 3% of the fuel weight, the proportion of solution needed to prepare is 20%. Before using, boric acid solution was uniformly sprayed to the surface of biochar. After that, the passivated biochar was dried at 105℃ to make its moisture content less than15%.

Results and discussion

Characteristics of sintering with non-modified biomass fuel

As using coke breeze alone, the optimal amount of coke in mixtures was determined at 3.85% by sintering experiments. As biochar hasn't been modified, the influence of biochar replacing coke breeze on sintering was studied on heat equivalent basis. The results are given in Figure 1. With the replacing proportions increasing, sintering speed is accelerated, while productivity, yield and tumbler index are decreased. When the proportion is not too high, the change of sinter indexes is relatively small. However, when replacement proportion increases to critical value, the output and quality of sinter are worsened significantly. The inflexion can be considered as a criterion of appropriate value for biomass replacing coke breeze. For biochar, when replacement proportion exceeds 40%, the sintering indices are worsened significantly. Therefore, the appropriate proportion of biochar replacing coke breeze is 40%.

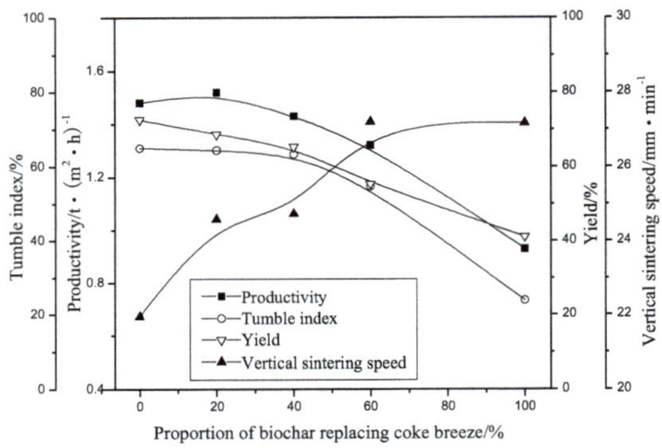

Figure 1. Effects of proportion of biochar(without modified) replacing coke breeze on sintering index

Mechanism of Biomass Fuel (without modified) Influencing Iron Ore Sintering

Since variances in physicochemical properties between biochar and coke are existed, combustion behavior and the gasification with CO_2 will be affected when coke breeze are replaced. Their combustibility and reactivity are shown in Figure 2.

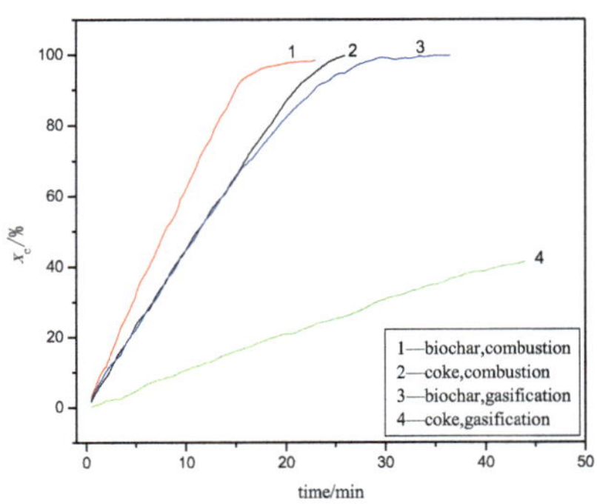

Figure 2. Isothermal combustion rate of fuels（1100℃）

Biochar can combust much faster, and the time needed to burn out is shorter than coke breeze. $R_{50\%}$ of biochar during combustion is 6.25 %/min, while that of coke breeze is 4.55%. Likely, the gasification speeds of biochar and coke breeze are 4.35%/min, 0.88%/min respectively. Biochar can be gasified much easier than coke breeze, which indicates that biochar has a higher reactivity. Biochar is easy to combust and gasify due to its large surface area and high porosity. Hence, biomass fuels can be burnt earlier and burnt out faster than coke breeze in sinter bed.

The influences of two fuels on sintering processing are studied. The results are given in Figure 3. The flame front speed is accelerated with the increase of the proportion of biochar. When the proportion increases from 0% to 40%, 100% respectively, the flame front speed rises from 34.1 mm/min to 41.7mm/min, 46.9 mm/min, while the heat front speed stays around 35mm/min, which indicates that the consistency of heat front speed and flame front speed is ruined remarkably when the replacing proportion reaches 40%. Meanwhile, the degree of fuels combustion decreases, the combustion efficiency ($CO_2/(CO+CO_2)$)decreases from 87.83% to 86.92%, 85.15% when replacement rises from 0% to 40%, 100% respectively, which indicates that the thermal utilizing efficiency of fuel is lowered as biochar replacing coke breeze.

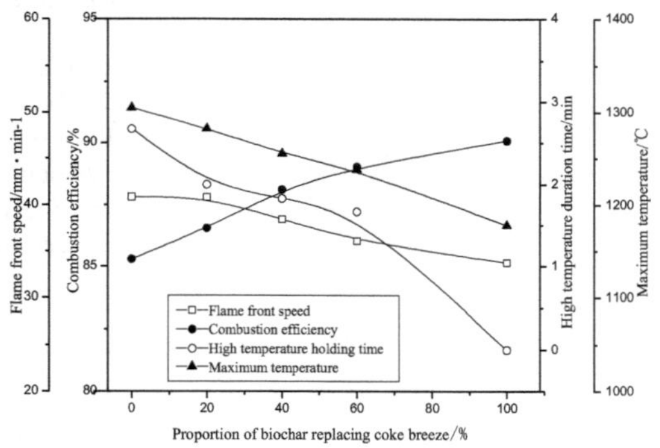

Figure 3. Effect of biochar (without modified) replacing coke breeze on the process of sintering

When replacing proportions increased from 0% to 40%, 100%, the maximum temperature decreases from 1305℃ to 1255℃, 1178℃ respectively, and the high temperature duration decreases from 2.67min to 1.83min, 0min. The maximum temperature is decreased and the high temperature duration is shortened since the coordination of flame front and heat front speed is damaged and the fuel combustion efficiency is reduced. As replacement ratio exceeds 40%, the maximum temperature of sinter bed is decreased and the high temperature duration is shortened significantly, all of which result in insufficient mineralization of sinter material, thus the yield and the tumbler strength are decreased.

Strengthening techniques of iron ore sintering with biomass fuels

Biochar has high porosity and large specific surface area. Therefore, it could burn at an excessively fast speed, which is the main reason for the decrease of sinter quality. Reducing the effective reactive porosity and specific surface area tends to be an important approach to strengthening sintering process. The effect of boric acid passivating the surface of biochar on sintering is studied in the paper. The effect of boric acid passivation on the structure of biochar is shown in Figure 4. The microstructure is tested by SEM, and it can be inferred that the surface of biochar is covered by the thin film of B_2O_3 after dehydration of boric acid. The passivation coating can reduce the contact between reaction gas and biochar.

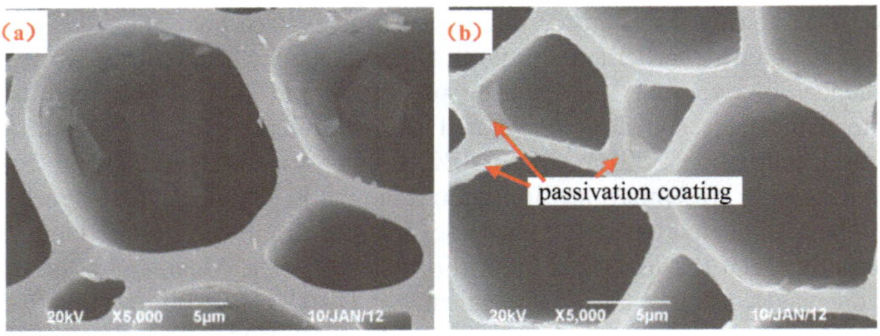

(a)No passivation; (b) Passivated by 3% boric acid
Figure 4. Effects of passivation on microstructure of biomass fuel

The effects of passivation on biochar combustion and gasification are investigated, and the results are shown in Figure 5. The burning rate and gasifying rate of biochar is decreased by with the help of boric acid. The $R_{50\%}$ of passivated biochar reduces from 6.25 %/min, 4.35%/min to 4.87 %/min, 2.68%/min respectively under 1100℃. The reaction velocities of passivated biochar are close to that of coke breeze. Especially, the burning rates of two fuels are roughly equal. The results show that the variance of combustibility and reactivity between biochar and coke breeze are reduced.

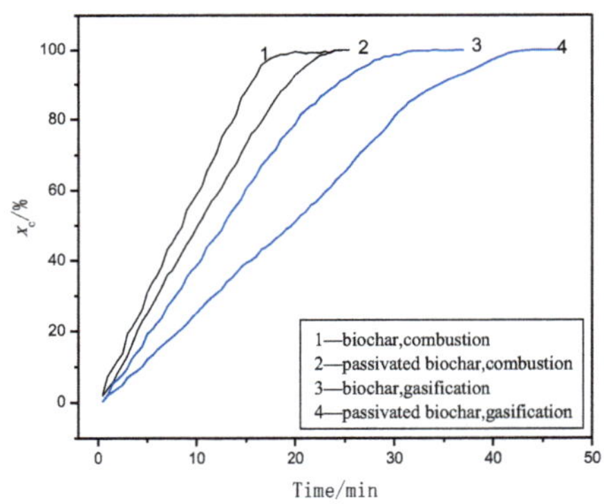

Figure 5. Modification on combustion and reactivity of biomass fuel（1100℃）

Furthermore, the behaviors of passivated biochar in bed during sintering are researched, and the results are given in Table III. As biochar replaces 40% coke, the flame front speed of biochar decreases from 41.7mm/min to 37.5mm/min after passivating, that the flame front speed and the heat front speed(35.1mm/min) tend to recover consistency again. And the combustion efficiency increases from 86.92% to 87.45%, which means that the heat utilization of biochar is enhanced. Those are advantageous to improve the thermal process during sintering. As Table Ⅳ shows that the maximum temperature increases from 1255℃ to 1280℃, and the high temperature duration extends from 1.83min to 2.17min. All those indicate that the characteristics of passivated biochar are more suitable to be applied into sintering, because it exhibits similar behaviors to coke.

Table III Effects of 3% boric acid passivating biochar on sintering process

Condition	Flame front speed/mm·min^{-1}	Combustion efficiency/%	Maximum temperature/℃	High temperature duration/min
Coke breeze alone	34.1	87.83	1305	2.67
Biochar replacing 40% coke	41.7	86.92	1255	1.83
Passivated biochar replacing 40% coke	37.5	87.45	1280	2.17

The effect of boric acid passivating the surface of biochar on sintering indices is shown in Table Ⅳ. The sintering speed is decreased through passivation, and the yield, productivity and tumbler index are enhanced, which means that the sintering with biochar is strengthened. With the increase of boric acid dosage, the strengthening effect is improved. However, when the dosage exceeds 3%, the level of increasing degree is marginal. When the dosage of boric acid is kept at 3%, the indices of sinter are comparable with that of using coke breeze alone.

Table Ⅳ Effects of passivating biochar on sintering indices

Boric acid dosage/%	Proportion of biochar replacing coke/%	Sintering speed /mm·min^{-1}	Yield (+5mm) /%	Tumble index/%	Productivity /t·m^{-2}·h^{-1}
0	0	21.94	72.66	65.00	1.48
0	40	24.73	65.30	63.27	1.43
1.5	40	24.11	67.25	63.80	1.45
3.0	40	23.57	71.23	65.20	1.49
4.5	40	23.22	71.89	65.13	1.50

Effects of passivated biochar on microstructure of sinter are shown in Figure 6. When fuel is coke breeze alone (Figure (a)), a large amount of acicular calcium ferrite appeared in sinter, and the interleaving structure with excellent strength is formed by calcium ferrite and magnetite.

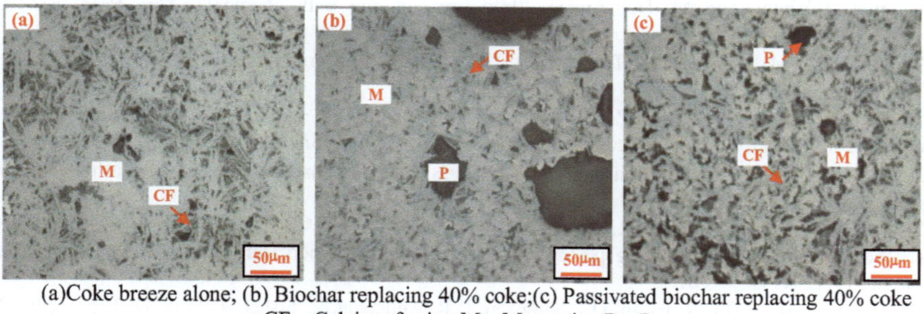

(a)Coke breeze alone; (b) Biochar replacing 40% coke;(c) Passivated biochar replacing 40% coke
CF—Calcium ferrite; M—Magnetite; P—Pores
Figure 6. Effects of passivating biochar on microstructure of sinter

However, when biochar replacing coke breeze by 40%, the content of calcium ferrite in sinter is decreased, which forms the structure of large pore and thin rampart, which is adverse to sinter strength. After being passivated by 3% boric acid, more calcium ferrite and less porosity in sinter can be observed when biochar replacing coke breeze by 40%. Compared with no passivation, the content of calcium ferrite increases from 33.99% to 35.20% if biochar passivated by boric acid. Because the maximum temperature of sinter bed is increased and the high temperature duration is extended significantly as using passivating biochar, the mineralization of sinter material is improved, thus the strength of sinter is increased.

Conclusions

(1) With the proportions of biomass fuel replacing coke breeze increasing, sintering speed is accelerated, while yield and tumbler index are decreased. The replacement proportion should be appropriate to satisfy the productivity and the quality of sinter, and it should not exceed 40%.

(2) Non-modified biochar can start burning earlier and be burnt out faster than coke breeze, which ruins the consistency of heat front speed and flame front speed, and decreases the combustion efficiency of fuels. Those also decrease the temperature of sintering layer, which is disadvantage to mineralize of sintering mixture, and eventually degrades sintering indices.

(3) Passivation coating are formed through adding boric acid solution into biochar, which can reduce the contact between reaction gas and biochar, and decrease combustion rate of biomass fuel properly, Sintering indices therefore are improved. Meanwhile, there is almost no effect on quality of pig iron because the additive of boric acid is a trace(<0.05%) in sintering mixture.

Acknowledgement

The authors are grateful to the National Natural Science Foundation of China(No.51304245) and the China Postdoctoral Science Foundation (No. 2013M540639 and No.2014T70691) for supporting this research.

References

[1] P. R.Dawson, "Recent developments in iron ore sintering", *Ironmaking Steelmaking*, 20(1993), 135-143.

[2] S.Peter, L. H.Bodo and E.Gerhard, "Measures to Reduce CO_2 and Other Emissions in the Steel Industry in Germany and Europe", *J. Iron Steel Res. Int*, 16(2009), 42.

[3] R.Lovel, K.Vining and M. Dell'Amico, "Iron ore sintering with charcoal", *Min. Proc. Ext. Met*, 116(2007), 85-92.

[4] M. Zandi, M. M. Pacheco, and T.A.T. Fray, "Biomass for iron ore sintering," *Miner. Eng.*, 23(2010), 1139-1145.

[5] M. Gan, X. H. Fan, X. L. Chen, Z. Y. Ji, W. Lv, Y. Wang, Z. Y. Yu, and T. Jiang, "Reduction of pollutant emission in iron sintering process by applying biomass fuels," *ISIJ Int.*, 52(2012), 1580-1584.

[6] Z.W.Jiang, "Study on combustion process of biomass fuels and characteristics of their ash", *Journal of Material Review*, 24(2010), 66-80. (in Chinese)

[7] T. C. Ooi, D. Thompson, D. R. Anderson, R. Fisher, T. Fray, M. Zandi, "The effect of charcoal combustion on iron-ore sintering performance and emission of persistent organic pollutants", *Comb. Flame*, 158(2011), 979-987.

[8] T. C. Ooi, E. Aries, B.C.R. Ewan, D. Thompson, D.R. Anderson, R. Fisher, T. Fray, D. Tognarelli, "The study of sunflower biomaterial as a substitute in the iron ore sintering process," *Min. Eng.*, 21(2008), 167-177.

Drying, Roasting, and Calcining of Minerals
Edited by: Thomas P. Battle, Jerome P. Downey, Lawrence D. May, Boyd Davis, Neale R. Neelameggham, Sergio Sanchez-Segado, and P. Chris Pistorius
TMS (The Minerals, Metals & Materials Society), 2015

THE PHASE TRANSFORMATION OF BAIYUNEBO IRON ORE TREATED WITH REDUCTANT SINTERING

Bingyi Bai[1], Yanling Guo [1,2, *], Lifeng Yang[1], Guoding Gao [1], Jieyu Zhang[1,2]

1. College of Materials Science and Engineering, Shanghai University, Shanghai 200072;

2. Shanghai Key Laboratory of Modern Metallurgy & Materials Processing, Shanghai 200072

Keywords: Baiyuneboite; Reduction; Ore phase; XRD.

Abstract: In terms of special resources with iron, niobium and rare earth coexisting in Baiyuneboite, the new smelting technology of ferrocolumbite was studied. Reduction experients for the Baiyuneboite and explores ore phase change of Baiyuneboite in the process of reduction was conducted with graphitic carbon as the main reducer and a temperature below 1000 ℃. Moreover, it studies the influence of reducer, roasting temperature and grind fineness on the reduction as well as focuses on the evolution course of iron phase. Through the analysis on in-situ XRD of the raw ore and carbon addition ore, it also develops microstructure change rules of Fe ore phase and ore phase of paragenetic elements for composite ore(Ca, Si, Nb, F) in the process of reduction roasting. In addition, the impacts of temperature on the reduction process are also discussed. With regard to minerals of different particle sizes, the thesis will explore their ore phase changes and microstructures after 10 hours' reduction under 900 ℃.

1. Introduction

Currently, with regard to analyses on RE-Nb-Fe paragenetic ore, there are beneficiating method, ferrocolumbium smelting method and direct reduction method. Baiyuneboite of Inner Mongolia is a typical RE-Nb paragenetic ore. Due to the limitation of beneficiation technology, people just make comprehensive recycling utilization for iron and rare earth, while the niobium has not been applied well [1-5]. Thus, direct reduction method is utilized to process RE-Nb paragenetic ore. In recent years, it has become a research hotspot. For example, Gao Peng et al[6] adopt coal-base to make direct reduction processing toward Baiyuneboite. However, the reduction temperature is as high as 1225 ℃ with a higher energy consumption. While, Chen Hong [7] utilizes gas-based direct reduction technology to process gravity concentrate of No. 2 ore in eastern contact belt of Baiyun Obo. Moreover, they have achieved good results. Nonetheless, the required ratio of mixed gas in this technology is 90%CO+10%CO_2. Obviously, it has a high demand for reducing gas composition.

By roasting Baiyuneboite with medium temperature, the thesis takes graphitic carbon as the main reducer and makes reduction experiment for ferrocolumbium of Baiyuneboite. Besides, the in-situ XRD technology is adopted to explore ore phase change of Baiyuneboite in the roasting reduction process as well as evolution process of iron phase and niobium phase.

2. Experiment

2.1 Experimental Materials

In this thesis, the experimental object is Baiyuneboite of Baotou in Inner Mongolia. Firstly, it conducts pretreatment including grinding and screening so as to make particle size of ore materials reach 50 meshes. Then, it makes drying treatment. To ensure the result, it carries out X-ray fluorescence analysis towards the raw ore. Moreover, the results are shown in table 1. At the same time, to study the effects of particle size on the ore roasting and reduction, it will conduct particle size classification for the pretreated ore materials. According to the standards of 50-100 mesh, 100-200 mesh and below 200 mesh, it makes separation and conducts X-ray fluorescence analysis towards ore samples of different particle sizes. Moreover, the results are shown in table 2.

In this experiment, choose reducer-graphite and broke it into 0.1 mm for later use. As a matter of fact, carbon is mainly and widely used in the ferroalloy smelting technology. Thus this is also the reason why graphite was chosen as the reducer [8, 9]. Moreover, the biggest characteristic of carbon is that it has great affinity for oxygen. Its reduction product is CO and CO_2 that can easily separated from the product.

Table 1 Element composition and content of Baiyuneboite

O	Fe	Ca	F	Ce	Si	La	P	Nd
33.79	27.06	21.17	3.98	3.68	3.35	1.62	1.46	1.44
S	Na	Mg	Pr	Al	Sr	Mn	Nb	Y
1.25	0.36	0.35	0.16	0.16	0.07	0.05	0.034	0.03

Table 2 Element composition and content of different particle sizes in Baiyuneboite

	Fe	O	Ca	Si	C	Ce	Nb
500-100 Mesh	36.02	31.10	13.86	5.49	5.29	2.43	0.027
100-200 Mesh	29.27	27.93	19.27	3.09	7.31	3.01	0.033
Below 200 mesh	15.17	30.43	23.96	2.75	9.85	4.33	0.034

2.2 Experiment Method

2.2.1 In-situ XRD Experiment

With in-situ XRD device, this experiment deposits and tiles Baiyuneboite raw and carbon addition ore on the platinum sheet respectively. Under the environment of high purity nitrogen, it will roast for the samples. In this atmosphere, the platinum strip won't react with the metallic product. As heating them to the expected temperature, it will make XRD test for the ore samples under heat preservation condition. As a result, it can analyze ore phase change towards ore samples in the whole roasting process.

2.2.2 Particle Classification Experiments

By adopting medium temperature roasting method, it conducts sintering experiments toward Baiyuneboite raw ore and carbon addition ore after particle classification. The equipment is tube-type lying muffle furnace. In the direct reduction experiments, take a certain amount of ore powder(about 5g), dry it at 100 ℃ drying oven for 2h. Then, put the sample into corundum crucible and take the crucible into the constant temperature area of box-type furnace with 5℃/min heating rate, the gas atmosphere in the muffle furnace is high purity nitrogen. Next, preserve heat for 10h after they reach the expected temperature so as to lead fully reaction. Later, take out the sample after the furnace cooling down to the room temperature and carry on research.

2.3 Research method

Firstly, the experiment makes in-situ XRD analysis towards Baiyuneboite raw ore and carbon addition ore. The XRD conditions are test voltage of 40kV, beam current of 100mA, a wavelength of 1.5406, step width for 0.02um, and the type of radiation is Cu Kα. In the XRD experiment, the speed of X-ray scanning is 8°/min. In this base, one can judge the ore phase composition and temperature after ore phase change of carbon addition ore in Baiyuneboite before and after roasting. Later, with regard to Baiyuneboite raw ore and carbon addition ore of different particle sizes, one can utilize muffle furnace for reduction roasting. Next, in terms of roasted ores, people can conduct XRD analysis, scanning electron microscopy as well as energy spectrum analysis so as to determine its ore phase change rules and explore influences of particle size on the ore phase change. From table 2, there are great differences among major element contents of Baiyuneboite under different particle sizes.

In the carbon addition process, one mainly uses Fe content to calculate Fe carbon fitting ratio. According to the principle that Fe element existed in the form of +3 valence, one can finally determine the carbon addition.

3. The experimental results and discussion

3.1 In-situ XRD Results Analysis on Raw Ore and Carbon Addition Ore

From the DSC/TG curve of raw ore (a of figure 1), one can find a peak at 100 ℃ which is generated from water absorption and gasification. At the same time, it leads to a certain amount of mass loss. In addition, with regard to 430℃, it is a peak caused by loss of crystal water in the process of ore phase change. As for interval from 800℃ to 1000℃, especially after 900℃, there is an obvious upward slope of DSC curve in Baiyuneboite samples. Moreover, the weight starts to reduce a lot.

Contrasting with DSC/TG curve after Baiyuneboite carbon addition ore (b of figure 1), one can find a peak after 900℃ in the curve. Actually, this is because Fe elements are reduced into zero-valent iron. Comparing DSC images of Baiyuneboite raw ore and carbon addition ore, one can found that there are bigger changes between them in the previous peak. However, with regard to weight change intervals of TG curve, they are almost the same. At interval from 800℃ to 1000℃, there are three obvious peaks in the DSC curve of Baiyuneboite carbon addition ore samples. Namely, one wave appears at 850℃. Besides, there are two temperature peaks of 930 ℃ and 960℃ which can be seen as the phase change temperature node. After 930℃, the weight of the ore samples begin to reduce significantly. Moreover, it can be judged that the interval after this is the area for reduction reaction.

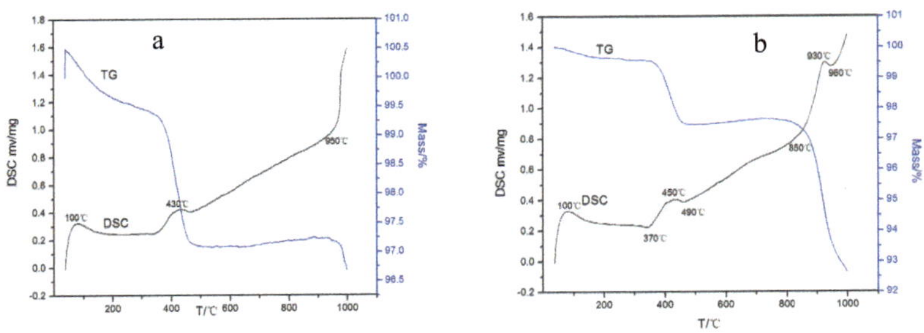

Figure 1 DSC/TG curve of Baiyuneboites raw ore and carbon addition ore
(a is raw ore, b is carbon addition ore)

According to ore phase change of ore samples and temperature points of weight change reflected in the DSC/TG curve, one can finally determine the temperature point of in-situ XRD for raw ore and carbon addition ore. As for raw ore, it selects 25 ℃, 450 ℃, 850 ℃ and 950 ℃ for testing temperature points. While for carbon addition ore, it selects 520℃, 850℃ and 950℃ as testing temperature points. Under these temperatures, one can conduct XRD test for ore samples to determine the ore phase change. The results are shown in figure 2(In figure 2(b), the line in 25℃ is raw ore). In situ XRD performed closed environment, high temperature product gas generated during the reaction will affect the result of the measurement.

In the raw ore, as temperatures increase, ore phase existing way of Fe element and Nb will have more changes. Generally speaking, Fe_3O_4 phase is still the main existing form. However, when temperatures increase, phases like $FeSiO_3$ and $FeNbO_4$ will appear. At the same time, ore phases of ferrite and iron as well as other element's intermetallic compound also begin to appear, such as $MgFe_2O_4$ and CeFe. In addition, at 950℃, one can find that Fe element exist in the form of +3 valence and +2 valence. Thus, it can be judged that Fe_3O_4 makes decomposition. In addition, there are no such big differences for the existing form of Nb elements. Nb elements basically exist in the form of +5 valence.With the increase of temperature, they will gradually change from Nb_2O to Nb_3O_7F with oxide into oxyfluoride.

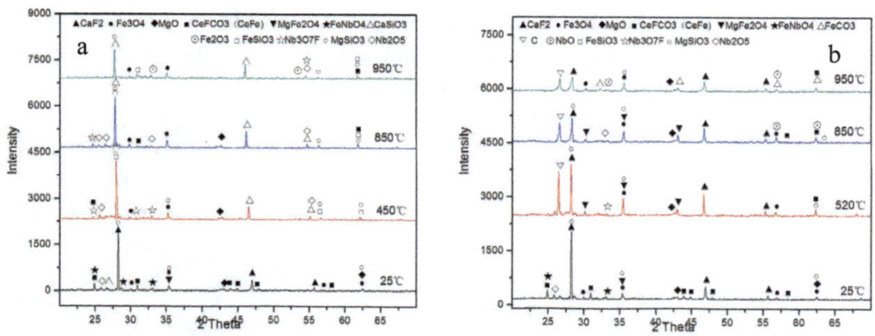

Figure 2 In-situ XRD spectrum of Baiyuneboites raw ore and carbon addition ore
(a is raw ore, b is carbon addition ore)

In the carbon addition ore, ore phase changes are various. After heated to 850℃, peak intensity of Fe_3O_4 weakens obviously and it will completely disappear until 950℃. As a result, Fe element ore phase like $FeSiO_3$ and $FeCO_3$ will take the place in the form of + 2 valence. Moreover, it shows that at 850 ℃, these Fe elements of +3 valence have reached reduction temperature. After 950 ℃, Fe elements of +3 valence will disappear completely. Combining with DSC/TG curve of carbon addition ore, it can be judged that the peak at 930℃ is a reduction process from +3 valence Fe elements of Fe_3O_4 to +2 valence. At the same time, with regard to Nb in the carbon addition ore heated to 850℃, it begin to exist in the form of NbO. And the carbon is depleted by the oxide, and the CaF_2 will discompose.

In the raw ore and carbon addition ore, there are ore phases such as CaF_2, $MgSiO_3$ and $CaSiO_3$, which have not made phase changes. At the same time, in the whole process of roasting, Ce always has ore phase changes. When the temperature getting higher, it will produce metal compounds with Fe elements.

3.2 XRD Analysis of Ore Samples after Particle Classification below 900℃

With regard to bigger differences of main elements for Baiyuneboite under distinct particle sizes in table 2, one can divide ore samples into three particle sizes. Namely, 50-100 mesh, 100-200 mesh and below 200 mesh. Together with ore samples without particle classification, there are totally four groups for comparative analysis. Put Baiyuneboite raw ore and carbon addition ore after particle classification in the box-type furnace for heating. Then set roasting temperature at 900 ℃ and preserve heat for 10 h. Make scanning electron microscope (SEM) and energy spectrum (EDS) analysis and so as to observe microstructure of ore samples and phase structure. Through surface scan, one can determine the elements distribution changes .Compare experimental results of raw ore and carbon addition ore under four particle sizes respectively, it can be shown in figure 3. Making the photos use the secondary electron imaging and the accelerating voltage is 25 kv. Make energy spectrum analysis on the basis of scanning electron microscopy, the results are shown in table 3.

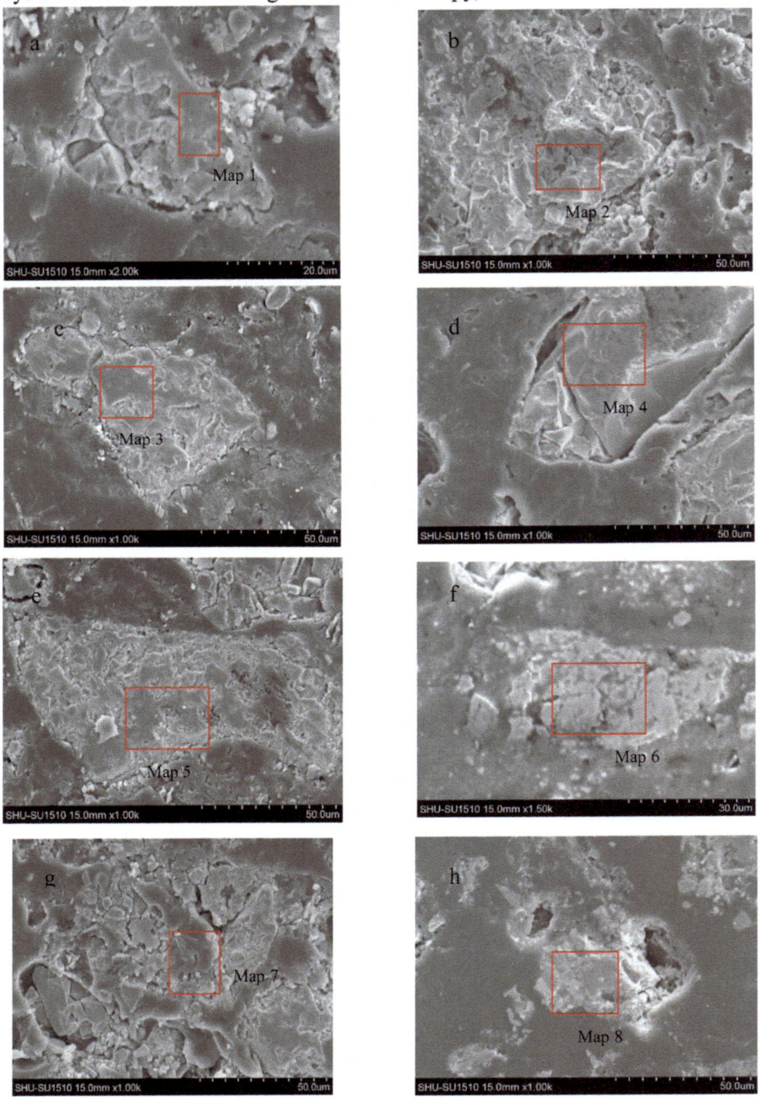

Figure 3 Scanning electron microscope (sem) images of raw ore and carbon addition ore under four particle sizes

(a and b figure: carbon addition ore without particle classification; c and d figure: raw ore and carbon addition ore of 50-100 mesh e and f figure: 100-200 mesh raw ore and carbon addition ore; g and h figure: below 200 mesh raw ore and carbon addition ore)

Table 3 Energy spectrum analysis of ore phase composition in raw ore and carbon addition ore under four particle sizes

	O	Fe	Ca	Si
Map 1	42.81	38.91	0.84	1.76
Map 2	32.06	55.97	0.49	0.99
Map 3	39.23	42.86	1.03	1.36
Map 4	27.40	61.49	0.30	2.69
Map 5	33.14	35.78	2.01	2.36
Map 6	24.45	60.97	0.31	1.26
Map 7	58.24	30.01	1.65	1.33
Map 8	36.12	54.23	0.24	0.99

With regard to microstructure of four group's ore samples, one can adopt tungsten filament sem for surface scan and energy spectrum analysis (each dot place in figure 3). Through scanning electron microscope images and energy spectrum data analysis, one can discover that obvious fractures appear on the surface of ore after 900 ℃ roasting of Baiyuneboite carbon addition ore. With regard to the raw ore after roasting, its surface becomes relatively smooth and some ore will stand out. Thus, it can be inferred that it is the result of reduction reaction. Compared with the shape and appearance of raw ore, carbon addition ore have obvious growth process in the reduction course of ore phase. This is because Fe elements gradually concentrate from original scattered distribution. In addition, through data changes before and after carbon addition for Fe and O atom ratio with energy spectrum analysis, it can also explain that iron oxide has reduction reaction. In terms of raw ore and carbon addition ore after roasting, their Fe/O has obvious differences. In general, after roasting, Fe/O of carbon addition ore is significantly higher than raw ore. At the same time, as particle size of ore sample becoming thinner and thinner, variation of Fe/O also increase correspondingly. Contrast Fe/O changes of map 3, 4, 5, 6, 7, 8, under the condition of different particle sizes, carbon addition ore and raw ore increase relatively. In brief, 50-100 mesh: from 1.09 to 2.24; 100-200 mesh: from 1.08 to 2.49, below 200 mesh: from 0.52 to 1.51. Thus, the thinner of particle size for ore samples become, the easier of Fe elements to be reduced by C. After particle classification, particle sizes of ore samples will become thin. Moreover, it is favorable for Fe to have reduction reaction.

4. Conclusions

(1) Through in-situ XRD test results, there are obvious changes for the existing method of elements in the heating and roasting process, though there are no valence state changes of element in the raw ore. Moreover, Fe elements will change from Fe_3O_4 to the state containing iron salt. In addition, in terms of carbon addition ore taking graphite as reducer, as roasting temperature reaching above 850 ℃, there are obvious reduction process for Baiyuneboite.By DSC/TG curve, it can be inferred that

the temperatures with phase changes are 850℃, 930℃ and 960℃. But no metallic iron formed. Baiyuneboite is too complex, it has many mineral phase, they will affect each other, so the 950℃ temperature is low. The metallic iron will form in 960℃ or more higher temperature. And in-situ XRD experiential, it has no holding time. The reducing process will take some time to react, so the reaction is not complete.

(2) As reduction temperature reaching above 900 ℃, through TG curve, one can clearly see the weight loss phenomenon. Besides, there is obvious exothermic process. Thus, it can be concluded that the process contains reduction of iron and oxygen discharge. As a matter of fact, ore phase composition of Baiyuneboite is very complicated. Therefore, there are significant changes of iron and niobium elements in the whole heating process. All in all, it can be inferred that medium temperature roasting under 1000℃ can change overall ore phase composition to some extent.

(3) As reduction roasting temperature of carbon addition reaching 900℃, by SEM and EDS analysis, there will be significant growth and concentrate phenomenon in the iron ore samples. In addition, on the surface of ore, iron will concentrate. At the same time, with the decrease of particle size, the reduction rate and concentrate degree of Fe elements will improve. Obviously, the thinner of ore samples, the easier for iron to have reduction reaction. However, under this temperature, there is no simple substance niobium precipitated. Thus, it lays a foundation for realizing concentrate and separation of iron and niobium as well as reducing beneficiating difficulty of Baiyuneboite.

References

[1] Jianzhong Cheng, Yunbing Hou, Liping Che. Rational development and utilization of resources in the Bayan Obo rare earth deposits [J].Rare Earth,2007,28(1):70-74.

[2] Yongfu Yu. China's rare earth ore beneficiation technology and its development [J]. China University of Mining Journal,2011,30(6):537-542.

[3] Jian Zhang. Bayan Obo ore beneficiation technology status symbiotic and Prospects [J]. BaoStell Technology,2005,31(4):1-5.

[4] Yongfu Yu, Chaoying Zhu. Baotou Rare Earth Mineral Processing Technology Progress [J]. Metal mines,1999(11):18-82

[5] Xiang Li, Yong Liu, Guanghui Li, Mudan liu, ZhenZhen Liu. New processes to separate and recover the iron and rare earth metals from low grade rougher concentrate of ferro-niobium [J]. Materials Research and Application,2014,8(1):67-72.

[6] Peng Gao, Yuexin Han, Yanjun Li. Bayan Obo ore deep oxidation - Magnetic experimental study [J]. Northeastern University, 2010,31(6):886-889.

[7] Hong Chen, Qiyong Han, Shoukun Wei. New methods of niobium and niobium-containing niobium iron ore extracted from the [J]. Steel,1999,34(3):13-19.

[8] Luo W, Feng Q M, Ou L M, et al. Kinetics of saprolitic laterite leaching by sulphuricacid at atmospheric pressure[J]. Minerals Engineering, 2010, 23(6):458-462

[9] Zengyu Li. Thermal Analysis [M]. Beijing: Tsinghua University Press, 1987: 220

DRYING, ROASTING, AND CALCINING OF MINERALS

SINTERING AND ENERGY USE

Session Chairs:
Mingming Zhang
Dean Gregurek

Drying, Roasting, and Calcining of Minerals
Edited by: Thomas P. Battle, Jerome P. Downey, Lawrence D. May, Boyd Davis, Neale R. Neelameggham, Sergio Sanchez-Segado, and P. Chris Pistorius
TMS (The Minerals, Metals & Materials Society), 2015

EFFECT OF SINTER BASICITY ON SINTER PRODUCTIVITY AND QUALITY WITH HIGH RATE OF RECYCLED MATERIALS

Mingming Zhang[1], Marshall S. Coe[2], Marcelo W. Andrade[3]

[1]Sr. Research Engineer, ArcelorMittal Global R&D, East Chicago, IN 46312, USA
[2]Plant Manager, ArcelorMittal Indiana Harbor, East Chicago, IN 46312, USA
[3]Manager, ArcelorMittal Global R&D, East Chicago, IN 46312, USA

Keywords: Ironmaking, Sinter, Pot-grate Sintering, Productivity, Quality

Abstract

To supply iron units and flux burden in blast furnaces, sinter basicity is one of the vital parameters for sinter plants. The sinter basicity also plays an important role in sinter quality since it imparts the formation of different mineral phases in sinter. Pot-grate sintering tests were carried out with low and high basicity to investigate the effects of sinter basicity on sinter mineralogy, properties and productivity. It was found that with an increase in sinter basicity, both sinter productivity and reducibility (RI) increased. Tumbler index (TI) and low temperature degradation index (LTD) of high basicity sinter were comparable to low basicity sinter. Softening and melting tests showed that sinter with 3.5 C/S basicity had slightly higher softening temperature comparing to lower basicity sinter with C/S 2.6, acid and fluxed pellets. XRD analyses confirmed the formation of silico-ferrite of calcium and alumina (SFCA) and calcium ferrite (CF) in the tested sinter samples.

Introduction

In the ironmaking process, sinter is an important burden materials for blast furnaces due to its desired physical, chemical and metallurgical properties. Depending on the quality of iron ore and sinter properties, an average of 20-70% sinter is used in the blast furnaces [1].

Sinter basicity is one of the most important process parameters which is typically controlled by dosing limestone into the sinter mix. Optimizing the basicity of the sinter can be a cost-effective way of improving the techno-economic indices of sinter plant operation [2], [3]. Flux addition often results in an increase in the rate of sintering due to two factors, i.e., a reduction in air volume required for sintering and an improvement in the average sintering permeability of the sinter mix.

Sinter plants in North America have been mainly iron unit recycling facilities due to the unavailability of local sinter feeds and high logistics cost for procuring sinter feeds. The sinter plants play an important role in an integrated steel mill for reducing costs and avoiding landfills. Among the raw materials used in the sinter plants, iron ore concentrates, ironmaking and steelmaking slags, dusts generated from gas cleaning system, coke breeze, water treatment plant sludge, and fluxes are major carbon and iron units. Sinter plants help reduce these solid waste by combusting waste products and capturing trace iron present in the mixture.

With an increasing amount of ironmaking and steelmaking by-products being recycled at sinter plant, sinter basicity is often affected by flux constituents which are typically pre-slagged (for example, C fines and BOF slags). These flux constituents may be of limited advantage for sintering since slag formation occurred already and they are difficult to mix with ores [2]. There is a vast amount of published work relating to the effects of flux additions and only a limited selection of literature is available discussing the effects of sinter basicity with high amount recycled materials.

This investigation was carried out at a laboratory scale pot-grade sintering test facility located at ArcelorMittal Global R&D East Chicago. The purpose of this study was to support sinter plant's effort to increase sinter production by increasing sinter basicity and to investigate the effects of sinter basicity on sinter mineralogy, sinter quality and productivity with high amount of recycled materials.

Experimental

Materials

The chemical compositions and size of the raw materials used in this study are shown in Table I and II, respectively. All materials were used as received except limestone which was screened at 6 mesh (3.36 mm) and the undersize fraction was used for sintering tests.

Table I. Chemical analyses of raw materials used for sintering tests

Raw Materials	Total Fe	FeO	SiO2	CaO	Al_2O_3	MgO	C	LOI
Sinter Blend[#]	44.51	23.1	6.61	17.32	3.00	4.50	4.26	/
Iron Ore Blend[&]	60.03	1.50	4.94	6.03	0.41	1.72	/	/
Return Fines*	52.84	6.86	4.67	14	1.62	3.27	/	0
Limestone	2.35	/	1.41	49.35	0.25	1.89	/	43.54
Solid Fuel (Flue Dust)	9.92	/	4.3	2.66	2.65	0.76	68.31	72.8

[#] Mainly recycled materials, 30% of total sinter raw mix

[&] Mainly fluxed pellet fines

* Chemistry for return fines may vary based on sinter basicity

Table II. Size distribution of major components in sinter blend

Size	Sinter Blend	Iron Ore Blend	Return Fines	Limestone	Solid Fuel (Flue Dust)
+12.70 mm	6.77	0.00	0.79	0.00	0.00
-12.70 + 6.35 mm	12.83	2.24	7.53	19.70	0.36
-6.35 + 1.68 mm	39.85	72.09	62.78	18.20	16.76
-1.68 + 0.84 mm	11.15	9.26	14.74	33.10	46.49
-0.84 + 0.25 mm	15.87	6.87	6.96	17.70	28.11
-0.25 + 0.15 mm	6.71	2.10	3.63	6.40	5.23
-0.15 mm	6.82	7.43	3.58	4.90	3.06

Methods

Sintering tests were conducted in a laboratory pot-grate sinter pot with a diameter of 400 mm and a height of 500 mm. A schematic diagram of the sintering apparatus is shown in Figure 1. The test was intended to simulate industrial sintering process with capability of high vacuum (up to 4800 mm W.G., or 47 kiloPascal) and high temperature ignition (up to 1350 °C).

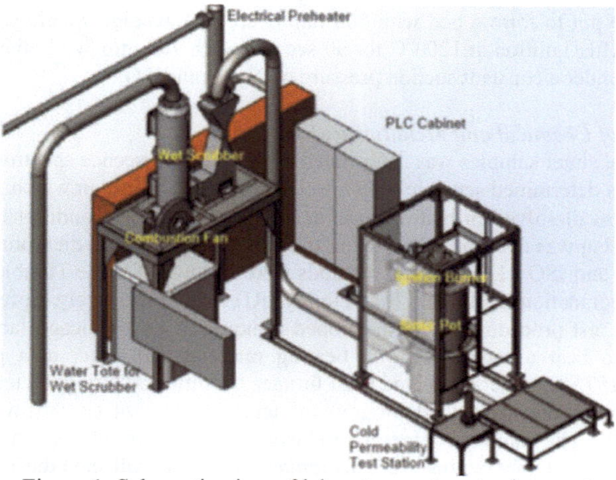

Figure 1. Schematic view of laboratory pot-grate sinter pot

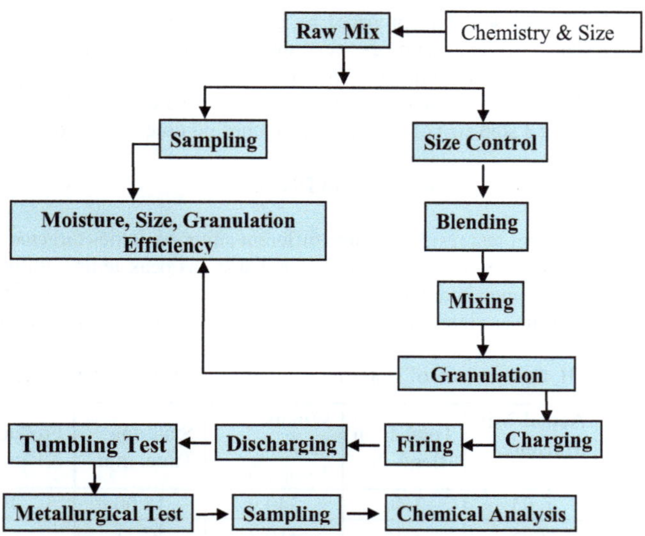

Figure 2. Flowchart of sinter pot-grate testing procedure

Sintering tests include proportioning, mixing, ignition, sintering, cooling and conditioning. The sinter mix was fixed as follows if not otherwise specified: sinter blend 30%; iron ore blend 30%; return fines 27%, solid fuel (flue dusts) 3.0%, basicity (CaO/SiO_2 ratio) ranged from 2.6 to 3.5 and the target moisture was set to 6.5%. The sinter mixture was proportioned and mixed manually and then charged in a cement mixer for dry mixing. Predetermined amount of water was then added into the mixer and the mixing was continued. After mixing the mixture was charged into the pot to form a bed with 500 mm deep. The average weight of each charge was about 150 kg. After ignition at 1200°C for 90 seconds with 700 mm W.G., the sintering process was conducted under a constant suction pressure of 2100 mm W.G.

Determination of Chemical and Metallurgical Properties

The chemistry of sinter samples was determined by X-ray fluorescence spectrometry (XRF). The FeO content was determined separately by a wet chemical procedure in which a portion of finely ground sinter was dissolved in hydrochloric acid in non-oxidizing conditions and the resulting iron (II) in solution was determined by redox titration with potassium dichromate. ASTM E279-97, ISO 4696-1 and ISO 11258 testing methods were followed for the Tumble Index (TI), Low Temperature Degradation (LTD) and Reducibility (RI) tests, respectively. Softening and melting test facility and test procedures were developed in-house at East Chicago laboratory. This test was designed to heat a sample of iron bearing material and coke, in a packed bed, to a temperature of 1575°C, under simulated blast furnace conditions. During the test, the sample was subjected to a computer-controlled program of temperature, gas composition and flow, and mechanical load. The rate and degree of reduction, contraction, change in permeability, and softening and melting temperatures, were determined from data collected during the test.

X-ray Diffraction (XRD) Analysis

The samples for XRD analyses were prepared by grinding the sinter samples down to 100% passing 100 mesh. The powdered samples were then spread over a flat sample holder. The diffraction patterns were obtained employing an X'pert Pro Diffractometer by PANalytical using Ni filtered Cu Kα radiation at 35kV and 50 mA. The samples were scanned at room temperature from 20 to 130 degrees at a step size of 0.10°, with a recording time of 4 seconds for each step.

Results and Discussion

Table II shows the sinter pot test results of three different sinter basicities targeted at 2.6, 3.0 and 3.5, respectively. It can be seen that both windbox exhaust gas peak temperatures and sintering time decreased with increasing sinter basicity. The calculated sinter pot productivity followed the same trend as sintering time as sinter yields were stable and not affected by sinter basicity.

Table II. Comparison of sintering parameters at different basicities

Sinter Basicity (C/S)	Actual Mix Moisture, %	Total C in Blend, %	Windbox Peak Temp., °C	Sintering Time, min	Productivity, $NT/m^2 \cdot 24hrs$
2.6	6.95	3.40	340.9	18.00	41.2
3.0	7.05	3.44	310.2	17.21	45.6
3.5	7.10	3.45	283.1	16.75	51.7

Chemical analyses of the sinter samples were performed by X-ray fluorescence (XRF) spectrometer. The results are summarized in Table III. Total Fe is the dominant element in all three samples with weight percentage ranging from 55.20 to 51.48. The decrease of total Fe content in sinter was caused by the increasing percentage of limestone as needed to reach target basicity level.

Table III. Chemical analyses of sinter samples with different basicities

	Low Basicity Sinter	**High Basicity Sinter**
C/S	2.6	3.5
Total Fe	55.20	51.48
FeO	10.07	8.60
SiO_2	4.44	5.08
Al_2O_3	1.46	1.50
CaO	11.62	17.77
MgO	2.33	2.64
K_2O	0.11	0.05
Na_2O	0.13	0.09
S	0.01	0.01
P	0.01	0.01
TiO_2	0.17	0.12
MnO	0.71	0.79

Table IV shows the sinter physical and metallurgical properties for two sinter samples with low (2.6) and high (3.5) basicity levels. It can be seen that high basicity sinter (C/S: 3.5) showed superior reducibility comparing to low basicity sinter, acid and fluxed pellets. The sinter tumble, abrasion indices and LTD for low and high basicity sinter were similar and there was no significant difference. The softening temperature of high basicity sinter was slightly higher than low basicity sinter and pellets but the difference was not significant either.

Table IV. Comparison of physical and metallurgical properties of sinters and pellets

Tests	Unit	Pellet		Sinter	
		Fluxed	Acid	Low Basicity	High Basicity
C/S		0.86	0.14	2.6	3.5
Tumble Index	%+6.35mm	-	-	76.2	77.5
Abrasion Index	%-0.6mm	-	-	4.98	4.52
ISO LTD	%wt. loss	2.59	3.28	0.06	0.06
500°C	%+6.35mm	93.68	79.25	80.54	78.44
	%+3.35mm	95.66	8.25	90.68	92.41
	%-0.5mm	3.56	5.31	3.18	1.47
Reducibility	dR/dt@ 0.90	0.97	0.75	0.89	1.09

(Fe/O Ratio)	dR/dt@ 0.75	0.74	0.65	0.74	0.93
	R40	1.04	1.02	1.12	1.48
% Red @75 min		80.83	74.50	78.67	83.20
Melting Test	Soft Temp (Ts)	1314	1328	1391	1404
max dP	Melt Temp (Te)	1469	1534	-	-
max dP	kPa (rel)	1.34	2.13	3.71	3.06
Range	Te – Ts	155	206	-	-

XRD analysis of the sinter samples revealed (Figure 3) that major mineral phases present in the sinter samples were Calcium Ferrites (CF), Hematite (H), Magnetite (M), Silico-ferrite of Calcium and Alumina (SFCA). Minor phases were Wustite (W), Silicates (S), Calcium Silicates (CS) and Iron Silicates (IS). Although SFCA and CF phases were identified in all three sinter samples with different basicities, their peaks were obviously stronger in higher basicity sinter sample comparing to lower basicity sinter sample. CF and SFCA are desirable phases for sinter because they are beneficial to form bonding matrix in sinter and also easily reducible. They are typically formed by a solid-liquid reaction between hematite and $Fe_2O_3{\cdot}CaO$ melt, with the subsequent assimilation of SiO_2 and Al_2O_3 in the melt during sintering. It has been reported that SFCA could only be present in iron ore sinter if the temperature remained below 1300°C, or the residence time at higher temperature was extremely short. If the temperature exceeded 1300°C, then SFCA would transform to Fe_2O_3 (at high pO2) or Fe_3O_4 (at low pO2) in a calcium silicate-rich melt [4]. Also acicular ferrite (<10 micron) formed at low temperature (<1300°C) is more reducible, while columnar ferrite (>10 micron) formed at high temperature (>1300°C, possibly coated with glass) is less reducible [5].

It should be noted that windbox gas peak temperature (indication of flame-front temperature) decreased with increasing sinter basicity (shown in Table II). This decrease in flame-front temperature was caused by endothermic reactions needed for calcining additional limestone in the sinter mix. More limestone addition could also accelerates the temperature cycle and reduce the extent of high temperature zone. As a result, more slag bond volume was generated and sinter yield was higher with high basicity sinter.

In the XRD patterns shown in Figure 3, calcium carbonate ($CaCO_3$) peaks were present most in higher basicity sinter. It could be residue limestone remained in sinter product which did not get calcined due to their coarser particle size. Shigaki et al. studied the effects of limestone particle size on sinter productivity and reduction disintegration index [6]. It was found that delayed decomposition of limestone and melt formation by increased size of limestone particles increases the permeability of the sintering bed and makes the heat pattern sharp, which leads to an increase in productivity and better quality sinter on the degradation characteristic during reduction.

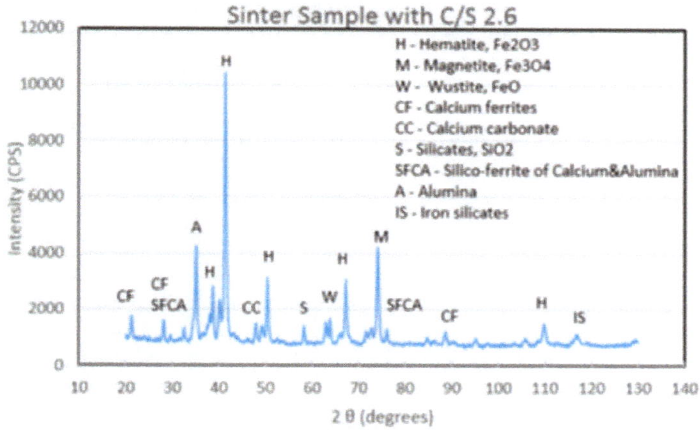
Sinter Sample with C/S 2.6
H - Hematite, Fe2O3
M - Magnetite, Fe3O4
W - Wustite, FeO
CF - Calcium ferrites
CC - Calcium carbonate
S - Silicates, SiO2
SFCA - Silico-ferrite of Calcium&Alumina
A - Alumina
IS - Iron silicates
Intensity (CPS)
2 θ (degrees)

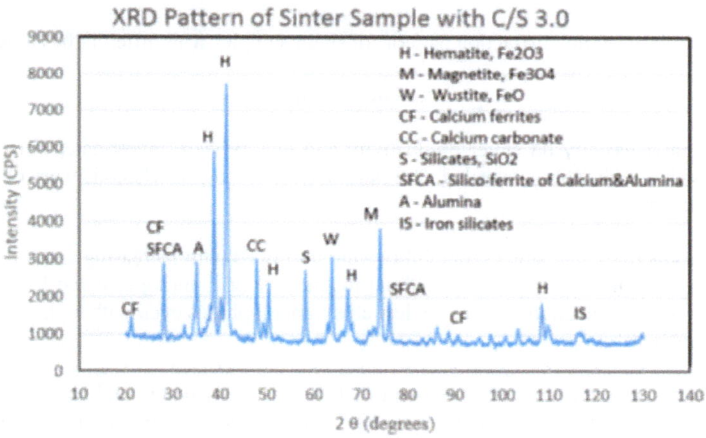
XRD Pattern of Sinter Sample with C/S 3.0
H - Hematite, Fe2O3
M - Magnetite, Fe3O4
W - Wustite, FeO
CF - Calcium ferrites
CC - Calcium carbonate
S - Silicates, SiO2
SFCA - Silico-ferrite of Calcium&Alumina
A - Alumina
IS - Iron silicates
Intensity (CPS)
2 θ (degrees)

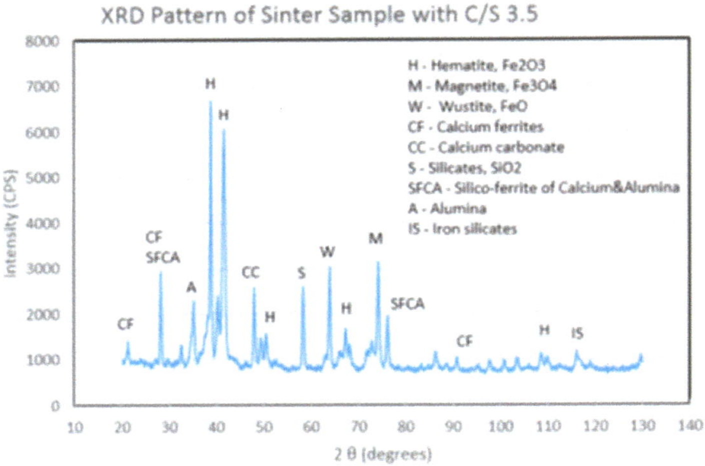

Figure 3. X-ray diffraction patterns of sinter samples with different basicities

Conclusions

Effects of sinter basicity on sinter productivity and quality were investigated for sinter plants operating with a high rate of recycled materials. The following conclusions were obtained from the pot-grate sintering tests:

1). Sinter productivity and sinter reducibility were improved with increasing sinter basicity up to 3.5. Increase of sinter basicity was beneficial for increasing sintering rate and lowering off-gas temperature, which helped to form more desirable mineral phases even with high rate of recycled materials (up to 30%).

2). XRD analyses confirmed the formation of silico-ferrite of calcium and alumina (SFCA) and calcium ferrite (CF) in the tested sinter samples. With increasing sinter basicity, the intensities of SFCA and CF peaks became stronger which indicated more SFCA and CF phases were present in higher basicity sinter.

Acknowledgements

The authors would like to thank ArcelorMittal Global R&D management for permissions to submit this paper for publication and local sinter plant personnel's support for supplying raw materials and sharing information.

References

[1]. Julius H. Strassburger, *Blast Furnace – Theory and Practice*, (Gordon and Breach Science Publishers, 1969), 211-215.

[2]. D. F. Ball, J. Dartnell, J. Davison, A. Grieve, and R. Wild, *Agglomeration of Iron Ores*, (Heinemann Educational Books Limited, London, 1973), 136-139.

[3]. M. Zhang, "Production of High Basicity Sinter" (Report RMIM0714, ArcelorMittal Global R&D, 2014).

[4]. N. A. S. Webster, et al., "Silico-ferrite of Calcium and Aluminum (SFCA) Iron Ore Sinter Bonding Phases: New Insights into Their Formation during Heating and Cooling," *Metallurgical and Materials Transactions B*, 43(6) (2012), 1344-1357.

[5]. M. Sasaki and Y. Hida, "Consideration on the Properties of Sinter from the Point of Sintering Reaction," *Tetsu-To-Hagane*, 68(6) (1982), 563–571.

[6]. I. Shigaki, et al., "Improvement of Productivity and Reduction Disintegration of Iron Ore Sinter by Increasing Size of Limestone Particles," *Tetsu-to-Hagane*, 71(16) (1985), 1880-1887.

Drying, Roasting, and Calcining of Minerals
Edited by: Thomas P. Battle, Jerome P. Downey, Lawrence D. May, Boyd Davis, Neale R. Neelameggham, Sergio Sanchez-Segado, and P. Chris Pistorius
TMS (The Minerals, Metals & Materials Society), 2015

RESEARCH ON NOx REDUCTION BY APPLYING COKE BREEZE PRETREATED WITH UREA ADDITIVE IN IRON ORE SINTERING PROCESS

Xiao-hui Fan, Zhi-yuan Yu[1], Min Gan, Xu-ling Chen, Yun-song Huang, Tao Jiang, Guanghui Li

(School of Minerals Processing and Bioengineering, Central South University, Changsha, Hunan 410083, China)

Keywords: NOx reduction; pretreated coke breeze; additives; iron ore sintering

Abstract

Compared with the traditional de-NOx processes by cleaning the waste gas, a new approach for reducing the formation of NOx during sintering process is proposed. Since more than 90vol% of NOx comes from the fuel combustion, the influence of sintering coke breeze pretreated by $CO(NH_2)_2$ additive on the NOx emission is researched. As adding $CO(NH_2)_2$, the reduction agents (NH_i) are generated during the pyrolysis of $CO(NH_2)_2$, which reinforce the NOx reduction, And the appropriate content of $CO(NH_2)_2$ is 1.0 mass% of coke breeze.

Introduction

The NOx reduction of industry has been one of the main interests since it can cause acid rain and photochemical fog, which has a harmful effect on human livings. The annual NOx emission from sintering machine, representing the major resource of NOx emission in iron and steel industry [1]. Therefore, it has a great significance by taking de-NOx technology in iron ore sintering process.

The traditional de-NOx processes in iron ore sinter plants covers process-integrated techniques and end-of-pipe measures. The process-integrated techniques are applied barely except the waste gas recycling techniques since its efficiency is less significant[2]. The end-of-pipe measures including the Selective Catalytic Reduction (SCR) process or Selective Non-Catalytic Reduction (SNCR) process treat the waste gas by removing the nitrogen oxides from the flue gas[3]. But the huge capital investment and the operational cost will strongly reduce the competitiveness of steelworks. Therefore, a more attractive countermeasure is to depress the formation of these pollutants within the sinter bed.

As reported by references, more than 90vol% of NOx comes from the fuel in iron ore sintering, which is formed as fuel-NOx[4,5]. So a route to depress the formation of fuel-NOx from the beginning of combustion is more significant instead of just reducing

---

[1] **Foundation item:** Project(2013bjjxj015) supported by the Outstanding and Creative Doctor Scholarship of Central South University, China; Project supported by the Hunan Provincial Innovation Foundation for Postgraduate(CX2014B094), China
Corresponding author: Zhi-yuan Yu, PhD; Tel: +86−13975104962; E-mail: y13975104962@126.com

the NOx emission by absorbing agent [6]. Therefore, the aim of this research is to reduce the formation of NOx by applying the coke breeze pretreated with urea additive during sintering.

Experimental

Properties of raw materials

Iron ore blending, fuel and fluxes (including dolomite, limestone and quicklime) were utilized to produce sinter with TFe 58.3%, SiO_2 4.83%, basicity ($R=CaO/SiO_2$) 1.85 and MgO 1.6%. The chemical compositions of raw materials and their mass fractions are given in **Table 1**.

Table 1. Chemical composition of raw materials and their percents in mixture / mass%

Raw materials	TFe	SiO_2	CaO	MgO	Al_2O_3	FeO	LOI	Ratio
Blending ore	60.01	4.39	2.84	0.61	1.37	4.59	5.16	58.96
Dolomite	0.32	0.41	31.66	21.38	0.01	0.39	45.56	2.48
Limestone	0.22	0.88	55.21	0.25	0.24	0.24	42.88	3.16
Burnt lime	0.30	3.45	70.77	0.39	0.68	0.36	23.80	1.50
Coke breeze	0.91	6.40	1.74	0.26	4.05	0.01	85.64	3.90
Return fines 1	57.06	5.14	9.47	1.84	1.70	6.64	0.77	20.00
Return fines 2	57.89	4.63	9.71	1.53	1.70	7.00	0.32	10.00

The properties of coke breeze used in the experiment were given in **Table 2**. It can be seen that the FC_{ad} of coke breeze was 82.72 mass% while V_{daf} of that was 2.92 mass%, which was a typical kind of coke used in sintering process. Otherwise, the elements of S and N composition in the coke breeze were tested, which were respectively 0.78 mass% and 0.72 mass%.

Table 2. Properties of coke breeze

Proximate analysis*/mass%			Main chemical composition of ash/mass%					S, N and H composition/mass%		
A_d	V_{daf}	FC_{ad}	TFe	SiO_2	Al_2O_3	CaO	MgO	S_{total}	N_{total}	H_{total}
14.36	2.92	82.72	6.34	44.57	20.20	12.12	1.81	0.78	0.72	0.33

* A_d: ash content (dry basis); V_{daf}: volatile matter content (dry ash-free basis); FC_{ad}: fixed carbon content (air dried basis).

Preparation of pretreated coke samples

The coke breeze was pretreated by $CO(NH_2)_2$ additive, which was AR (analytical reagent) [5]. The experimental procedure was described as follows. The solution with $CO(NH_2)_2$ was prepared after the AR of $CO(NH_2)_2$ adding into beaker, where the solution was stirred by a rod for 10min with 20r/min speed. Then the coke breeze was charged into the beaker and soaked for 30min. The $CO(NH_2)_2$ powder were adsorbed and distributed on the surface or pore of coke breeze because of its adsorption property. The coke sample pretreated by $CO(NH_2)_2$ was obtained after drying at 110℃ for 4 hours. Meanwhile, a series of pretreated coke breeze with different amount of $CO(NH_2)_2$ were achieved through regulating the $CO(NH_2)_2$ concentration of solution.

Combustion tests

The combustion behavior of coke breeze in sintering process will be changed as well as the NOx emission after pretreated with additives. In order to reveal the influence mechanism, the combustion behavior of pretreated coke breeze was tested in the device of 30mm diameter that described in **Figure 1**. The experiments conducted as follows, 2g pretreated coke breeze which were mixed with silica sands in advance, were charged into the cup of 25mm diameter with 20 holes distributed evenly at the bottom. The particle size of charging material was controlled in 5~8mm while the height of packed bed was about 15mm. And the artificial air which was controlled as $0.7m^3/h$, was introduced into the quartz tube from the bottom after mixing, which was passed through the holes of charging cup; meanwhile the temperature was regulated and controlled by the temperature controller. The emission rule of NOx in outlet gas was detected by gas analyzer along with the heating period. After doing research of combustion tests, the results were validated by using the sintering pot test which included proportioning, mixing, granulation, ignition, sintering, cooling and screening. And it was evaluated by vertical sintering speed, productivity, yield, tumble index and NOx reduction. Each group of experiments was conducted parallelly, and the average data was adopted in this paper. The Nitrogen Conversion Rate (***NCR***) was defined as the ratio of nitrogen to NOx while the reduction of NOx emission in combustion process was described as ***R*** (rate of NOx reduction), and it was calculated by the following formulas.

$$NCR = \frac{14\int_0^t V_t dt\upsilon}{22.4N_{total}} \times 100\% \tag{1}$$

$$R = \frac{\int_0^t V_{t0}dt - \int_0^t V_t^* dt}{\int_0^t V_{t0}dt} \times 100\% \tag{2}$$

Where: "V_t"-the NOx emission volume fraction at "t" time; "υ"-flow rate; "N_{total}"-the total nitrogen mass in coke breeze; "V_{t0}"-the NOx emission volume fraction of non-pretreated coke breeze at "t" time; "V_t^*"-the NOx emission volume fraction of pretreated coke breeze at "t" time.

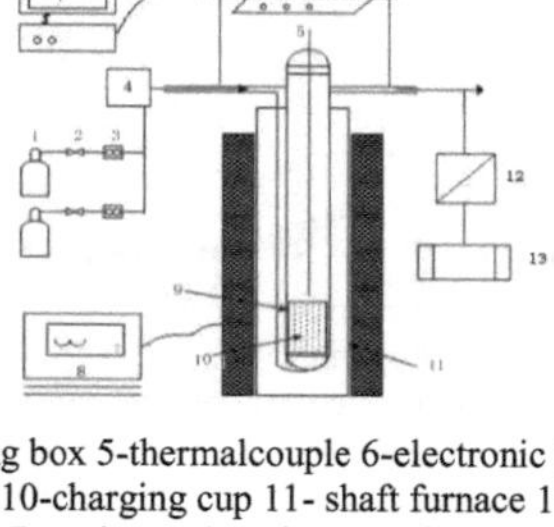

1-gas 2-valve 3-flowmeter 4-mixing box 5-thermalcouple 6-electronic balance 7-computer 8-temperature controller 9-quartz tube 10-charging cup 11- shaft furnace 12-dryer 13-gas analyzer

Figure 1. Experimental equipment of combustion tests

Results and Discussion

Effect of $CO(NH_2)_2$ additive on fuel combustion

There are three most widely used nitrogenous reduction agents: ammonia, urea as well as ammonium bicarbonate. Because urea exists as stable solid, it has advantages in transportation and storage. The effect of $CO(NH_2)_2$ additive content on fuel combustion behavior under different temperatures was investigated, which was given in **Figure 2**. It can be seen that with the increase of temperature, the burnout time (time that CO_2 content in off-gas is lower than 1.0%) of coke breeze were significantly decreased, which meant the fuel emission was enhanced; however, the CO_2 content in off-gas and the CO_2 emission velocity were increased in case of applying coke breeze pretreated with $CO(NH_2)_2$ additive, especially under 700℃. It is mainly because some active units (such as H, OH and O etc.) are produced during the pyrolysis of urea, which can promote the coke combustion.

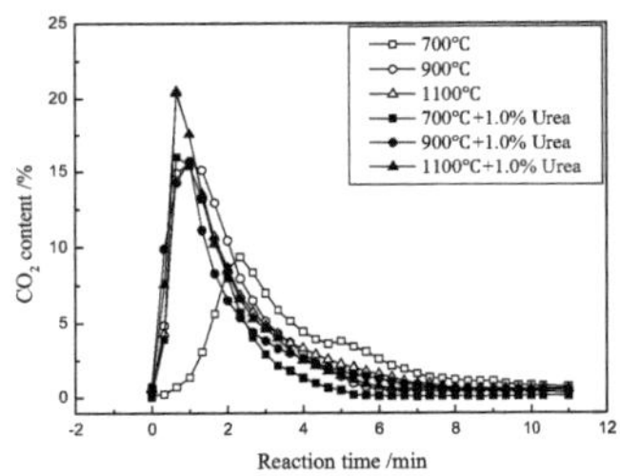

Figure 2. Effect of urea additive under different temperatures on fuel combustion behavior

The effect of $CO(NH_2)_2$ additive content in coke breeze on fuel combustion behavior under 900℃ was researched. As described in **Figure 3**, the CO_2 emission rules were matched with each other when the combustion temperature was 900℃. However, the burnout time of fuel was decreased with the increase of urea additive content in coke breeze, which meant the fuel combustion was enhanced in part.

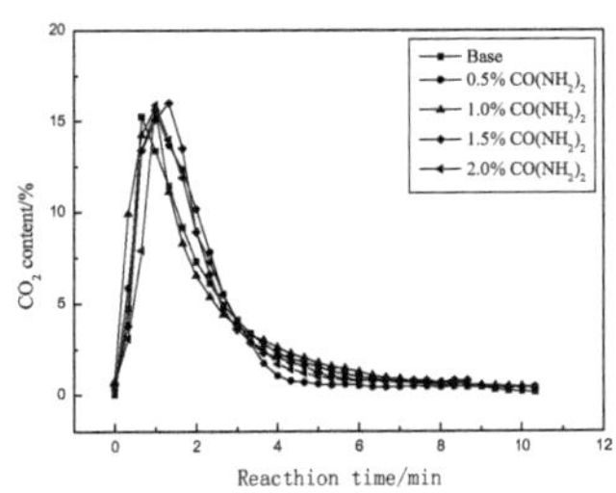

Figure 3. Effect of $CO(NH_2)_2$ additive content on fuel combustion behavior under 900℃

Effect of $CO(NH_2)_2$ additive on NOx emission

The effect of $CO(NH_2)_2$ additive content on NOx emission was researched, and the result was given in **Figure 4**. It can be seen that the NOx emission was reduced when the urea additive content increased to 1.0%. However, when the urea additive content was 1.5% or 2.0%, the NOx emission peak was significantly increased, as well as the total amount of NOx emission in off-gas. It is mainly because the pyrolysis products of urea are NH_i active units, which can not only act as the reducing agents of NOx reduction but also be oxidized to NOx. Thus, the NOx emission during fuel combustion can be reduced by applying coke breeze pretreated with an appropriate content urea additive.

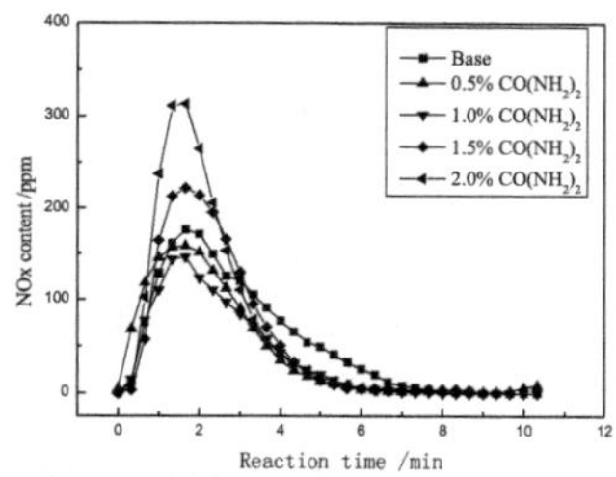

Figure 4. Effect of $CO(NH_2)_2$ content on NOx emission

As shown in **Figure 5**, the ***NCR*** was decreased from 43.34% to 30.15% as the total NOx emission decreased from 70 mg/m^{-3} to 48 mg/m^{-3} when the urea additive content increased from 0% to 1.0%, and the same time the ***R*** was increased to 30.43%; while continued to increase the urea additive content, the ***NCR*** increased while the ***R*** began to decrease since the total amount of NOx emission increased. Therefore, the appropriate urea additive content is about 1.0%.

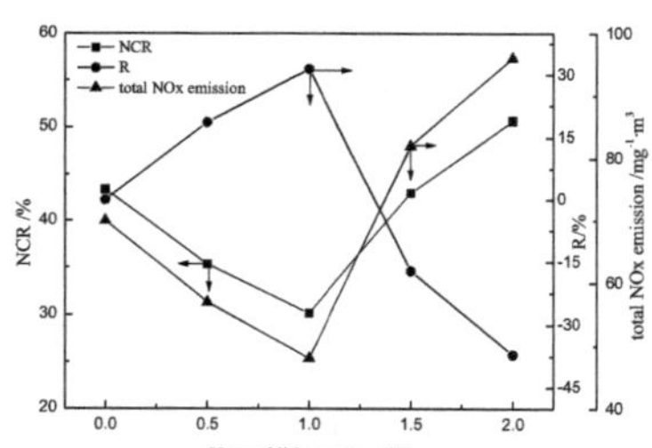

Figure 5 Effect of $CO(NH_2)_2$ content on NCR and NOx reduction rate

The pyrolysis of urea will be conducted as the reaction equation (3) and (4) during the heating process. The nitrogenous products of pyrolysis mainly include: NH_2, HNCO and NH_3, which are the reducing agents for the NO reduction during the combustion process.

$$CO(NH_2)_2 \xrightarrow{\Delta} NH_2 + H + HNCO \tag{3}$$

$$CO(NH_2)_2 + H_2O = 2NH_3 + CO_2 \tag{4}$$

The NO reduction reactions will conduct as the following reaction equations:

$$NH_3 + OH \leftrightarrow NH_2 + H_2O \tag{5}$$

$$NH_2 + NO \leftrightarrow NNH + OH \tag{6}$$

$$NH_2 + NO \leftrightarrow N_2 + H_2O \tag{7}$$

$$NNH + NO \leftrightarrow N_2 + HNO \tag{8}$$

$$HNCO + OH \leftrightarrow NCO + H_2O \tag{9}$$

$$NCO + NO \leftrightarrow N_2O + CO \tag{10}$$

$$NCO + NO \leftrightarrow N_2 + CO_2 \tag{11}$$

As presented in the above equations, the active units (like OH, O etc.) are the necessary for the NOx reduction. Therefore, the effect of NOx reduction will reach a high rate at 900℃ since the active units generation starts to conduct under 900℃. The generation reactions of active units include:

$$H + O_2 \leftrightarrow OH + O \tag{12}$$

$$O + H_2O \leftrightarrow OH + OH \tag{13}$$

The active units used to induce the prophase NOx reduction reaction are supplied by the reaction (12) and (13), and then the later active units are mainly from the reaction (6). The NOx emission of coke breeze pretreated with 1.0% urea under different temperature was researched, and it was described as **Figure 6**. Compared with base, when the temperature increased from 700℃ to 900℃, the total NOx emission was decreased with the existence of 1.0% urea. The NOx reduction under 900℃ was enhanced significantly even though the NOx generation was risen with the increase of temperature. However, the NOx emission continued to increase when the temperature was 1100℃, which meant a higher temperature was adverse to the NOx reduction.

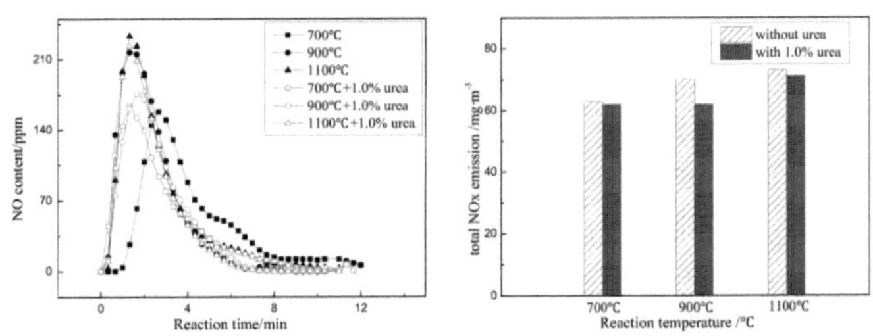

Figure 6. Effect of urea additive on NO emission under different temperatures

Effect of $CO(NH_2)_2$ additive on NOx emission in sintering process

The effect of applying coke breeze pretreated with $CO(NH_2)_2$ additive on sintering process was researched. **Table 3** was described the comparison of sinter indexes with different coke breeze samples. The sinter indexes were changed a little since the coke breeze applying in sintering process was pretreated with different content additives. However, the variation was incredibly small because the additives dosage was just 0.5~2.0 mass% of coke breeze. It can be assumed that the sinter indexes were merely changed by applying the coke breeze pretreated with 1.0 mass% of urea additive.

Table 3. Comparison of sinter indexes with three different coke breeze samples

$CO(NH_2)_2$ content/%	Vertical velocity/mm·min^{-1}	Yield /%	TI/%	Productivity /t·(m^2·h)$^{-1}$
0.0	25.26	62.38	49.20	1.36
0.5	25.56	61.62	50.00	1.32
1.0	25.61	61.68	50.00	1.34
1.5	25.50	61.96	49.60	1.34
2.0	25.22	61.62	50.00	1.32

The effect of additives on NOx emission in sintering process was studied, and it was presented in **Figure 7** and **Table 4**. Compared with the base, the NOx emissions of applying coke breeze with additives in sintering process were reduced obviously except when the urea additive content was 2.0%; the ***NCR*** was decreased from 60.32% to 50.53% and the ***R*** arrived at 16.23% on the condition that the urea additive content was 1.0%. Meanwhile, as shown in **Table 4**, the NOx emission from sinter machine applying with base coke was 285 $g \cdot t^{-1}_{sinter}$ while it was reduced gradually with the increase of urea additive content. The NOx emission was 258 $g \cdot t^{-1}_{sinter}$ when the urea content was 1.0%. However, the ***NCR*** and ***R*** were increased when the urea content raised to 2.0%. Therefore, the NOx emission from sintering machine can be reduced by applying coke breeze pretreated with 1.0% urea additive in iron ore sintering process.

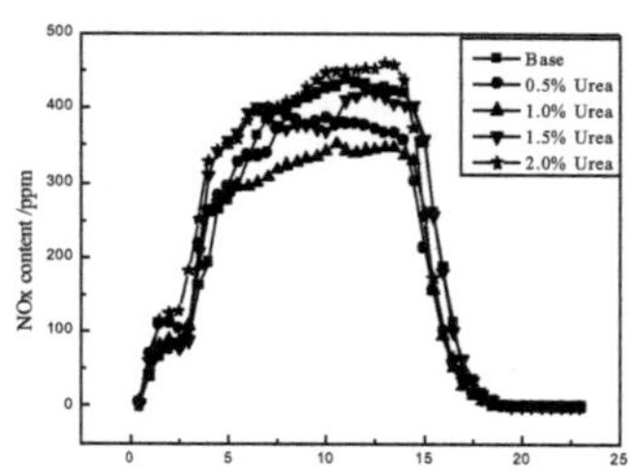

Figure 7. Effect of additives on NOx emission in sintering process

Table 4. Comparison of nitrogen conversion with three different coke breeze samples

$CO(NH_2)_2$ content/%	***NCR***/%	Total NOx emission/$g \cdot t^{-1}_{sinter}$	***R***/%
0.0	60.32	308	-
0.5	55.01	281	8.77
1.0	50.53	258	16.23
1.5	61.26	313	-1.62
2.0	64.45	329	-6.85

Conclusion

(1) When the sintering coke breeze was pretreated by 1% urea additive, the NCR was decreased from 43.34% to 30.15% ,and The NOx emission is reduced from 308 $g{\bullet}t^{-1}$sinter to 258 $g{\bullet}t^{-1}$sinter .

(2) When the the content of urea was over 1%, the total NOx emission increased. The reason is that the nitrogenous (including NH_2, HNCO and NH_3) products during the pyrolysis of urea, which are used as the reducing agents of NOx reduction and also can be oxidized to NOx.

References

1.Y. G. Chen, Z. C. Guo, G. S.Feng, "NOx reduction by coupling combustion with recycling flue gas in iron ore sintering process," *International Journal of Minerals, Metallurgy, and Materials,* 18(4)(2011), 7.

2.R. Remus, M. A. A. Monsonet, S. Roudier, L. D. Sancho, "*Best Available Techniques (BAT) Reference Document for Iron and Steel Production*"(Report EUR 25521 EN, Joint Research Centre of institute for propspective technological studies, 2013).

3.C. L. MO, C. S. TEO, I. HAMILTON, J. MORRISON, "Admixing Hydrocarbons lron Ore Sintering Process," *ISIJ International,* 37(4)(1997), 350-357.

4.K. Morioka et al., "Primary application of the In-Bed-deNOx process using Ca-Fe oxides in iron ore sintering machines," *ISIJ international,* 40(3) (2000), 280-285.

5.Y. G. CHEN, Z. WANG and Z. C. GUO, "Simulation of NOx Reduction by Using Modified Coke Breeze During Sintering Process," *Journal of Iron and Steel Research,* 21(5)(2011), 8-11. (in Chinese)

6.C. L. Mo, "A study of in-plant de-NOx and de-SOx in the iron ore sintering process,"(Ph.D. thesis, THE UNIVERSITY OF WOLLONGONG, 1997,172-186)

Drying, Roasting, and Calcining of Minerals
Edited by: Thomas P. Battle, Jerome P. Downey, Lawrence D. May, Boyd Davis, Neale R. Neelameggham, Sergio Sanchez-Segado, and P. Chris Pistorius
TMS (The Minerals, Metals & Materials Society), 2015

THE GENERATION ABILITY OF LIQUID PHASE FOR MIXTURE OF IRON ORE POWDERS AND LIME: PREDICTION, CHARACTERIZATION AND INFLUENCING FACTORS

Changyang Ji, Xuewei Lv*, Xiangwei Zheng, Chengyi Ding, Rende Zhang
College of Materials Science and Engineering, Chongqing University,
Chongqing 400044, China
*corresponding author, Email address: lvxuewei@163.com

Keywords: iron ores; roasting; theoretical calculation; the mass of liquid phase

Abstract

The fraction of liquid phase in sintered ore can be the most important factor to determine the sinter quality and yield in sintering process. The generation ability of liquid phase for mixture of iron ore powders is determined by basicity, roasting time, roasting temperature and chemical composition of iron ores. Five different iron ores were selected in this study and the dried powders of iron ore and CaO reagent (<200 mesh) were homogeneously mixed together at their required proportions, then 15g of the mixture was pressed under a pressure of 20MPa to make a cylindrical sample with a diameter of 20mm. The sintering temperature in this study was kept at 1280℃, the gas phase is air, the holding time was varied with its three levels of 5min,10min and 15min. After that, the sample was cooled under room temperature in the air, and the cooled sample was analyzed by using zeiss optical microscope. The mineral phase compositions were quantitatively determined with an open source image processing software, Image J whicn can calculate the fraction of liquid phase. Besides, the study on how to estimate the sintering behavior of the raw materials in different conditions by the theoretical calculation is of vital importance, and the development of this will greatly improve the sintering production. Factsage will be used to calculate the liquid phases formulation during the sintering. Comparing the experimental results and the theoretical calculation shows that the mass of liquid phase increases with increasing basicity during the sintering, besides, the mass of calcium ferrite increases with increasing basicity. The generation ability of liquid phase of iron ores can be predicted through the experiment verification.

Introduction

The liquid phase generated from the sinter mixture consisted of iron ore powders, fuel (coke breeze) and flux (lime, dolomite) at high temperature (1200~1400℃)[1] is greatly influenced by the properties of iron ores and the dosages of fuel and flux. The fraction of liquid phase in sintered ore can be the most important factor to determine the sinter quality and yield in sintering process. The physicochemical properties that influence the formation of liquid phase for iron ores included are mineral phase compositions, gangue compositions and their contents, particle sizes, micro-pore structure of the iron ore particles, etc. Therefore, the comprehensive studies on the generation capability of liquid phase for each single iron ore should be important on guiding the burdening procedure.

On this topic, considerable studies had been carried out. Kasai and Wu proposed a series concepts for characterization of iron ore sintering[2-8]: assimilation, fluidity of liquid phase, strength of binding phase and generation ability of the calcium ferrite. These concepts assistant the selection of iron ore in the plant practice. These four indexes can give a comprehensive estimation of the behavior of iron ore during sintering. However, these indexes can only be measured in a special equipments and the value of the index is a relative data. Hida and others[9] discovered that calcium ferrite was produced in 1205℃ in sintering process. With the increase of temperature, the mass of liquid phase increased gradually. The melt was formed as the mineralogical structure in cooling process. Zhang[10] researched the influence of basicity on producing calcium ferrite. The basicity increases from 1.6 to 2.8, the percent of calcium ferrites increased gradually and the reducibility of sinter was improved in the ironmaking process. The shape of calcium ferrites have acicular(or needle shaped), columnar and tabular, the acicular calcium ferrites can enhance the sinter property, while the columnar and tabular calcium ferrites can lead to bond phase strength reduction.

However, there are also some questions about the previous research: the generation of liquid phase only relied on the experimental which is very complex and random, the generation ability of SFCA was also lack of quantitative description. Therefore, we suggested an estimation of liquid phase percentage by the thermodynamic software, some kind of iron ores were selected in present study, the mixture between iron ore and the lime were baked, polished and observed under the microscope for verification the theoretical calculation.

Experimental

Raw materials

Five different iron ore powders were selected in this study, with their chemical compositions shown in Table 1.The total iron (TFe) for ore E is the lowest (only 44.82%), on the contrary, ore B is a magnetite ore with the highest content of TFe (65.27%). The mass percent of Al_2O_3 for ores A and E is high (>2.5%), while the mass percent of SiO2 fo ores C and D is high (>6.0%).

Table 1 Chemical composition of raw materials (mass %)

Iron ore	TFe	FeO	CaO	SiO_2	MgO	Al_2O_3	P	LOI	Crystal water
A	59.55	1.80	0.21	4.54	0.15	2.63	0.098	5.64	4.87
B	65.27	23.53	0.27	5.92	0.39	0.70	0.014	1.43	0.88
C	61.30	0.90	0.03	8.00	0.05	1.30	0.048	2.00	1.21
D	60.32	0.86	0.03	9.34	0.18	1.38	0.033	1.31	0.79
E	44.82	5.75	3.68	15.83	1.48	2.96	0.092	5.49	4.62

The particle sizes of the raw materials have a significant impact on granulation process and sintering process. Table 2 shows the particle sizes of five sinter raw materials, with their cumulative particle size distributions plotted in Figure 1.

Table 2 Particle sizes of sinter raw materials(mass %)

Iron ore	Size range (mm)							
	>10	10～7	7～5	5～3	3～1	1～0.7	0.7～0.2	<0.2
A	3.17	7.61	10.37	23.68	23.58	10.21	15.50	5.88
B	0.00	0.00	1.57	6.48	10.98	6.64	42.00	32.33
C	4.56	6.74	7.84	14.38	13.59	7.33	17.11	28.44
D	3.12	7.66	9.30	14.95	16.29	8.48	20.70	19.51
E	30.91	12.00	7.96	10.47	8.73	5.15	11.48	13.31

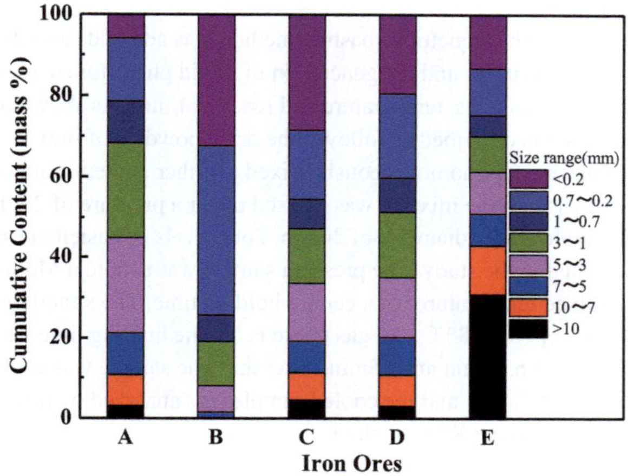

Fig.1 The percent of cumulative particle of sinter raw materials

Fig.2 shows the phases of the five different iron ores determined by using XRD method. Table 3 shows the phase composition for the five different iron ores. Except for ore B, the main phase for ores A, C, D and E is Fe_2O_3 (hematite). Ores A, C and D contain some hydroxyl phases. The crystal water of ores D and E is high and decomposes at 400~500℃, so the LOI for them is higher than other ores.

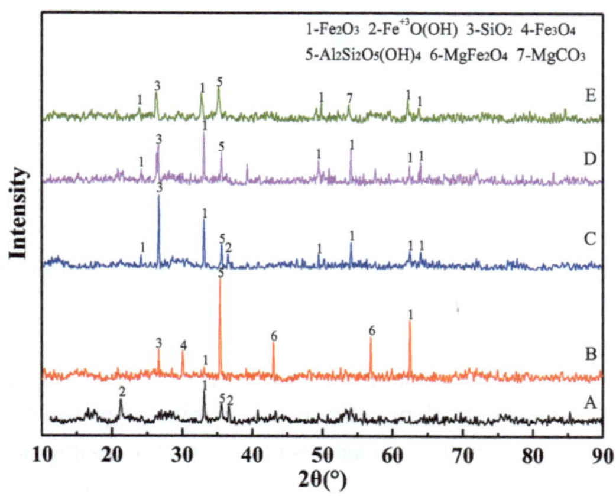

Fig.2 Phase of raw materials

Table 3 Mineral phase of iron ores

Iron ore	Mineral Phase
A	Fe_2O_3、$Fe^{+3}O(OH)$、$Al_2Si_2O_5(OH)_4$
B	Fe_2O_3、Fe_3O_4、SiO_2、$Al_2Si_2O_5(OH)_4$、$MgFe_2O_4$
C	Fe_2O_3、$Fe^{+3}O(OH)$、SiO_2、$Al_2Si_2O_5(OH)_4$
D	Fe_2O_3、SiO_2、$Al_2Si_2O_5(OH)_4$
E	Fe_2O_3、SiO_2、$Al_2Si_2O_5(OH)_4$、$MgCO_3$

Laboratory procedure

A sinter process with briquetting-roasting method was adopted to study the shift of mineral phase compositions and the generation of liquid phase for each iron ore roasted at different basicities, temperature and roasting times. As shown in Figure 3, the sinter procedure is described as follows: the dried powders of iron ore and CaO reagent (<200 mesh) were homogeneously mixed together at their required proportions, then 15g of the mixture was pressed under a pressure of 20MPa to make a cylindrical sample with a diameter of 20mm. Four levels of basicity (1.6, 1.8, 2.0, 2.3) was researched in the study. The pressing sample was sent to a MoSi2 furnace to be roasted at a fixed temperature for a certain holding time. The sintering temperature in this study was kept at 1280℃, the gas phase is air, the holding time was varied with its three levels of 5min,10min and 15min. After that, the sample was cooled under room temperature in the air, and the cooled sample was analyzed by using zeiss optical microscope and the XRD method.

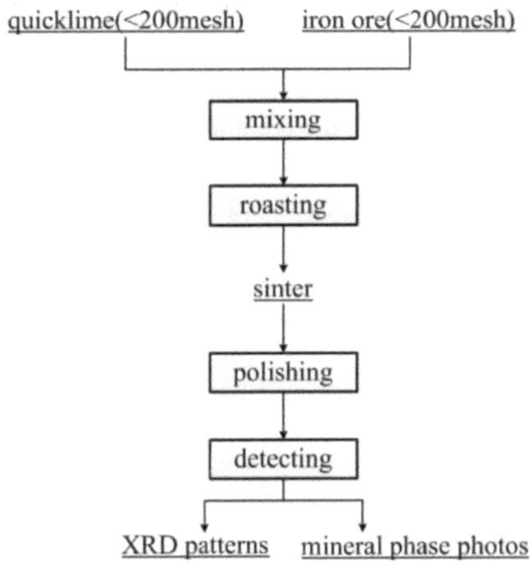

Fig.3 The sinter process with briquetting-roasting method for each ore used

Results and discussion

The optical images of the mineral phase structure for each sintered sample were obtained with the zeiss optical microscope. The mineral phase compositions were quantitatively determined with an open source image processing software, Image J. The following results and discussion are the analyses on the optical images for each ore and the discussions on the content of mineral phases calculated from each optical image.

Optical images of mineral phases

Figure 4 to Figure 8 show the mineral phase photos of iron ore in different basicity and different roasting time, from top to bottom, the sintering time were 5min, 10min and 15min, from left to right, the basicity were 1.6, 1.8, 2.0, 2.3. There are several mineral phases in the optical image of the sintered samples[11]: (1) hematite (Fe_2O_3)—the gray part, has an obvious interface with the calcium ferrite, and its refractive index is approximately 22~25%. (2) calcium ferrite ($CaFe_2O_4$)—the shape is granular or columnar, the reflected light is gray with a light cyan, and its refractive index is about 18~18.5%. (3) magnetite (Fe_3O_4)— granular shaped and anhedron magnetite are existed which was corroded with calcium ferrites, and its refractive index is approximately 20%. (4) silicate— olivine, pyroxene and other silicon-oxygen compounds, and its refractive index is about 5~8%, reflected light is deep grey. (5) holes—black, the light is absorbed completely. Mineral phase images of every group show that the percent of hematite decreased and the percent of calcium ferrites increased with the rise of basicity and sintering time. The mineral phase photos of ore A shows that the mass of liquid phase increase with the increase basicity and roasting time, the pictures also show that the hole of ore A is more than other ores because ore A has more crystal water than other ores, besides the shape of calcium ferrites were presented acicular (or needle shaped). To ore B, the mass of calcium ferrites was generated mostly in basicity 2.3, To ore C, the generation ability of producing calcium ferrites is stronger than other ores, and in basicity 2.0, the mass of calcium ferrites were reached most and in basicity 2.3 ore C had occurred over melting. To ore D, there is almost no calcium ferrites in basicity 1.6, because the percent of SiO_2 is high relatively, silicate was generated firstly. To ore E, the mineral phase photos show that the main phase is silicate, because the percent of SiO_2 is highest than other ores.

5

10

15

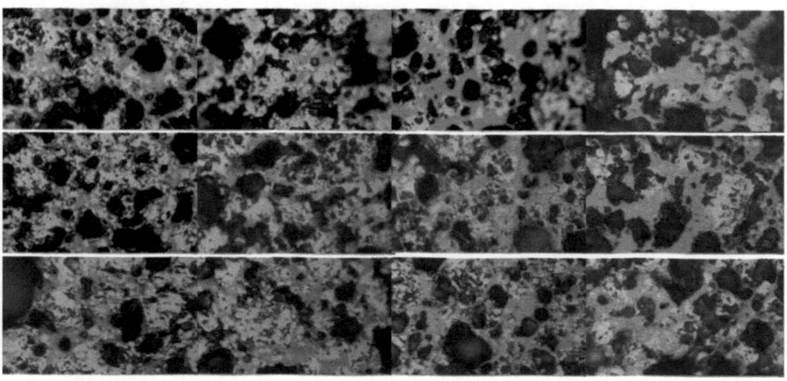

1.6 1.8 2.0 2.3

Fig.4 Ore microscopys for various sinter times and various basicities (reflected light×500 A)

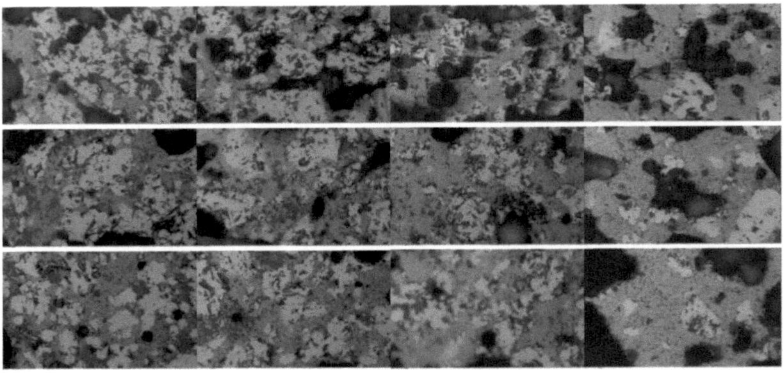

Fig.5 Ore microscopys for various sinter times and various basicities (reflected light×500 B)

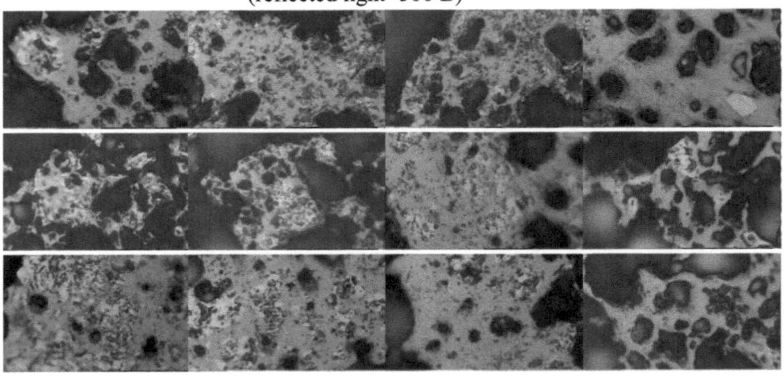

Fig.6 Ore microscopys for various sinter times and various basicities (reflected light×500 C)

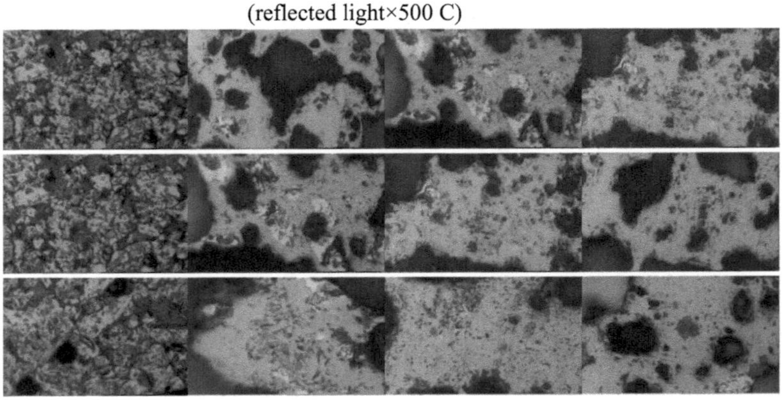

Fig.7 Ore microscopys for various sinter times and various basicities (reflected light×500 D)

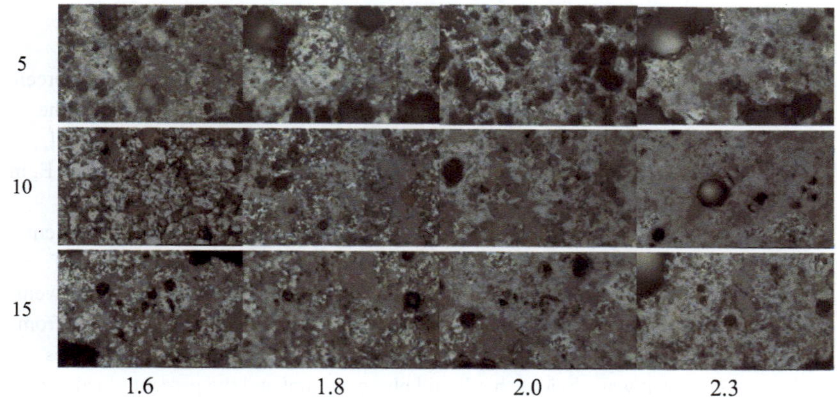

Fig.8 Ore microscopys for various sinter times and various basicities (reflected light×500 E)

Quantitative analysis of mineral phases

The mass of liquid pahse was calculated by the ImageJ which is a public image processing software. Hematite, calcium ferrites and silicate have different reflectivity index under the ore microscopys pictures which can count the mass of liquid phase through ImageJ.

Figure 9 shows the area fraction of each phases versus basicity under different sinter times (holding time) for each sample.

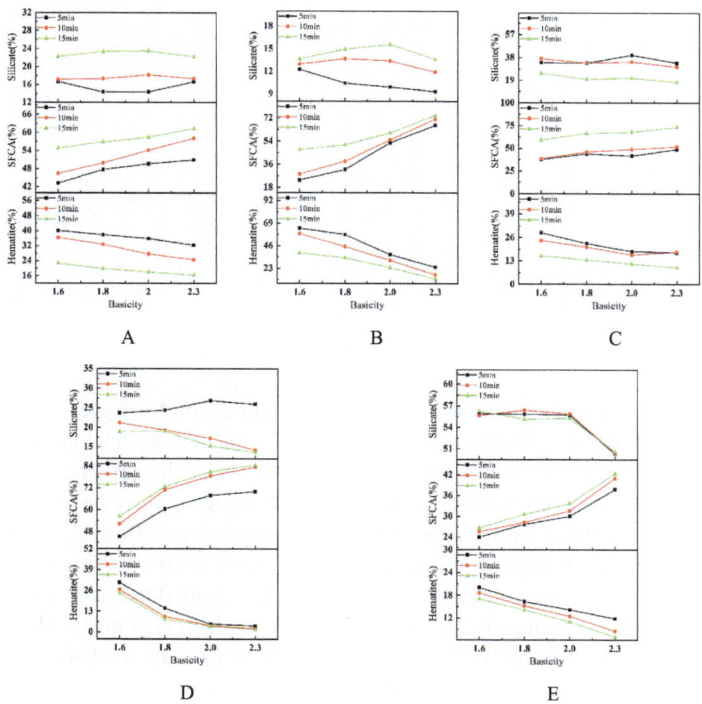

Fig.9 Area fraction of phases versus basicity for different sinter times

Figure10 shows the area fraction of phases versus basicity for five ores, the percent of calcium ferrites and liquid phase amount of the sinter sample increased with the basicity from 1.6 to 2.3, the variation of the percent of calcium ferrites and liquid phase amount of every raw material were approximately same. To ores A, C and E, in different sintering time the percent of calcium ferrites and liquid phase amount increased with the increase of basicity, the variation of liquid phase amount of them were same basicly that liquid phase amount increased gently with the increase of basicity. To ore B, liquid phase amount and the percent of calcium ferrites were went through a gentle-rapid-gentle process with the increase of basicity. The basicity from 1.6 to 1.8, the variation of liquid phase amount and the percent of calcium ferrites changed slowly, it was obvious that liquid phase amount and the percent of calcium ferrites increased quickly when basicity was 2.0 while it increased slowly with the basicity increasing to 2.3. To ore D, the basicity from 1.6 to 2.0, liquid phase amount and the percent of calcium ferrites increased obviously, the basicity from 2.0 to 2.3, the variation of liquid phase amount and the percent of calcium ferrites were almost constant. Contrasting the five kinds of iron ore known that the ability of producing liquid phase amount in basicity 1.6: E>C>D>A>B, in basicity 1.8 and 2.0: D>E>C>A>B, in basicity 2.3: D>E>C>B> A. Summing up the appeal, the generation ability of liquid phase of ores E and D are stronger than other ores while ores A and B are poor in producing liquid phase.

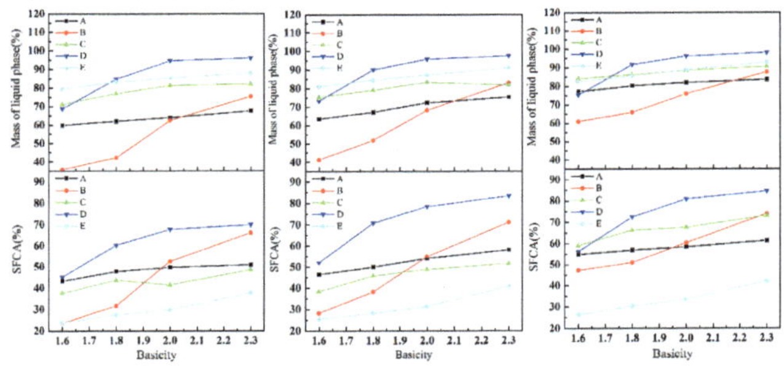

5min 10min 15min

Fig.10 Area fraction of phases versus basicity for five iron ores

Thermodynamics theoretical calculation with Factsage

Up to now, the most effective method on studying the iron ore sintering peocess is the experiments in the laboratory which is carried out in the sintering pot[12]. No mathematical model or experience equations were developed. Therefore, the study on how to estimate the sintering behavior of the raw materials in different conditions by the theoretical calculation is of vital importance, and the development of this will greatly improve the sintering production. Factsage will be used to calculate the liquid phases formulation during the sintering. The Equilib model is used in the calcaulation, the gas phase is included in the calculation. It is helpful to combine theoretical calculation with the experimental results in sintering process. Factsage can give some good explanations on the phenomenon of the melting behavior of different samples and reduce the extra experiments[13].

The mass of liquid phase created with increasing the temperature was shown in Figure 11. It shows that the mass of liquid phase increases with increasing the temperature. In 1.6, ores A, B, C, D and E generated liquid phase at 1200℃; in 1.8, ores A and E generated liquid phase at 1150℃, ores B, C and D generated liquid phase at 1200℃; in 2.0 and 2.3, ores A, B, C, D and E generated liquid phase at 1100℃, because the percent of CaO in 2.0 and 2.3 is more than in 1.6 and 1.8, it is helpful to generate liquid phase obviously at low temperature. The mass of liquid phase generated in ore E increases dramatically after 1200℃ in 1.6, 1.8 and 2.0, while in 2.3 the mass of liquid phase generated in ore E increases dramatically after 1100℃, it reaches more than 95% at 1300℃, and 100% in mass at 1300℃. The mass of liquid phpase generated in ores A, B, C and D increase obviously after 1200℃ in 1.6, 1.8, 2.0, and 2.3, and increase with the increase of temperature. Besides, it is obviously that the generation ability of liquid phase of ore E is better than other ores in 1.6, 1.8, 2.0, and 2.3. In 1.6 and 1.8, the generation ability of liquid phase of five iron ores is E>D>C>B>A in general, but ores A and B have a little difference at 1300-1400℃, because the percent of SiO_2 of ore B is more than ore A, and the initial liquid phase is silicate at 1200-1250℃, the generation ability of liquid phase is B>A; over 1250℃, the main liquid phase is calcium ferrite, and the percent of Fe_2O_3 of ore B is less than ore A, so the mass of liquid phase of ore A is more than ore B in 1300-1400℃; with increasing temperature, the generation ability of liquid phase is B>A. In 2.0, the generation ability of liquid phase of five iron ores is E>D>C>A>B in 1100-1400℃, and the generation ability of liquid phase of five iron ores is E>D>C>B>A in 1400-1500℃. In 1100-1400℃, the generation ability of liquid phase is A>B, because the percent of CaO is more in high basicity, the binding capacity of CaO and Fe_2O_3 is stronger than CaO and SiO_2, with increasing temperature the high melting point phase is generated and the generation ability of liquid phase is B>A. In 2.3, the generation ability of liquid phase of five iron ores is E>D>C>A>B in 1100-1200℃, and at 1250℃, the generation ability of liquid phase of five iron ores is E>C>A>D>B, the most calcium ferrites are generated at 1250℃ because the mass of CaO is enough in 2.3, and the percent of TFe of ores C and A is more than other ores. The generation ability of liquid phase of five iron ores is E>D>C>A>B in 1250-1350℃, and the generation ability of liquid phase of five iron ores is E>D>C>B>A in 1350-1500℃, with increasing temperature the high melting point phase is generated and the generation ability of liquid phase of ores is B>A.

Compared the four figures, the generation ability of liquid phase of ores E, D and C is stronger than ores A and B. The initial liquid phase is silicate in low temperature and with increasing temperature, the calcium ferrites are generated, then the aluminate is generated in high temperature. Because the chemical composition of five ores show that the percent of SiO_2 of ores E, D and C are more than other ores, the mass of liquid phase of ores E, D and C is more than ores A and B in spite of high basicity.

The thermodynamics theoretical calculation can indicate the experiment. Although the results of theoretical calculation had a little difference with the results of experiment, it could be basically the same.

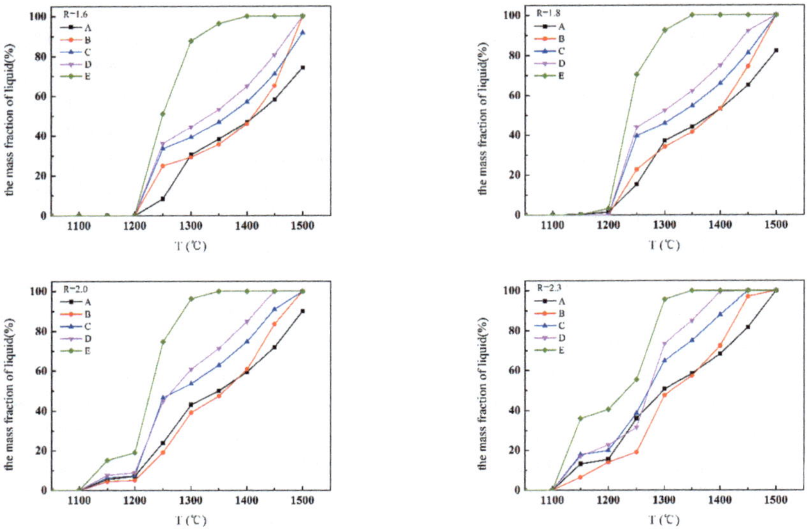

Fig.11 Mass of liquid phase verse temperature for five ores

Conclusions

The liquid phase formulation during the iron ore sintering were studied by using the FactSage software, and the experiments were carried out for the validation. The conclusions can be summarized as followings:

(1)Comparing the experimental results and the theoretical calculation shows that the mass of liquid phase increases with increasing basicity during the sintering, besides, the mass of calcium ferrite increases with increasing basicity.

(2)The experimental results shows the generation ability of liquid phase of five iron ores in basicity 1.6: E>C>D>A>B, in basicity 1.8: D>E>C>A>B, in basicity 2.0: D>E>C>A>B and in basicity 2.3: D>E>C>B>A at 1280℃.The theoretical calculation shows the generation ability of liquid phase of five iron ores in basicity 1.6: E>D>C>B>A , in basicity 1.8: E>D>C>B>A, in basicity 2.0: E>D>C>A>B and in basicity 2.3: E>D>C>A>B. Although the results of theoretical calculation had a little difference with the results of experiment, it could be basically the same.

(3)The generation ability of liquid phase of iron ores can be predicted through the theoretical calculation and experiment verification.

Acknowledgements

The authors are especially grateful to the Nature Science Foundation of China (Grant No. 51104192) for the financial support on this work.

References

[1] Zhongyu Liang, *Ironmaking science* (Beijing, BJ: Metallurgical Industry Press, 2009), 57-65.

[2] Kasai E., and Y. Omori, "Subjects on the Evaluation of the Process and Products in the Sintering of Iron Ore," *Tetsu to Hagane*, 79 (11) (1993), 1217-1223.

[3] Kasai Eiki, Yoshio Waseda, and Maximiano V Ramos, "Thermal analyses of the sintering reactions of iron ores," *Tetsu to Hagane*, 83 (9) (1997), 539-544.

[4] Loo CE, Penny GC, and Witchard D, "Effective sintering of iron ore blends containing porous ores at low moisture contents," *Transactions of the Institution of Mining and Metallurgy (Section C: Mineral Processing and Extractive Metallurgy)*, 105 (1996), C22-C36.

[5] S. Wu et al., "The new concepts about Iron ore sintering characteristics," *Journal of University of Sicience and Technology Beijing*, 24 (3) (2002), 254-257.

[6] S. Wu et al., "The flow characteristics of iron ore sintering liquid phase," *Journal of University of Sicience and Technology Beijing*, 27 (3) (2005), 291-291.

[7] S. Wu et al., "The research about the assimilative capacity of iron ore with CaO," *Journal of University of Sicience and Technology Beijing*, 24 (3) (2002), 258-258.

[8] S. Wu et al., "The evaluation about flow characteristics of iron ore sintering liquid phase," *Journal of University of Sicience and Technology Beijing*, 30 (10) (2008), 1095-1100.

[9] Hida Y. et al., "Study on sintering of iron ore with advanced analyzers (production of high-reducibility sinter with low fuel consumption)," *Nippon steel technical report*, 35 (1987), 59-67.

[10] Zhang, F. et al., "Effect of Basicity and Alumina-Silica Ratio on Formation of Silico-Ferrite of Calcium and Aluminum," *Journal of Iron and Steel Research*, 19 (4) (2012), 1-5.

[11] Fan X. et al., "Influence factors of calcium ferrite formation in iron ore sintering," *J. Cent. South Univ.(Science and Technology)*, 06 (2008), 1125-1131.

[12] Kongoli F. et al., "Effect of oxygen potential and fluxing components on phase relations during sintering of iron ore," *Journal of Mining and Metallurgy B: Metallurgy*, 46 (2) (2010), 123-130.

[13] Xuewei Lv et al., "Behavior of liquid phase formation during iron ores sintering," *ISIJ international*, 51 (5) (2011), 722-727.

AUTHOR INDEX
Drying, Roasting, and Calcining of Minerals

SUBJECT INDEX
Drying, Roasting, and Calcining of Minerals

MIX
Papier aus verantwortungsvollen Quellen
Paper from responsible sources
FSC® C105338
FSC
www.fsc.org

If you have any concerns about our products,
you can contact us on
ProductSafety@springernature.com

In case Publisher is established outside the EU,
the EU authorized representative is:
Springer Nature Customer Service Center GmbH
Europaplatz 3, 69115 Heidelberg, Germany

Printed by Libri Plureos GmbH
in Hamburg, Germany